The M68HC11 Microcontroller

Applications in Control, Instrumentation, and Communication

MICHAEL KHEIR

**Real-Time Systems Design
and Control Consultant**

Prentice Hall

Upper Saddle River, New Jersey   Columbus, Ohio

Library of Congress Cataloging-in-Publication Data

Kheir, Michael.
    The M68HC11 microcontroller: applications in control, instrumentation, and communication / by Michael Kheir.
       p. cm.
    Includes index.
    ISBN 0-13-205550-3
    1. Programmable controllers. 2. Microprocessors. I. Title.
TJ223.P76K47   1997
629.8'9—dc21
                                  96-36941
                                      CIP

Cover Photo: Phil Matt
Editor: Charles E. Stewart, Jr.
Production Editor: Christine Harrington
Text Designer and Production Coordinator: Custom Editorial Productions, Inc.
Cover Designer: Jill Bonar
Production Manager: Laura Messerly
Marketing Manager: Debbie Yarnell

This book was set in Times Roman by Custom Editorial Productions, Inc., and was printed and bound by Quebecor Printing Book Press, Inc. The cover was printed by Phoenix Color Corp.

© 1997 by Prentice-Hall, Inc.
A Pearson Education Company
Upper Saddle River, NJ 07458

Notice to the Reader: The publisher and the author do not warrant or guarantee any of the products and/or equipment described herein nor has the publisher or the author made any independent analysis in connection with any of the products, equipment, or information used herein. The reader is directed to the manufacturer for any warranty or guarantee for any claim, loss, damages, costs, or expense arising out of or incurred by the reader in connection with the use or operation of the products or equipment.

    The reader is expressly advised to adopt all safety precautions that might be indicated by the activities and experiments described herein. The reader assumes all risks in connection with such instructions.

Printed in the United States of America

10 9 8 7 6 5 4 3 2 1

ISBN 0-13-205550-3

Prentice-Hall International (UK) Limited, London
Prentice-Hall of Australia Pty. Limited, Sydney
Prentice-Hall Canada Inc., Toronto
Prentice-Hall Hispanoamericana, S.A., Mexico
Prentice-Hall of India Private Limited, New Delhi
Prentice-Hall of Japan, Inc., Tokyo
Pearson Education Asia Pte. Ltd., Singapore
Editoria Prentice-Hall do Brasil, Ltda., Rio De Janeiro

A special thanks to my wife Eman and my sons Michael Jr. and Matthew for their patience and support

Contents

List of Example Programs

Preface

This book is intended to be used in a one- or two-term course toward a two- or four-year engineering technology degree, a computer science degree, or a computer engineering degree. It can also be used by professionals and hobbyists in the field of microprocessor/microcontroller programming and interfacing.

The book differs in scope from many other books on microcontrollers in that it combines in-depth coverage of the M68HC11's applications in control, instrumentation, and communication in a single source. It is designed to help readers develop hands-on experience in both software and hardware. We selected the M68HC11 because it is one of the fastest growing, best selling lines worldwide for embedded control applications.

This book offers a unique approach because it is the first microcontroller textbook to include:

- Real-world applications on power control, switching circuits, LED and LCD displays, data acquisition, communications between microcontrollers, and many others
- Case studies on robot control, engine control, and barometric pressure control
- PID algorithms and programs on speed control of DC motors, control of stepper motors speed and direction, and process control techniques
- A whole chapter on real-time programming using C
- Other advanced MCUs such as M68HC11K4, M68HC16, and M68HC332

The text includes many application programs and real-world problems and projects that make them useful to implement in the laboratory as well as in real life. It also includes many basic hardware design examples and problems. It reflects the assumption that students in a science or engineering computer degree program should master the basics of design with microcontrollers.

In "crash courses" or entry level courses on microcontrollers, Chapters 1–6 and 11 are most essential. For advanced and more intensive courses, all chapters are recommended.

The book includes an entire chapter (Chapter 10) on programming microcontrollers in C. Students must have a basic knowledge of C before working through this chapter.

Finally, the text features three case studies: two at the end of Chapter 7 and one at the end of Chapter 8. The first involves an industrial machine application: a robot motion control program. In this case study, the pneumatic system, solenoid valves, and robot motion are explained. The second case study involves an automotive application: an engine control program. This case study clarifies the air/fuel ratio, open-/closed-loop control of delivered fuel, and exhaust gas sensors. The third case study involves an instrumentation application: a barometric pressure gauge program. Motorola IC pressure sensors, circuitry design, and LCD interfacing are explained.

Appendixes include other industrial and automotive projects that require students to go to technical libraries and search for answers.

The following is a more detailed summary of the chapter contents.

Chapter Descriptions

Chapter 1 explains the basic microcontroller architecture and its internal buses and number systems. It also introduces the addressing modes of the MCU M68HC11.

Chapter 2 explains the instruction set of MCU M68HC11 and provides examples.

Chapter 3 introduces the program development stages (design and coding phases) and the basics of structured programming. It also introduces compilers, assemblers, assembly directives, and object and listing files.

Chapter 4 illustrates in detail how to develop many algorithms using pseudocode, flowcharts, and assembly language to perform arithmetic operations and write logic code (using AND and OR terms).

Chapter 5 presents in detail the on-chip I/O resources such as clock generators, I/O ports, SCI, SPI, timers, ADC, and address decoding, with many programming examples.

Chapter 6 explains basic interfacing requirements of analog and digital devices. Examples and problems explain how to interface LEDs, switches, keypads, DC motors, and zero-voltage switching (ZVS) devices to the MCU.

Chapter 7 introduces the modes of digital control, PID algorithms, servosystems, programs on speed control and speed measurement of DC motors, programs to implement the control of stepper motor speed and direction, and process control techniques. The chapter also includes case studies on both robot motion control and engine control.

Chapter 8 introduces several programs to interface 7-segment, dot-matrix, and alphanumeric displays (LEDs and LCDs) to the MCU. It also presents many transducers and how to use them in data acquisition systems. The chapter includes a case study on barometric pressure gauge design.

Chapter 9 explains in detail the serial transmission modes, synchronous and asynchronous protocols, GPIB and VME buses, and the communications between microcontrollers. Programs exemplify most of these topics.

Chapter 10 briefly covers some topics pertaining to C language such as functions, arrays, pointers, and so on. It also includes programming examples and problems on PWM generation, period measurement, and DC motor control.

Chapter 11 introduces development and debugging tools such as emulators, simulators, and logic analyzers. It introduces the debugging techniques in the testing phase. This chapter also briefly covers some other advanced members of the M68HC11 family and other Motorola 16-bit and 32-bit MCUs.

Practical Projects, at the end of the book, cover many subjects, such as mobile robot control, automotive design, data acquisition, communications, and more. Students are encouraged to develop and add more features to these projects.

Appendix A lists the details of the M68HC11 instruction set used throughout this book.

Appendix B includes data sheets of some of the transducers or drivers used in Chapters 7 and 8.

Appendix C lists the MC68HC11E9 header files used by some C program examples presented in Chapter 10.

An instructor's solutions manual with transparency masters and a floppy disk with selected labs accompany this book.

Acknowledgments

I would like to thank Professor V. S. Anandu of Southwest Texas State University and Professor Subra Ganesan of Oakland University in Michigan for their helpful advice during review of the manuscript, and the people at Custom Editorial Productions, Inc., who provided excellent support during the editorial and production phases.

1 Introduction to Microcomputers

A computer revolution in the last 15 years has produced computers with very high speeds and computing power while keeping their sizes compact. This revolution has occurred as a result of the development of *Large-Scale Integration* (LSI) and *Very Large-Scale Integration* (VLSI) technologies, which put tens of thousands of transistors on a single chip. This has made it possible to fabricate the heart of a microcomputer as a single chip called a *microprocessor* (MPU). This chip, with additional auxiliary chips called *peripherals,* constitutes "a microcomputer." Such peripherals are I/O ports, memories, timers, and so on.

The new technology has also made it possible to integrate this microprocessor and its peripherals in a single chip called a *microcontroller* (MCU). That is the reason a microcontroller is called a *single-chip microcomputer.*

1.1   Microcontroller Structure

A microcontroller consists of four basic parts: central processing unit (CPU), internal memory, registers, and I/O subsystem. These parts are connected internally by an internal bus. The I/O subsystem allows the MPU to exchange information with the outside world. The I/O subsystem is grouped into units called *I/O ports*. Each I/O port has I/O lines (usually eight lines) to transfer information between the external devices and the ports. These lines can be input only, output only, or programmable to be either. Each port also has its own I/O registers. The register types are control, status, and data. (See Figure 1.1.)

I/O Ports

Control Register

The CPU loads the control register with a control word specifying to the I/O device one or more commands. Usually each bit of the control word indicates a specific command. The information stored by the CPU in this register determines operational characteristics such as port direction and the ability of a port to cause interrupts.

Status Register

The status register reflects the status of I/O device at any time. The CPU reads the contents of this register and determines the operational conditions of the device.

Data Register

The data register could be a data input register or a data output register. If the outside device is ready to provide the I/O port with new data, such data are stored in the data input

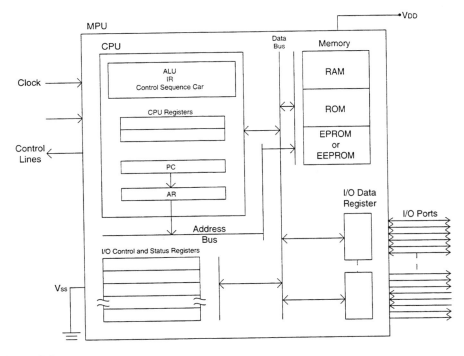

Figure 1.1

register. At the same time, it sets the appropriate bits of the status register to indicate that that has been done. If the outside device is ready to accept data from the I/O port, the new data are placed in the data output register.

Note that not all of the above registers are always needed.

Central Processing Unit

The central processing unit (CPU) consists of three main sections:

- Arithmetic logic unit (ALU)
- Registers and accumulators
- Control unit

Arithmetic Logic Unit

The ALU performs arithmetic and logic operations on operands; these operations include addition, subtraction, multiplication, division, ANDing, ORing, and so on. The operands are held temporarily in registers. The width of the ALU corresponds to the basic word length of the system. Inputs to the ALU may come from accumulators or temporary registers, and the results are placed back in accumulators. The type of operation is determined by the control unit.

Registers and Accumulators

Registers and accumulators are high-speed temporary memory locations during CPU operations. They exchange information through one or more internal buses. The length of each register is equal to the width of the internal data bus. Registers are usually divided into general-purpose registers and special-purpose registers.

General-Purpose Registers. A general-purpose register may be used as a data register for arithmetic and logic operations or as an accumulator. An accumulator is a register that stores the results of an arithmetic operation performed by the ALU. The general-purpose register may also be used as an address register that points to a memory location.

Special-Purpose Registers. All microcontrollers include special-purpose registers, each of which is dedicated to a specific function. Some of these registers are:

Program Counter (PC). At the beginning of a program execution, the PC is loaded with the starting address of that program. Thereafter, it always points to the location of the next instruction. When the CPU executes an instruction telling it to jump or branch to another part of the program (such as a subroutine), the new address is loaded into the PC and the sequential order resumes.

Instruction Register (IR). The IR extracts the operation code (opcode) from an instruction. An instruction consists of an opcode part and operand(s). The control unit decodes the contents of the IR and generates the necessary control signals in order to carry out the actions specified by that instruction.

Stack Pointer (SP). The stack is a specially reserved area in memory where information items are added or removed in last-in-first-out (LIFO) form. The stack pointer register points to the top of the stack—that is, it holds the address of the top of the stack. When the data are written into (pushed onto) the stack, the stack moves downward and SP is decremented.

On the other hand, when the data are read from (pulled off or popped off) the stack, the stack moves upward and the stack pointer is incremented. Stacks are used mainly during subroutines or interrupt handling.

Flag or Status Register. The status register consists of *flag bits* and *control bits*. Flags are set automatically during arithmetic and logic operations. Such flags include N,Z,V,C, and so on. Control bits are set by the program to enable certain modes of CPU operation. More details later.

Address Register (AR). The PC register sends "the pointed to" address to the AR register. The AR register then sends this address to the address bus to select this address in memory.

Control Unit

The control unit includes the instruction decoding, timing, and control circuitry. The control unit generates two groups of signals:

1. Internal control signals for activation of the ALU and the opening and closing of the data paths between registers.
2. External control signals concerning the memory and I/O. These signals are sent either for activation of data transfers or as a response to an interrupt.

A *microinstruction* is a binary pattern used to encode the control signals for each step of execution. A sequence of microinstructions makes a *microprogram.* A microprogram is stored in ROM.

Memory Unit

The memory unit stores the program to be executed and data that are to be operated on by the program. The memory unit operation is controlled by CPU signals (read and write). When the CPU sends data to the memory, it is called a *write* operation, and when the CPU receives data from the memory, it is called a *read* operation. Memory is divided into internal and external memory.

Internal memory usually means memory within the chip. Examples of memory include RAM, ROM, PROM, EPROM and EEPROM. External memory generally means memory outside the chip. This type of memory includes the semiconductor type and serial memory such as magnetic disks, magnetic tapes, and bubble memory. Internal memory is a semiconductor type with low capacity and high speeds, while serial memory with high capacity and low speeds, is not a semiconductor type.

Types of Memory ICs

Memory ICs may be volatile or nonvolatile. Volatile memory loses its data after the power is removed. With nonvolatile memory, these ICs store data permanently or at least semi-permanently (10 years or more) even when power is removed from the chip. Volatile memory includes RAM, while nonvolatile memory includes battery-backed RAM, PROM, EPROM, and EEPROM.

RAM. There are two types of random access memory (RAM): static ram (SRAM) and dynamic ram (DRAM). SRAM uses an integrated flip-flop for each of its storage cells, as shown in Figure 1.2. These cells are arranged in matrices. Column and row select lines are used to select a specific cell for read and write operations. SRAMs are fast, need no refresh circuitry, and are ideal for small programs (less than 16K).

DRAMs are usually organized for storing a single bit for location which is referred to as "×1" organization. For example, some DRAMs are 64K × 1 or 256K × 1 in size. A DRAM storage cell consists of one MOSFET and a capacitor. If the capacitor is charged, a logic 1 is stored in the cell. If the capacitor has no charge, it indicates a logic 0. It is addressed by means of a row address and a column address.

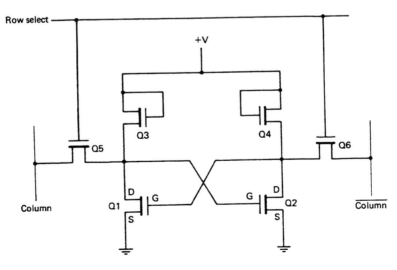

Figure 1.2
From J. Uffenbeck, *Microcomputers and Microprocessors: The 8080, 8085, and z-80.* Copyright 1985 by Prentice Hall. Reprinted by permission of Prentice Hall, Upper Saddle River, New Jersey.

To reduce the number of external interface lines, the row and column address lines are multiplexed on the same input lines (A0–AN). To access a cell, the row address is placed over these lines, and then the row address strobe ($\overline{RAS}$) is activated. Next, the column address is placed and the column address strobe ($\overline{CAS}$) is activated. The DRAM decodes the two stored addresses to locate the addressed cell.

The write enable ($\overline{WE}$) line indicates whether this access is for reading or writing. A logic 1 indicates a read operation, while a logic 0 indicates a write operation. Figure 1.3 shows a block diagram and a pin configuration of a DRAM.

The charge stored in the small capacitors of a DRAM array dissipates fast because of leakage. Therefore, a refresh cycle is needed periodically. DRAMs allow refreshing of the cells of an entire row in a single operation. Usually they require refreshing of 128 rows every 2 mS or 256 rows every 4 mS.

A dynamic memory controller (DMC) simplifies a DRAM interface circuit. It has three functions:

1. Multiplex the address
2. Arbitrate memory access
3. Generate the control cycles

The advantages of DRAMs over SRAMs include their high density, lower power dissipation, and greater cost-effectiveness in large programs (more than 16K).

A third type of RAM is nonvolatile RAM or NVRAM. It consists of a SRAM and battery backup circuitry. The backup circuitry contains a control circuit and a 3 V lithium source. If Vcc should fall lower than 3 V, the controller switches automatically to the lithium source, preventing the RAM's data from getting lost.

ROM. The read-only memory (ROM) storage cell consists of a MOSFET or a bipolar transistor. If the gate is connected to the row line, it is considered a logic 1. If the gate is not connected, it is considered a logic 0.

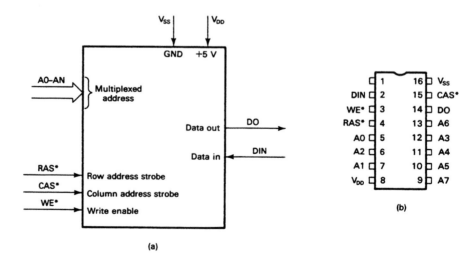

(a)

(b)

Figure 1.3

From D. A. Protopapas, *Microcomputer Hardware Design.* Copyright 1988 by Prentice Hall. Reprinted by permission of Prentice Hall, Upper Saddle River, New Jersey.

Manufacturers program ROMs at the manufacturing stage using a mask process; thus they are called *mask-programmed ROMs*. Once the ROMs are programmed, they cannot be changed like correction of software errors. Therefore, ROMs are used in mass production after the software has been debugged completely.

PROM. The programmable ROM consists of cell arrays interconnected by means of fusible links. When the PROM is manufactured, all fusible links are connected, indicating that all cells contain logic 1. When the user blows a fusible link using a PROM burner, the cell has a logic 0. Like ROMs, PROMS are one-shot programmable.

EPROM. The erasable programmable read-only memory is a nonvolatile memory and can be recognized by the quartz window above its integrated circuit. In an erasing operation, this window is exposed to ultraviolet (UV) light for between 30 to 50 minutes.

An EPROM contains a matrix of metal-oxide semiconductor (MOS) transistors with floating polysilicon gates. When programming a memory location, a higher voltage (ranging from 12.5 to 25 V) is applied that causes electrons to collect on a floating gate and remain even after the programming voltage is removed. The programmed location is read as logic 0. When an EPROM is erased when exposed to UV light, its contents are changed to 1s. The programming process then changes all 1s that should be 0s to 0s.

All EPROMs have the quartz windows except the *one-time-programmable* (OTP) EPROM, which has no window and thus is not erasable. The number of erase/write cycles of EEPROMs is a little over 100.

EEPROM. One of the drawbacks of EPROMs is their need for external programming or erasing devices. The entire chip must be erased before programming even a single byte. Electrically erasable PROMs (EEPROMs) use a floating-gate technology called Fowler-Nordheim tunneling that allows them to be erased electrically.

Modern EEPROMs automatically erase each byte before programming. They are powered from a single +5 V source and can generate their own programming voltages on-chip.

Writing a byte to an EEPROM can take as long as 10 mS, which is considered too long for MPU/MCU operations. However, in modern EEPROMs, once the write cycle has started, the EEPROM itself takes over and completes the process on its own. This will allow the MPU/MCU to process other tasks. The number of erase/write cycles is about 10,000.

Memory Characteristics

Speed. Memory speed is defined in terms of *access time* and *cycle time*. Access time is defined as the time elapsing from the moment we present a stable address until memory responds with stable data. The cycle time determines how fast we can access memory. Figure 1.4 illustrates both times.

Density. Density is defined as the number of bits per memory chip.

Power Dissipation. Power dissipation is specified in terms of operating power and standby power. Low standby power represents a very important advantage.

Cost. The price of memory devices affects the cost of the overall memory subsystem.

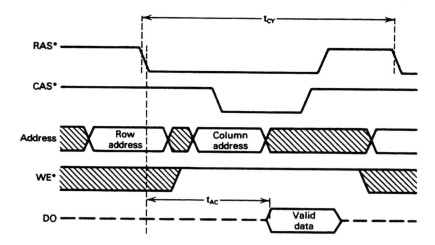

Figure 1.4
From D. A. Protopapas, *Microcomputer Hardware Design.* Copyright 1988 by Prentice Hall. Reprinted by permission of Prentice Hall, Upper Saddle River, New Jersey.

1.2   Microcomputer Internal Buses

The diagram in Figure 1.5 shows the communication between the heart of the microcomputer (CPU) and memory through three buses: address bus, data bus, and control bus. A bus is a group of lines carrying specific data.

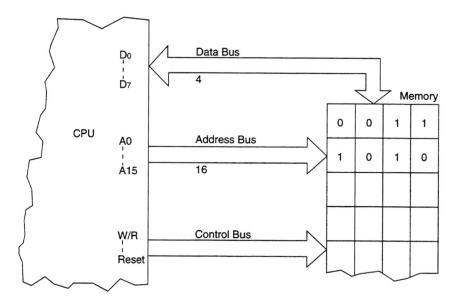

Figure 1.5

Address Bus

This bus carries the address of a location in memory or I/O, generated by the CPU, and it is *unidirectional,* as shown in Figure 1.5. If the CPU needs to access data at a certain location, this address in binary form is placed on the bus to fetch its data. The number of address lines determines the number of addresses or locations that can be accessed. For example, our CPU in Figure 1.5 has 16 address lines. Then the number of accessed locations is equal to $2^{16} = 65,535$ addresses = 64K, where K = 1024.

Generally, the number of accessed addresses = 2^n, where n = number of address lines (bits). The lowest address is 0 and the highest is 2^{n-1}.

Data Bus

This bus is *bidirectional* and is used to send or receive data to or from the memory or the I/O peripherals. A data bus could be 8, 16, or 32 bits (lines). The data bus size (beside the internal registers of the CPU) determines the size of the microcomputer. For example, if we say that a certain microcomputer is an 8-bit microcomputer, it means that its data bus has 8 bits.

Control Bus

The number of control bus lines varies from one microcomputer to another. Some of these lines—such as: R/W, IRQ, Reset, and so on—are common. The control signals synchronize the operation of the CPU with the memory and peripherals.

EXAMPLE 1
A memory chip has 8 data lines and 16 address lines. What will its size be?

Solution
First, we calculate how many addresses or locations it contains. As we mentioned before, the number of addresses = $2^{16} = 65,535$ locations = 64K. Second, the number of data lines = number of bits per location = 8 bits. So the size of this memory chip is:

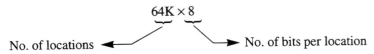

$$64K \times 8$$

No. of locations ← / ↘ → No. of bits per location

Generally, the size of a memory chip is expressed by two numbers. The first indicates the number of locations in this chip. The second indicates the number of bits per location. Usually the data lines are called D0, D1, . . . , DN and the address lines are called A0, A1, . . . , AN.

Exercises

1. Draw the block diagram of a basic microcontroller and explain the function of each section.
2. State two differences between internal and external memory.

3. A memory chip size is 16M × 16. How many address and data bits does it have?

4. Explain the advantages and disadvantages of SRAM, DRAM, NVRAM, ROM, PROM, EPROM, and EEPROM.

1.3   Instruction Cycle

The sequence of steps involved in fetching and executing an instruction from the memory is known as an *instruction cycle*. An instruction cycle consists of a few machine cycles. Figure 1.6 shows an instruction cycle with five machine cycles. Types of machine cycles include the following:

- Read memory
- Write memory
- Interrupt operation
- Interrupt acknowledge

The machine cycles are synchronized by the internal clock cycles. A machine cycle consists of three to four clock cycles.

In Figure 1.6, the instruction cycle consists of five machine cycles. Here is how the CPU fetches and executes an instruction:

1. In machine cycle 1, the CPU reads the first word of the instruction using the contents of the program counter (PC). This is accomplished by the end of clock cycle T3. This machine cycle is extended by one additional clock cycle (T4), to give the CPU a chance to decode the opcode and the address mode fields of the instruction. At the end of T4, the CPU knows that the instruction includes one more word and also knows what type of operation is required.

2. In machine cycle 2, the CPU reads the address of the memory.

3. In machine cycle 3, knowing the address, the CPU fetches the operand from memory.

4. In machine cycle 4, the CPU adds 1 to the operand.

5. In machine cycle 5, the CPU stores the results back into memory.

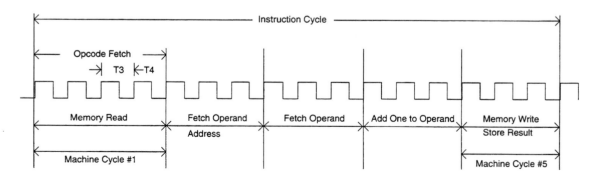

Figure 1.6

1.4   Number Systems

Each number system has a base (or radix) that defines a number of symbols used to represent numbers in that system. The value given to each digit is determined by its position in relation to a reference point called a *radix point*. For example, decimal numbers are base 10, and the radix point is referred to as the *decimal point*. The decimal digits are $0, 1, 2, \ldots, 9$. Binary numbers are base 2, and the radix point is referred as the *binary point*. The binary digits are 0 and 1. The hexadecimal (called hex) numbers have a base of 16; their digits are: $0, 1, \ldots, 9, A, \ldots, F$.

Signed and Unsigned Numbers

Unsigned Numbers

For m-bit unsigned integers the range is usually from 00 to $2^m - 1$. For example, in 8-bit unsigned integers, the range is from 0 to 255_{10} in decimal or from 00 to FF_{16} in hex. In 16-bit number systems, the range is from 0 to $65,535_{10}$ or from 00 to $FFFF_{16}$.

Signed Numbers

There are two representations of signed integers: *sign magnitude* (S&M) notation and *complement* notation. In both notations, the most significant bit (MSB) of a number represents the sign bit. If this bit is 0, the number is positive; if it is 1, the number is negative.

Sign-Magnitude Representation. In this representation, the positive version of a number differs from the negative only in the sign digit. The other digits indicate the magnitude of this number. For example, the number 7 in an 8-bit binary number is 0000 0111 and the number –7 is 1000 0111.

Complement Representation. The two most common complement systems are the *radix-complement* and the *diminished radix-complement* systems. The 1s and 9s complements are examples of diminished radix-complement systems for binary and decimal numbers respectively.

   The 1s complement of a number is accomplished by converting 0s to 1s and 1s to 0s. For instance, the 1s complement of 00111000 is 11000111.

   The two common radix-complement systems are the 2s complement and the 10s complement systems. Most microprocessors and microcontrollers have arithmetic instructions that operate on negative numbers represented in radix-complement form.

   The 2s complement of a binary signed integer is accomplished using two steps:

- Obtain the 1s complement.
- Add 1 to the 1s complement.

In this system, the sign bit (0 or 1) is a part of the number system. For example, the number 7 in an 8-bit binary number is 0000 0111 and the number –7 is

$$
\begin{array}{ll}
1111\ 1000 & \text{1s complement} \\
\underline{+1} & \\
1111\ 1001 & \text{2s complement}
\end{array}
$$

or F9 in hex.

Conversion of Positive Numbers from Binary to Decimal

The total decimal value of an integer binary number is the sum of the values designated by each bit, namely, $b_0 \times 2^0 + b_1 \times 2^1 + \ldots + b_n \times 2^n$.

EXAMPLE 2

Convert 1011 0110 to decimal form.

Solution

$$1011\ 0110 = 0 \times 2^0 + 1 \times 2^1 + 1 \times 2^2 + 0 \times 2^3 + 1 \times 2^4 + 1 \times 2^5 + 0 \times 2^6 + 1 \times 2^7 = 182\ \text{dec}$$

EXAMPLE 3

Convert the binary number 10110.110 to decimal form.

Solution

The integer part $= 0 \times 2^0 + 1 \times 2^1 + 1 \times 2^2 + 0 \times 2^3 + 1 \times 2^4 = 22$
The fraction part $= 1 \times 2^{-1} + 1 \times 2^{-2} + 0 \times 2^{-3} = 0.75$
Total value $= 22.75$ dec

Conversion of Negative Numbers from Binary to Decimal

The decimal value of an 8-bit negative number in 2s complement form is

$$-2^7 + b_6 \times 2^6 + b_5 \times 2^5 + \ldots + b_0 \times 2^0$$

EXAMPLE 4

Convert the 2s complement value 1100 1100 to decimal form.

Solution

Decimal value $= -2^7 + 1 \times 2^6 + 0 \times 2^5 + 0 \times 2^4 + 1 \times 2^3 + 1 \times 2^2$
$\qquad\qquad + 0 \times 2^1 + 0 \times 2^0 = -128 + 76 = -52$

Binary Scaling and Binary Point

Some processors do not contain floating-point hardware, nor does the cross-assembler simulate a floating-point data type. For values stored in RAM or ROM, which require the accuracy of a fractional number, users should use scaled arithmetic to retain the resolution of a fractional part. Scaling requires that the programmer maintain a record of the magnitudes of the variables throughout the problem.

Binary point (BP) is a value used by the MCU for binary scaling to convert unscaled values into engineering units. BP can be zero, positive, or negative. BP = 0 means that the BP location is at the right side of the binary number. Negative BP means that the BP location is shifted n times to the right (called *upscaling*). Positive BP means that this BP location is shifted n times to the left (called *downscaling*), where n is the absolute binary point number value.

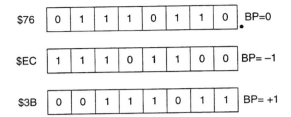

Figure 1.7

For BP = n, scaled value = unscaled value $\times 2^{-n}$ and for BP = $-n$, scaled value = unscaled value $\times 2^{+n}$.

EXAMPLE 5

An 8-bit hex number has a value of 76 in hex and the BP is 0. What is the scaled value if:

1. BP = 1.
2. BP = -1.

Solution

As you can see from Figure 1.7, for a BP = 0 of any number, the period must be to the right of this number. According to the BP rules mentioned above, for a negative BP, we move the BP one digit to the right. For a positive BP, we move the BP one digit to the left. Thus:

1. For BP = -1, scaled = $76 \times 2^1 = $ EC.
2. For BP = $+1$, scaled = $76 \times 2^{-1} = $ 3B.

EXAMPLE 6

The fraction binary 1011 1.110 has BP = 3. What is its value if

1. BP = 2.
2. BP = -3.

Solution

1. For BP = 2, number = 1011 11.10
2. For BP = -3, number = 101 1111 0000

Some Rules on Scalars

Addition and Subtraction

Added or subtracted scalars must have the same binary point. For example,

$$Y = X1 \times 2^{BP1} + X2 \times 2^{BP1}$$

and

$$Y = X1 \times 2^{BP2} - X2 \times 2^{BP2}$$

Multiplication and Division

The binary points are added in multiplication and subtracted in division. For instance,

$$Y = (X1 \times 2^{BP1}) \times (X2 \times 2^{BP2}) = X1X2 \times 2^{(BP1+BP2)}$$

and

$$Y = (X1 \times 2^{BP1}) / (X2 \times 2^{BP2}) = (X1/X2) \times 2^{(BP1-BP2)}$$

EXAMPLE 7

Calculate the multiplication and division of these numbers:

$$X1 = 9 \text{ hex with } BP = 4$$

$$X2 = 2 \text{ hex with } BP = 3$$

Solution

$$Y1 = 9 \times 2^4 \times 2 \times 2^3 = 12 \times 2^7 = 2304 \text{ decimal}$$
$$Y2 = (9 \times 2^4)/(2 \times 2^3) = 4.5 \times 2 = 9 \text{ decimal}$$

Exercises

1. What is the difference between sign-magnitude and complement systems for negative numbers?
2. Convert unsigned CFA1 to decimal notation.
3. Convert 1010.1010 to decimal notation.
4. Convert signed 8E to decimal notation.
5. Calculate the value in hex of the following terms:
   a. 10_{10}, $BP = 4$
   b. 48_{16}, $BP = -3$
   c. 11.010, $BP = 2$
   d. $(A, BP = -2) \times (4, BP = 3)$
   e. $(C, BP = 4)/(3, BP = -4)$

1.5 MC68HC11 Family

The M68HC11 family is an 8-bit microcontroller manufactured by Motorola. The HCMOS technology used combines smaller size and higher speed with the low power and high noise immunity of CMOS. The M68HC11 family is one of the fastest-growing, best-selling lines worldwide for embedded control applications. There are many versions of this product, and they are software compatible. The difference between these versions resides in their hardware features such as RAM, ROM, EPROM, EEPROM, and CONFIG register contents. The basic microcontroller is the MC68HC11A8. At the time of writing, there are about 16 versions of the 68HC11. Table 1.1 shows these versions and the differences between them.

The alphanumeric code for the 68HC11 family is shown in Figure 1.8.

Table 1.1 M68HC11 Family Members

Part Number	EPROM	ROM	EEPROM	RAM	CONFIG[2]	Comments
*MC68HC11A8	—		512	256	$0F	Family built around this device
MC68HC11A1	—	—	512	256	$0D	'A8 with ROM disabled
MC68HC11A0	—	—	—	256	$0C	'A8 with ROM and EEPROM disabled
MC68HC811A8	—	—	8K+512	256	$0F	EEPROM emulator for 'A8
MC68HC11E9	—	12K	512	512	$0F	Fourinput capture/bigger RAM 12K ROM
MC68HC11E1	—	—	512	512	$0D	'E9 with ROM disabled
MC68HC11E0	—	—	—	512	$0C	'E9 with ROM and EEPROM disabled
MC68HC811E2	—	—	2K[1]	256	$FF[3]	No ROM part for expanded systems
MC68HC711E9	12K	—	512	512	$0F	One-time programmable version of 'E9
MC68HC11D3	—	4K	—	192	N/A	Low-cost 40-pin version
MC68HC711E9	4K	—	—	192	N/A	One-time programmable version of 'D3
MC68HC11F1	—	—	512[1]	1K	$FF[3]	High-performance nonmultiplexed 68-pin
MC68HC11K4	—	24K	640	768	$FF	>1 Meg memory space, PWM, CS, 84-pin
MC68HC711K4	24K	—	640	768	$FF	One-time programmable version of 'K4
MC68HC11L6	—	16K	512	512	$0F	LIke 'E9 with more ROM and more I/O, 64/68
MC68HC711L6	16K	—	512	512	$0F	One-time programmable version of 'L4

*Basic microcontroller
Notes:
1. The EEPROM is relocatable to the top of any 4K memory page. Relocation is done with the upper four bits of the CONFIG register.
2. CONFIG register values in this table reflect the value programmed prior to shipment from Motorola (Internal MCU register).
3. At the time of this printing, a change was being considered that would make this value $0F.
(Copyright of Motorola. Used by permission.)

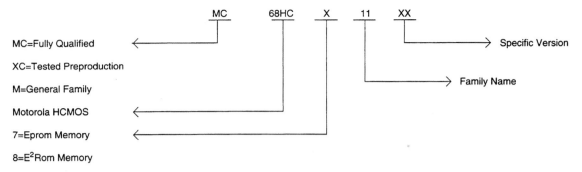

MC   68HC   X   11   XX

MC=Fully Qualified

XC=Tested Preproduction

M=General Family

Motorola HCMOS

7=Eprom Memory

8=E^2Rom Memory

→ Specific Version

→ Family Name

Figure 1.8

The MC68HC11E9 Microcontroller

Throughout this book, we focus on the 68HC11E9 version. As a rule, if you became familiar with one version of the 68HC11 family, you have no problem working with any other version because, as mentioned before, all versions are compatible. As shown in Table 1.1, the MCU 68HC11E9 includes 12K of ROM, 512 bytes of EEPROM, and 512 bytes of RAM.

Features

- MC68HC11 CPU
- Power-saving instructions such as Stop and Wait
- 12K bytes of on-chip EPROM
- 512 bytes of on-chip EEROM
- 512 bytes of on-chip RAM
- 16-bit timer system:
    - 4 or 5 output compare channels
    - 3 or 4 input capture channels
    - 8-bit pulse accumulator
    - Real-time interrupt feature
- COP watchdog feature
- Synchronous serial peripheral interface (SPI)
- Asynchronous non-return to zero (NRZ) serial communication interface (SCI)
- 8 channel 8-bit analog-to-digital (A/D) converter
- 38 General-purpose I/O lines
    - 16 bidirectional I/O lines
    - 11 input-only lines and 11 output-only lines
- Available in a 52-pin plastic leaded chip carrier (PLCC) and 52-pin ceramic cerquad

The block diagram of MC68HC11E9 is shown in Figure 1.9.

1.6   Programming Model of M68HC11

The programming model (sometimes called a software model) of the 68HC11 includes the registers and accumulators accessible to the programmer, as shown in Figure 1.10.

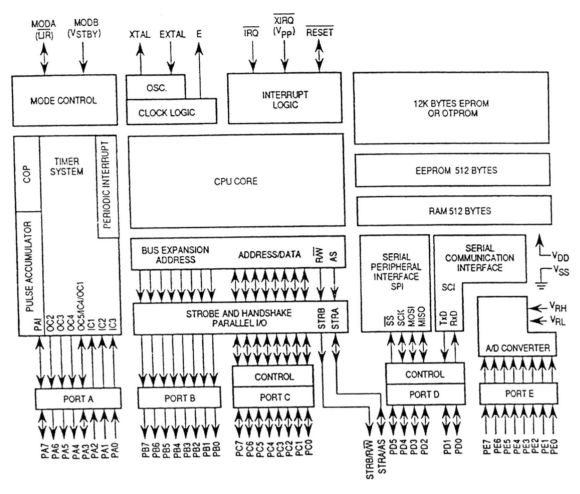

Figure 1.9
Copyright of Motorola. Used by permission.

Accumulators (A, B, and D)

Accumulators A and B are general-purpose 8-bit registers used to communicate with memory or I/O devices and hold the results of 8-bit arithmetic operations. Most operations can use accumulator A or B interchangeably. However, some instructions treat the combination of these two accumulators as a 16-bit double accumulator called accumulator D. Accumulator D is used for 16-bit arithmetic and data manipulation operations.

Index Registers (X and Y)

The 16-bit index registers X and Y are used for the indexed addressing mode, as we will see in the next section. They are also very useful in cases where X and Y can be used as pointers to data tables to manipulate these data.

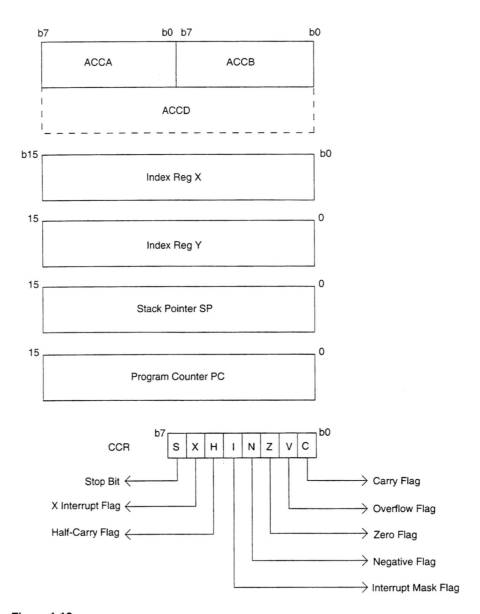

Figure 1.10

Stack Pointer (SP)

The stack is an area of RAM used for temporary storage of data. It may be located anywhere in the 64K address space and can be any size up to the amount of memory available in the system. The stack pointer is a 16-bit register that points to the next free location on the stack. When a subroutine is called, the address of the next instruction in the calling program is pushed onto the stack. The return from subroutine (RTS) instruction causes the saved address on the stack to be pulled off, and the execution of the main program continues.

Also, in case of an interrupt, a return from interrupt (RTI) instruction causes the saved registers to be pulled out of the stack and the execution of the main program continues. More details will be provided later.

Program Counter (PC)

The program counter is a 16-bit register that holds the address of the next instruction to be executed.

Condition Code Register (CCR)

The CCR contains 8 bits: 5 status bits (called flags), 2 interrupt mask bits, and a stop disable bit. The 5 status flags are as follows: (carry/borrow) C, overflow V, zero Z, negative N, and half-carry H. The two interrupt flags are I and X bits. The S bit is the stop bit.

Carry Flag (C)

The C flag is set (logic 1) if there is a carry from bit 7 (MSB) in 8-bit numbers or bit 15 (MSB) in 16-bit numbers in the case of addition or if there is a borrow from bit 7 in 8-bit numbers or bit 15 in 16-bit numbers in the case of subtraction. Figure 1.11 provides an example.

The subtraction is performed using the 2s complement of negative numbers. For example, if we would like to subtract $15 from $20, then the negative number (–15) is converted internally first to its 2s complement form, then added to the positive 20. The 2s complement of a number is the 1s complement plus 1. In Figure 1.12, $EB = 2s complement of (–$15), ignore the carry.

If a borrow is 1, it is complemented internally to 0 to indicate that carry flag = 0 and the result is positive. The sign bit in 8-bit numbers is the MSB bit. If this bit is zero, it means that the number is positive; and if it is 1, it means that the number is negative. Keeping this in mind, this result is positive. In this example, $20 – $15 = $0B = 11 dec, and the carry flag is 0. Now, let's subtract $20 from $15, as in Figure 1.13, where $E0 = 2s complement of (– $20).

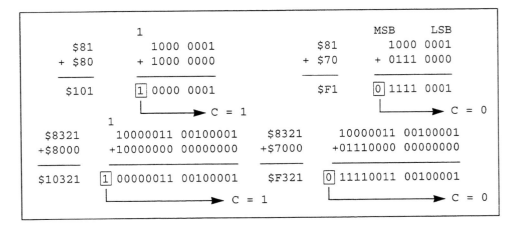

Figure 1.11

```
            -15 H       = 0 0 0 1  0 1 0 1

            1s comp  = 1 1 1 0  1 0 1 0
            Add 1      = 0 0 0 0  0 0 0 1

            2s comp  = 1 1 1 0  1 0 1 1
                       =      E       B

    $20      0010 0000                 $20       0010 0000
  + $EB    + 1110 1011    ◄────      - $15     - 0001 0101

    $0B    ⎡1⎤ 0000 1011
            |Complemented
            └───────► C = 0
```

Figure 1.12

```
            -20 H       = 0 0 1 0  0 0 0 0

            1s comp = 1 1 0 1  1 1 1 1
            Add 1      = 0 0 0 0  0 0 0 1

            2s comp = 1 1 1 0  0 0 0 0
                       =      E       0
    $15      0001 0101                 $15       0001 0101
  + $E0    + 1110 0000    ◄────      - $20     - 0010 0000

    $F5    ⎡0⎤ 1111 0101
            |Complemented
            └───────► C = 1
```

Figure 1.13

The carry C = 0 is then complemented to 1 internally to indicate that there is a borrow, since the sign bit is 1. The results are in 2s complement form and will be stored as such in the CPU. To know the actual results outside the computer, we have to get the 2s complement of this result manually.

The 2s complement of $F5 is $05. Since there is a borrow, the result of this subtraction is –05. The same technique is used for 16-bit subtraction. Note that the range of signed 8-bit numbers is –128 to +127 decimal and the range of signed 16-bit numbers is –32,768 to +32,767 decimal.

Overflow Flag (V)

The V flag is used to indicate *overflow* when numbers are added or subtracted. For example, if we add two 8-bit signed numbers and the addition results were outside the

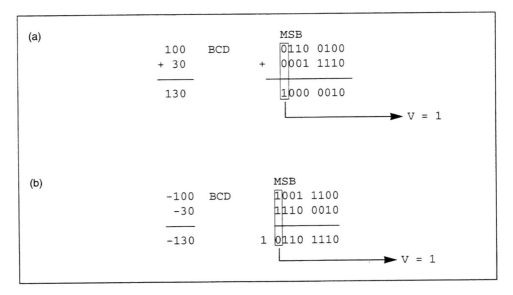

Figure 1.14

range −128 and 127, then the overflow flag is set. In the example shown, we added two signed numbers +100 BCD (binary coded decimal) and +30 BCD and the result is +130 BCD which is higher than the positive range +127. Thus V is set. (See Figure 1.14a.)

The computer will also look at the MSB of signed numbers (the sign bit) and will find that by adding two positive numbers (MSB = 0), the result should be positive. But in our case, the sign bit is 1. So the computer knows that the result is not negative but there is an overflow. The V flag is generated from a two-input exclusive OR gate inside the CPU. These inputs are the N and C flags. Generally overflow occurs in the following cases:

1. Positive plus positive equals negative.
2. Negative plus negative equals positive.
3. Positive minus negative equals negative.
4. Negative minus positive equals positive.

Figure 1.14b provides another example of adding two negative numbers; the addition is a positive number.

Zero Flag (Z)

The zero flag is set if the result of an arithmetic operation on 8-bit or 16-bit number is zero. For example, if we add $80 (hex) plus $80 we notice that the 8-bit result is all zeros. Thus Z = 1. Also notice that C = 1. (See Figure 1.15.)

Negative Flag (N)

In signed numbers, if the sign bit (MSB) of an arithmetic result is 1, this means that it is a negative, as mentioned before, and the N flag is set. If the MSB of this result is 0, it means that this number is positive and N = 0.

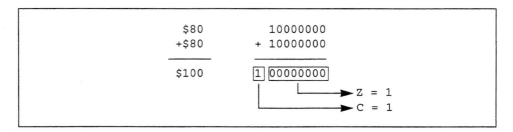

Figure 1.15

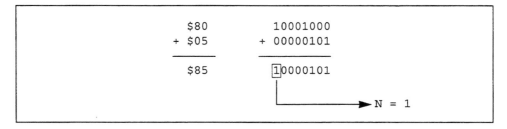

Figure 1.16

As an example, if we add the numbers $80 and $05, the result is $85 and its sign bit is 1. Thus N = 1. (See Figure 1.16.)

Half-Carry Flag (H)

The H flag indicates a carry from bit 3 to bit 4 during an arithmetic operation. This status indicator allows the CPU to adjust the result of an 8-bit BCD addition so it is in the correct BCD format. This H flag is updated by ABA, ADD, and ADC instructions. The H flag is used by the DAA instruction to compensate the result in accumulator A to correct the BCD format. For example, if we add $08 and $08, then a carry from b3 to b4 occurs. Thus H = 1. (See Figure 1.17.)

Interrupt Mask Flag (I)

The I flag is the interrupt request (IRQ) bit. The IRQ is a maskable interrupt—that is, it can be prevented by the programmer. If the programmer wishes to ignore an interrupt, I is set, but to recognize a certain interrupt, I is cleared. I is set or cleared by software instruction SEI or CLI respectively. As the CPU recognizes the interrupt and branches to the interrupt service routine (ISR), the I flag is automatically set to 1. More about interrupts is explained in Chapter 5.

Interrupt Mask Flag (X)

The X flag is the nonmaskable interrupt request (XIRQ). After reset, the X bit can be cleared to enable a nonmaskable interrupt using TAP instruction. The X bit is set only (after reset) by hardware and not by a software instruction. Thus, the XIRQ is a nonmaskable interrupt.

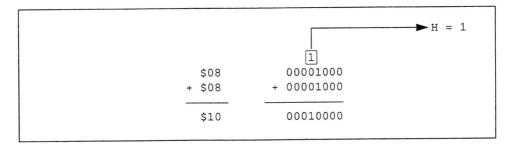

Figure 1.17

Stop Flag (S)

The stop disable bit (S) is used to allow or disallow the stop instruction. If the stop bit is set to 1 and the stop instruction is executed by the CPU, this instruction is ignored and is treated as a NOP (no operation) instruction. But if the stop bit is cleared to zero and stop is executed, all system clocks halt, and the system is placed in power standby condition.

Exercises

1. A crystal frequency of a certain MCU is 12 MHZ. If the machine cycle has four clock cycles, calculate the machine cycle time.

2. Which flag(s) is set when the following arithmetic operations are performed? (*Note:* Numbers are in Hex.)

a.	8000 + 8000	b.	FF + FF	c.	4C −6A

d.	34 + 57	e.	C001 + 3C40

3. If the contents of the CCR register are $90, indicate whether each of the following statements is true or false:
   a. The $\overline{\text{IRQ}}$ interrupt is ignored.
   b. The $\overline{\text{XIRQ}}$ interrupt is ignored.
   c. The stop instruction is allowed.

Instruction Format

A program is a sequence of instructions (commands) that perform a certain task. The instruction could be 1, 2, 3, or 4 bytes long, as shown in Figure 1.18.

Each instruction has an operation code (opcode) and operand(s) (data). The opcode is 1 or 2 bytes. The operand could be 0, 1, 2, or 3 bytes long.

Figure 1.18

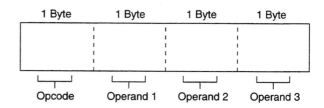

1.7   Addressing Modes

The addressing modes define the way an instruction is to obtain the data required for its execution. The power of any MCU lies in the capability of its addressing modes to access memory. For the 68HC11, there are six addressing modes:

1. Immediate
2. Direct
3. Extended
4. Indexed
5. Relative
6. Inherent

More examples on the instruction set are presented in Chapter 2.

Immediate Mode

Instructions that use this mode are 2, 3, or 4 bytes long. The pound symbol (#) is used to indicate an immediate addressing mode and is written just before the data. Other symbols that may precede the data and be recognized by the assembler are shown in Table 1.2.

EXAMPLE 8

The instruction LDAA #15 loads accumulator A with the decimal value 15. If we check the contents of ACCA, we find it is 0F in hex, which is equivalent to 15 in decimal. The microprocessors and microcontrollers deal with hex numbers only.

EXAMPLE 9

In the instruction LDAB #$15, ACCB is loaded with hex value 15. If we check the contents of ACCB, we find it is 15 hex. When you load or store values in the accumulator, registers, or memory, the old data are lost and replaced by the new data.

Table 1.2

Symbol	Type of Data
#	Decimal
$	Hex
@	Octal
%	Binary
,	ASCII Character

EXAMPLE 10
The instruction LDAA #%01011111 uses the binary format to show the user that specific bits have to be set or cleared. When ACCA is checked, we find its contents are 5F in hex.

Direct Mode

The range of this addressing mode is from $0000 to $00FF or 256 bytes. This range is on page zero. Most instructions that use this mode are 2 bytes long: one is the opcode and the other is the least significant byte of the effective address.

EXAMPLE 11
The instruction LDAA $60 means to load ACCA with contents of address $60 (30 hex). As shown in Figure 1.19, the direct addressing mode allows the programmer to use instructions that take one less byte of program memory space than the equivalent instruction using the extended addressing mode.

Extended Mode

Most instructions that use this mode are 3 bytes long. The first byte is the opcode, the second is the high address byte, and the third is the low address byte. The address range is from $0000 to $FFFF.

EXAMPLE 12
The instruction STAA $1234 stores the contents of ACCA (45 hex) into the address (location) $1234. (See Figure 1.20.)

Indexed Mode

This mode is powerful because it comes in very handy to access tables or blocks of data in memory. It uses the contents of either index register X or Y and adds it to an offset contained in the instruction to calculate the effective address.

Figure 1.19

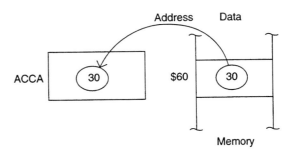

Figure 1.20

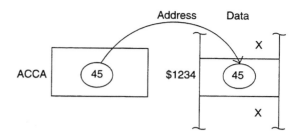

EXAMPLE 13

In the instruction LDAB $6,X, the number before the comma is the offset. If X contains $2000, the effective address = $2000 + $06 = $2006. Now the instruction loads the "contents" of the effective address $2006, which is "7E" in this case, to accumulator B. The offset is unsigned and could be 8-bit or 16-bit. (See Figure 1.21.)

The same technique is used for index register Y.

Relative Mode

This mode is used for branch instructions. The instruction consists of 2 bytes: the first byte is the opcode and the second is the relative address. The relative address byte is a "signed" number that is added to the present program counter (PC) to obtain the new address to be branched to. If the relative address is negative (MSB = 1), the branching is *backward*, but if it is positive (MSB = 0), the branching is *forward* as shown in Figure 1.22.

The address from where branching starts is called the *origination address,* and the address to where branching ends is called the *destination address.*

The branching range is between −128 and +127. The CPU calculates the relative address automatically. The relative address = destination − (origination + 2). More details on branching will be presented in the next chapter.

Inherent Mode

The instruction in this mode is 1 or 2 bytes only; it is an opcode with no operands. For example, CLRA is an opcode with no operands.

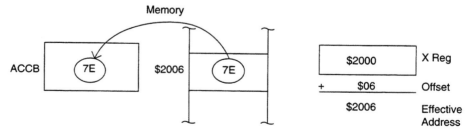

Figure 1.21

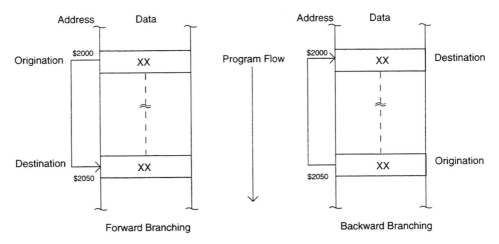

Figure 1.22

Exercise

Find the addressing mode of each line in the following program.

```
LDX   #$4000
LDY   #$5000
LDAA #$66
LDAB #100
STAA $1A
STAB $2250
CLRA
LDAA 0,X
LDAB 40,Y
```

1.8 Selecting the Right Microcontroller

Selecting the proper MCU for your application is one of the critical decisions determining the success or failure of your project. The main criteria in selecting an MCU are listed below, in order of importance:

1. Suitability for the application system
   a. Does it have the required number of I/O ports? If too few, it cannot do the job and if too many, the cost will be excessive.
   b. Does it have all the other required peripherals, such as serial ports, I/O, RAM, ROM, A/D, D/A, and so on?
   c. Does the CPU core have the correct computing power to handle the system requirements over the life of the system for the chosen implementation language? Too much is wasteful and too little will never work.
   d. Does the project budget allocate enough funds to permit using this MCU?

2. Availability
   a. Is the device available in sufficient quantities?

    b. Is the device in production today?

    c. What about the future?

3. Development support availability

    a. Assemblers

    b. Compilers and linkers

    c. Debugging tools such as evaluation modules, in-circuit emulators, logic analyzers, debug monitors, and so forth

    d. Online bulletin board service (BBS) such as real-time, executives, application examples, utility software, and so on

    e. Applications support

       1. Application engineers, technicians, or marketers

       2. Telephone/fax support

4. Manufacturer's history

    a. Demonstrated competence in design

    b. On-time delivery performance

    c. How long in business?

    d. Financial report

2 The Instruction Set

The power of an instruction set is judged on the basis of the number of instructions needed to perform a specific complex task. Fewer instructions generally mean less memory space for the program and shorter task execution times. Hence, the instruction set is considered more powerful.

Instruction sets may differ not only in the type of operations they specify, but also with respect to the types of data they operate on. For example:

- In 8-bit MCUs, instructions operate on bytes and sometimes on 16-bit words.
- In 16-bit MCUs, instructions operate on bytes, 16-bit words, and sometimes 32-bit words.
- In 32-bit MCUs, instructions operate on bytes, 16-bit words, 32-bit words, and sometimes 64-bit words.

Instructions can be divided into four major categories:

1. Data transfer (movement) instructions

2. Data manipulation instructions
   a. Arithmetic
   b. Shift and rotate
   c. Logical
   d. Bit manipulation

3. Control instructions
   a. Program control
   b. CPU control

4. Miscellaneous instructions

2.1   Data Transfer (Movement) Instructions

These instructions involve transferring data from memory or input/output (I/O) devices to the CPU's internal registers and vice versa.

Table 2.1 gives the "mnemonic" and the types of addressing modes of the data movement instructions. Some examples will show how these instructions are executed.

Table 2.1

Function	Mnemonic	IMM	DIR	EXT	INDX	INDY	INH
Clear Memory Byte	CLR			X	X	X	
Clear Accumulator A	CLRA						X
Clear Accumulator B	CLRB						X
Load Accumulator A	LDAA	X	X	X	X	X	
Load Accumulator B	LDAB	X	X	X	X	X	
Load Double Accumulator D	LDD	X	X	X	X	X	
Pull A from Stack	PULA						X
Pull B from Stack	PULB						X
Push A onto Stack	PSHA						X
Push B onto Stack	PSHB						X
Store Accumulator A	STAA		X	X	X	X	
Store Accumulator B	STAB		X	X	X	X	
Store Double Accumulator D	STD		X	X	X	X	
Transfer A to B	TAB						X
Transfer A to CCR	TAP						X
Transfer B to A	TBA						X
Transfer CCR to A	TPA						X
Exchange D with X	XGDX						X
Exchange D with Y	XGDY						X

Copyright of Motorola. Used by permission.

EXAMPLE 1   LOAD, STORE Instructions

In the following program, what are the contents of the accumulators and memory after execution?

```
LDAA $0000
LDAB $0001
STD  $0005
```

Solution

The first instruction is an extended addressing mode. It states: load ACCA with contents of address $0000. The contents of address $0000 (91 hex) copied to ACCA and the address $0000 still keeps its contents, 91 hex. The old contents of ACCA are lost and replaced by the new data. This operation is similar to recording a song on a cassette from the original cassette. The original cassette is not affected by the copying process, though any previously recorded material on the second cassette is lost and is replaced by the new song. The second instruction does the same thing with ACCB.

The third instruction states: store contents of ACCD (combination of ACCA and ACCB) into the address $0005. Since ACCD is 2 bytes long and location $0005 is 1 byte long, the high byte of ACCD (ACCA) is stored into location $0005 and the low byte of ACCD (ACCB) is stored into location $0006 as shown in Figure 2.1.

Now the contents of address $0005 have become 91; AA is in $0006.

EXAMPLE 2   PUSH, PULL Instructions

Describe the operation performed by the following instruction:

```
LDS  #$FF
PSHA
PULB
```

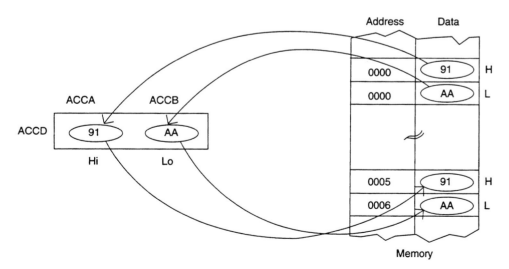

Figure 2.1

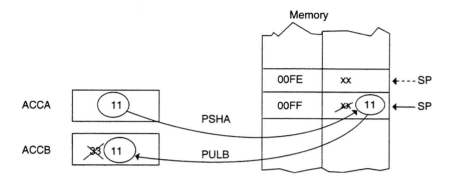

Figure 2.2

Solution

The stack pointer has the contents of $00FF—that is, it points to the address $00FF. The rules of push and pull are as follows:

1. PSHA (or PSHB): It stores the contents of ACCA or ACCB into the stack location indicated by the stack pointer (SP), then decrements SP by 1.

2. PULA (or PULB): First the SP is incremented by 1 and then ACCA or B is loaded with the contents of the stack location pointed to by SP.

   To implement these rules in our three-instruction program, the contents of ACCA are stored to location $00FF, then the contents of SP are decremented to $00FE. SP now points to the address $00FE. In the second instruction, PULB, SP is first incremented by 1 to gain the address $00FF again, then the contents of this location (11 hex) are stored into ACCB, as shown in Figure 2.2.

EXAMPLE 3   EXCHANGE Instructions

Describe the operation performed by the following instruction:

XGDX

XGDY

Solution

Assume X = $4630, Y = $E1F0, and ACCD = $00AB. These two instructions are very powerful. The first instruction means exchange ACCD with the X register, or in other words, swap D and X. ACCD now has $4630, and index X has $00AB. The second instruction means exchange ACCD with the Y index register. Now ACCD and register Y are swapped. The new contents of ACCD are $E1F0 and of the Y register are $4630.

2.2   Data Manipulation Instructions

Arithmetic Instructions

Table 2.2 summarizes the entire list of arithmetic operations and addressing modes supported by each instruction.

Table 2.2

Function	Mnemonic	IMM	DIR	EXT	INDX	INDY	INH
Add Accumulators	ABA						X
Add Accumulator B to X	ABX						X
Add Accumulator B to Y	ABY						X
Add with Carry to A	ADCA	X	X	X	X	X	
Add with Carry to B	ADCB	X	X	X	X	X	
Add Memory to A	ADDA	X	X	X	X	X	
Add Memory to B	ADDB	X	X	X	X	X	
Add Memory to D (16-Bit)	ADDD	X	X	X	X	X	
Compare A to B	CBA						X
Compare A to Memory	CMPA	X	X	X	X	X	
Compare B to Memory	CMPB	X	X	X	X	X	
Compare D to Memory (16-Bit)	CPD	X	X	X	X	X	
Decimal Adjust A (for BCD)	DAA						X
Decrement Memory Byte	DEC			X	X	X	
Decrement Accumulator A	DECA						X
Decrement Accumulator B	DECB						X
Increment Memory Byte	INC			X	X	X	
Increment Accumulator A	INCA						X
Increment Accumulator B	INCB						X
Two's Complement Memory Byte	NEG			X	X	X	
Two's Complement Accumulator A	NEGA						X
Two's Complement Accumulator B	NEGB						X
Subtract B from A	SBA						X
Subtract with Carry from A	SBCA	X	X	X	X	X	
Subtract with Carry from B	SBCB	X	X	X	X	X	
Subtract Memory from A	SUBA	X	X	X	X	X	
Subtract Memory from B	SUBB	X	X	X	X	X	
Subtract Memory from D (16-Bit)	SUBD	X	X	X	X	X	
Test for Xero or Minus	TST			X	X	X	
Test for Zero or Minus A	TSTA						X
Test for Zero or Minus B	TSTB						X

EXAMPLE 4   ADD, NEGATE Instructions

Describe the operation performed by the following program:

```
LDAA   $0000  ......1
LDAB   $0001  ......2
ABA           ......3
ADDA   #05    ......4
NEGB          ......5
STD    $0005  ......6
```

where

```
0000   48 21 XX XX XX ...
```

Solution

The first and second instructions have already been explained. The third instruction adds the contents of ACCB to the contents of ACCA and stores the result in ACCA. So ACCA now has 69 hex. The fourth instruction adds 05 hex to 69 and stores the total in ACCA. ACCA has 6E hex. The fifth instruction, "negate," is to perform 2s complement of the contents of ACCB. Since ACCB originally has 21 hex, its 2s complement is DE hex. The sixth instruction is to store the contents of ACCD to locations $0005 and $0006. The high byte of ACCD (ACCA) is stored to $0005 and the low byte (ACCB) is stored to $0006.

EXAMPLE 5   ADDD Instruction

The above addition program is done using 8-bit addition since we used ACCA, ACCB. The 16-bit addition can also be done using the 16-bit ACCD. Consider the following instructions:

```
LDD   $0020
ADDD  $0022
STD   $0024
```

```
0020    70 10 20 30 FF A9 XX XX
```

Solution

The first instruction loads ACCD with the contents of locations $0020 and $0021. The second instruction adds the contents of locations $0022 and $0023 to ACCD and the result is stored in ACCD. Notice that with a proper carryover of bit 7 if it occurs, the high byte of memory is added to the high byte of ACCD and the low byte of memory is added to the low byte of ACCD. The third instruction stores the new contents of ACCD to locations $0024 (90 hex) and $0025 (40 hex). The subtraction instruction uses the same techniques as the addition instruction.

EXAMPLE 6   COMPARE Instructions

This example illustrates the compare instructions that are used to compare given data with known data. Compare instructions perform subtraction but do not alter any data. The Condition Code Register flags affected by these instructions are the N, Z, V, and C flags (the same as in the subtraction process). The result of a comparison tells the CPU either to continue executing the program in sequence or to branch (jump from a location to another, skipping some instructions in between). The branch instructions include BEQ (branch if equal zero) and BNE (branch if not equal zero). Consider the following program and explain each instruction:

```
LDAA $0000
LDAB $0001
CBA
```

```
0000    A4 A1 XX XX
```

Solution

The first instruction loads ACCA with A4. The second loads ACCB with A1. The third compares the contents of ACCA to the contents of ACCB—that is, ACCA – ACCB—but none of the accumulators are changed. There is usually a branch instruction after this comparison, as we will see. The result here is positive.

EXAMPLE 7   TEST Instructions
The test instructions allow the CPU to test for zero or minus. The instruction subtracts 00 from the contents of accumulators A, B, or memory and sets the CCR flags N and Z accordingly. The subtraction is accomplished internally without modifying either ACCA, ACCB, or memory.

```
LDAA $0010
TSTA
TST  $0011
TST  $0012
```

```
0010    81 00 79 XX .. ..
```

Solution
The second line tests ACCA for 00 or minus. As you can see, the contents of ACCA (81 hex) is a negative number. Thus the N flag is equal to 1. Usually there is a branch instruction after each test instruction. The third line tests the contents of $0011, which is 00 for zero or minus. Again it modifies the N and Z flags accordingly, and in this case $Z = 1$, $N = 0$. The fourth line tests the contents of $0012, which is 79 hex, for zero or minus. In this case, $N = 0$ and $Z = 0$.

DAA Instruction

Decimal Adjust for Addition (DAA) is used to add BCD numbers and adjust the result in BCD format. The instruction adds 06 or 60 to the result of addition in the following cases:

1. If $H = 1 \rightarrow$ add 06 to the low nibble
2. If $C = 1 \rightarrow$ add 60 to the high nibble
3. If the BCD number $> 9 \rightarrow$ add 06 either to the low or high nibble as explained later

EXAMPLE 8
First Case: Add the numbers 38 and 49 in BCD form and see how the DAA instruction adjusts the results.

```
LDAA #$38
ADDA #$49
DAA
```

```
                              1
           38 BCD        0011 1000
          + 49         + 0100 1001
Correct → 87 BCD         1000 0001
                         8    1 ◄── Incorrect
```

Solution

By adding the decimal numbers, the result is (87), while in adding the BCD equivalent, the result is (81). Here H = 1, and DAA must add (06) to the result, as follows:

$$
\begin{array}{r}
\text{Result} \longrightarrow \quad 1000\ 0001 \\
+ \quad 06 \qquad +\ 0000\ 0110 \\
\hline
1000\ 0111 \\
\end{array}
$$

$$\qquad\qquad 8 \qquad 7 \longleftarrow \text{Correct}$$

The correct answer is 87.

Second Case: Now let us add (87) and (81) in BCD form and find out how DAA adjusts the addition:

```
LDAA #$87
ADDA #$81
DAA
```

$$
\begin{array}{rr}
 & 1 \\
87\ \text{BCD} & 1000\ 0111 \\
+\ 81 & +\ 1000\ 0001 \\
\hline
\text{Correct} \longrightarrow 168\ \text{BCD} & 1\ 0000\ 1000 \\
\end{array}
$$

$$\qquad\qquad 1 \quad 0 \quad 8 \longleftarrow \text{Incorrect}$$

By adding these two numbers, we found that the result is not correct and C = 1. If C = 1, DAA adds 60 to the results as shown; and it is adjusted to the right answer.

$$
\begin{array}{r}
\text{Result} \longrightarrow \quad 1\ 0000\ 1000 \\
+\ 60 \qquad +\ 0110\ 0000 \\
\hline
1\ 0110\ 1000 \\
\end{array}
$$

$$\qquad\qquad 1 \quad 6 \quad 8 \longleftarrow \text{Correct}$$

Third Case: If we would like to add 87,08 in BCD form and see how the DAA adjusts the addition:

```
LDAA #$87
ADDA #$08
DAA
```

$$
\begin{array}{rr}
87\ \text{BCD} & 1000\ 0111 \\
+\ 08 & +\ 0000\ 1000 \\
\hline
\text{Correct} \qquad 95\ \text{BCD} & 1000\ 1111 \\
\end{array}
$$

$$\qquad\qquad\qquad\qquad\qquad \longrightarrow \text{Not BCD}$$

The result is not in BCD form, because the low nibble is greater than 9. In this case, we have to add 06 to the results as shown:

```
Result   ────►   1000 1111
+  06          + 0000 0110
               ──────────
                 1001 0101

                   9    5  ◄── Correct
```

The correct result is 95.

Using the same technique, if the number in the higher nibble is greater than 9, then the DAA will add (60) to adjust the results.

Multiply and Divide Instructions

The M68HC11 has one multiply and two divide instructions. The multiply instruction (MUL), multiplies two unsigned 8-bit values and produces an unsigned 16-bit value. If we need to multiply 16-bit by 16-bit values or more, we have to write our own routine to accomplish that. The integer divide (IDIV) instruction performs a 16-bit by 16-bit divide, producing a 16-bit result and a 16-bit remainder. The fractional divide (FDIV) divides a 16-bit numerator by a larger 16-bit denominator, producing a 16-bit result and a 16-bit remainder. If we want to perform division using more than 16-bit values, we have to write our own routine to accomplish that. See Chapter 4 for multiplication and division routines.

EXAMPLE 9 MUL Instruction

Figure 2.3 shows how multiplication of two unsigned 8-bit numbers is accomplished in the following program:

```
LDAA #$4C
LDAB #$3A
MUL
```

The third instruction, MUL, will multiply the contents of ACCA by contents of ACCB and store the result (16-bit) in ACCD, as shown in Figure 2.4.

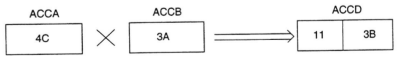

Figure 2.3

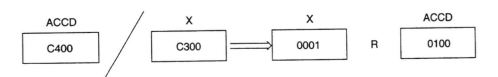

Figure 2.4

EXAMPLE 10 IDIV Instruction

Explain how this division operation works:

```
LDD #$C400
LDX #$C300
IDIV
```

D/X = XRD where R = remainder.

Solution

The IDIV instruction divides the contents of ACCD by the contents of index register X. The quotient is placed in index register X and the remainder is placed in ACCD.

In our program, we divide $C400 by $C300. The quotient (0001) is placed in index register X and the remainder (0100) in ACCD. What if we let the numerator be less than the denominator, as in the following program?

```
LDD #$C300
LDX #$C400
IDIV
```

We find that the quotient of 0000 is placed in index register X and the remainder of $C300 in ACCD.

EXAMPLE 11 FDIV Instruction

Using the fractional divide (FDIV), the quotient is a binary weighted fraction between 0 and 0.999998 that is stored in X. A 16-bit remainder is stored in ACCD. The same rule as IDIV applies:

D/X = XRD

```
LDD #05
LDX #07
FDIV
```

Solution

The quotient in X is B6BD and the remainder in ACCD is 3 because the quotient is a binary weighted fraction that, after being converted to decimal form, should be divided by 2^{16} or 65,536. So the result (quotient) in X is 46811 (B6BD hex)/65536 = 0.714279.

If we divide 5 by 7 using a calculator, the result is 0.714285. The difference (.000006) is expressed as a remainder of 3 in ACCD. Why 3? The difference is usually divided by 0.000002 and stored in ACCD as an integer remainder.

Using the FDIV, let's make the numerator larger than the denominator. Check the contents of X and D.

```
LDD #07
LDX #05
FDIV
```

If the numerator is larger than the denominator, this is an overflow. Index register X always has $ FFFF, and ACCD has a remainder of 07.

Shift and Rotate Instructions

Table 2.3 summarizes all shift and rotate instructions. Notice that all these instructions involve the C flag bit. Also, the ASL and LSL instructions have the same performance. The ASR instruction always keeps the MSB bit the same and shifts the other bits to the right. This is useful in dealing with signed numbers, when we should keep the sign bit (MSB) the same. Now let's see how these instructions operate, as shown in Figure 2.5.

EXAMPLE 12 LSLA Instruction

```
LDAA #$46
LSLA
LSRA
LSRA
```

The second instruction shifts ACCA logically one bit to the left and places 0 in LSB. The MSB bit is placed in carry flag C. (See Figure 2.6.) What are the contents of ACCA now? The answer is $8C. What is the relation between $8C and $46? The first number is twice the second. So by shifting any contents (values) to the left n number of shifts, the contents are multiplied by 2^n. In this case, since we shifted the contents of ACCA once to the left, $n = 1$. The contents are multiplied by $2^1 = 2$—that is, $46 \times 2 = \$8C$. If we shift the contents twice to the left, we multiply this number by 4, and so on.

Table 2.3

Function	Mnemonic	IMM	DIR	EXT	INDX	INDY	INH
Arithmetic Shift Left Memory	ASL			X	X	X	
Arithmetic Shift Left A	ASLA						X
Arithmetic Shift Left B	ASLB						X
Arithmetic Shift Left Double	ASLD						X
Arithmetic Shift Right Memory	ASR			X	X	X	
Arithmetic Shift Right A	ASRA						X
Arithmetic Shift Right B	ASRB						X
(Logical Shift Left Memory)	(LSL)			X	X	X	
(Logical Shift Left A)	(LSLA)						X
(Logical Shift Left B)	(LSLB)						X
(Logical Shift Left Double)	(LSLD)						X
Logical Shift Right Memory	LSR			X	X	X	
Logical Shift Right A	LSRA						X
Logical Shift Right B	LSRB						X
Logical Shift Right D	LSRD						X
Rotate Left Memory	ROL			X	X	X	
Rotate Left A	ROLA						X
Rotate Left B	ROLB						X
Rotate Right Memory	ROR			X	X	X	
Rotate Right A	RORA						X
Rotate Right B	RORB						X

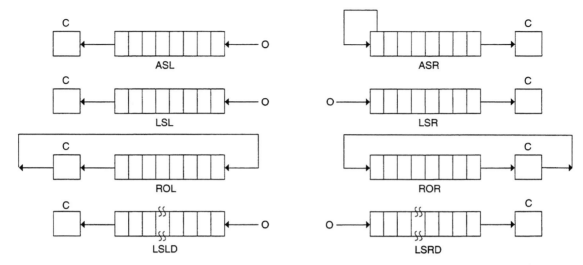

Figure 2.5

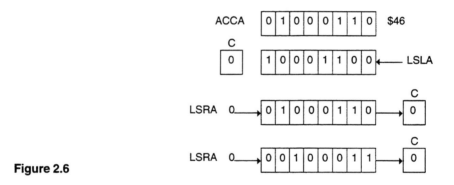

Figure 2.6

The third instruction shifts ACCA to the right one bit logically. Now the contents have returned to the original value of $46. By shifting the value to the right logically, we divide the contents by 2^n where n = number of shifts. The fourth instruction shifts these contents one more time to the right to become $23, which is $46/2.

In the real world, constants are "scaled up" and "scaled down" to make arithmetic operations easier. Scaling up means multiplying a constant by 2^n. Scaling down means dividing a constant by 2^n.

EXAMPLE 13 ASR Instruction

Assume we wish to shift a signed number $81 to the right. With a signed number, we want to keep the sign bit the same, as Figure 2.7 shows. In this case the number is negative and its sign bit is 1.

```
LDAB #$82
ASRB
```

The second instruction shifts $82 to the right and still keeps its sign bit as a negative number.

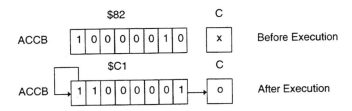

Figure 2.7

Logical Operation Instructions

Table 2.4 summarizes the logical operations of each instruction. The affected flags are N and Z. This group of instructions is used to perform the boolean logical AND, OR, exclusive OR (EOR), and 1s complement. The AND instruction is used to selectively "clear" specific bits of a data word while not affecting the other bits. This is called a *mask operation*. The masking technique is also often used to isolate the value of a single bit of a data word. The OR instruction can be used to selectively "set" a specific bit of a data word while not affecting the other bits. The EOR instruction is often used to selectively "toggle" specific bits of a data word while not affecting the other bits.

EXAMPLE 14 AND, OR, EOR Instructions

Describe the logical operations performed by the following program:

```
LDAA #$74
ANDA #$0F
ORAA #$32
EORA #$B3
```

The second instruction performs the logical AND between each bit of ACCA and its corresponding bit of memory contents, as shown in Figure 2.8a. Here we multiply each bit of ACCA by its corresponding bit of memory M. The result is stored back to ACCA.

Table 2.4

Function	Mnemonic	IMM	DIR	EXT	INDX	INDY	INH
AND A with Memory	ANDA	X	X	X	X	X	
AND B with Memory	ANDB	X	X	X	X	X	
Bit(s) Test A with Memory	BITA	X	X	X	X	X	
Bit(s) Test B with Memory	BITB	X	X	X	X	X	
One's Complement Memory Byte	COM			X	X	X	
One's Complement A	COMA						X
One's Complement B	COMB						X
OR A with Memory (Exclusive)	EORA	X	X	X	X	X	
OR B with Memory (Exclusive)	EORB	X	X	X	X	X	
OR A with Memory (Inclusive)	ORAA	X	X	X	X	X	
OR B with Memory (Inclusive)	ORAB	X	X	X	X	X	

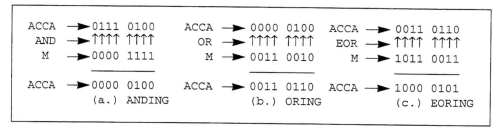

Figure 2.8

The third instruction performs the logical OR between each bit of ACCA and the corresponding bit of memory. The result is stored back to ACCA.

The fourth instruction performs the logical EOR between each bit of ACCA and its corresponding bit M. The result is stored back to ACCA.

Data Testing and Bit Manipulation Instructions

Table 2.5 lists these instructions. The BIT test instructions perform the AND operation, as explained earlier, without changing the data. The affected flags are N and Z. The clear bit(s) instruction BCLR and the set bit(s) instruction BSET read the operand, manipulate selected bits within the operand, and write the result back to the operand address. The bit(s) to be cleared or set must be specified by logic 1 in the mask byte. All other bits at the memory location are not affected.

EXAMPLE 15 BCLR Instruction

To clear bit 6 of location $0070, the following instruction does the job:

BCLR $0070,$40

address–| ◄──┘   └──► |→mask byte

The format of BCLR and BSET is that the opcode is followed by the location to be affected, followed by a comma and the mask byte. As mentioned earlier, to clear bit 6, it

Table 2.5

Function	Mnemonic	IMM	DIR	EXT	INDX	INDY
Bit(s) Test A with Memory	BITA	X	X	X	X	X
Bit(s) Test B with Memory	BITB	X	X	X	X	X
Clear Bit(s) in Memory	BCLR		X		X	X
Set Bit(s) in Memory	BSET		X		X	X
Branch if Bit(s) Clear	BRCLR		X		X	X
Branch if Bit(s) Set	BRSET		X		X	X

has to be set to 1 while the other values in the mask byte are 0. Thus, the mask byte is 01000000 or $40. The same is valid for BSET instruction.

2.3 Control Instructions

Control instructions are divided into program control instructions and CPU control instructions.

Program Control Instructions

Branch Instructions

Conditional Branch Instructions. Table 2.6 lists all the branch instructions. The branch range is from −128 to 127. For every branch condition, there is a branch for the opposite condition. The CCR flags affected are C, N, Z, and V. These branches are called *conditional branches* because their branching depends on the status of the CCR flags mentioned. As noted in Chapter 1, branching could be forward or backward.

EXAMPLE 16   BLS,.. Instructions
Describe the following branch instruction performance:

```
LDAA #$65
CMPA $C060
```

Table 2.6

Function	Mnemonic	REL	DIR	INDX	INDY	Comments
Branch if Carry Clear	BCC	X				C = 0?
Branch if Carry Set	BCS	X				C = 1?
Branch if Equal Zero	BEQ	X				Z = 1?
Branch if Greater Than or Equal	BGE	X				Signed ≥
Branch if Greater Than	BGT	X				Signed >
Branch if Higher	BHI	X				Unsigned >
Branch if Higher or Same (same as BCC)	BHS	X				Unsigned ≥
Branch if Less Than or Equal	BLE	X				Signed ≤
Branch if Lower (same as BCS)	BLO	X				Unsigned <
Branch if Lower or Same	BLS	X				Unsigned ≤
Branch if Less Than	BLT	X				Signed <
Branch if Minus	BMI	X				N = 1?
Branch if Not Equal	BNE	X				Z = 0?
Branch if Plus	BPL	X				N = 0?
Branch if Bit(s) Clear in Memory Byte	BRCLR		X	X	X	Bit Manipulation
Branch Never	BRN	X				3-cycle NOP
Branch if Bit(s) Set in Memory Byte	BRSET		X	X	X	Bit Manipulation
Branch if Overflow Clear	BVC	X				V = 0?
Branch if Overflow Set	BVS	X				V = 1?

```
                              BLS   NEXT
                              ADDA  #$40
                               |
                               |
NEXT                          LDAB  #$88


C060    90 XX XX
```

Solution

BLS and some other branch instructions are used when comparing unsigned numbers. See Table 2.6. The contents of ACCA are compared to the contents of location $C060, which is 90 hex. By now, you are familiar with the COMPARE instruction. If the contents of the reference (ACCA) is less than or the same as the second data value ($C060 contents), the branch to a label called "NEXT" will occur. Otherwise this branch does not occur and the following instruction (BLS NEXT) is executed. In our case, the contents of ACCA ($65) is less than $90 and the branch occurs. If the contents of ACCA are changed to $90, does the branch occur? Yes, because both values are the same. If ACCA contents are increased to $95, the branch does not occur and instruction "ADDA" will be executed.

EXAMPLE 17   BLE,... Instructions

If we are dealing with signed numbers, we use branch instructions such as BGT, BGE, BLT, and BLE. For instance, the previous example would have to be modified as follows:

```
                              LDAA  #$65
                              CMPA  $C060
                              BLE   NEXT
                              ADDA  xx
                               |
                               |
NEXT                          LDAB  xx
                               |
```

Solution

The contents of location $C060 are $90, which is a negative number. Further, 90 hex (signed) = –70 H, since 65 H > –70 H; thus the branch never happens.

    If the branch range is higher than –128/+127, a "jump" instruction is used.

Unconditional Branch and JMP Instructions. These two instructions are shown in Table 2.7. BRA and JMP do not depend on the status of CCR flags. BRA is similar to the JMP instruction except that its range is limited to the range –128 to +127. The jump

Table 2.7

Function	Mnemonic	DIR	EXT	INDX	INDY	INH	REL
Jump	JMP	X	X	X	X		
Branch Always	BRA						X

instruction is the same as BRA because it is an unconditional jump, but the jump range is different. The relative address of the jump instruction is 2 bytes long and its range is from –32,768 to +32,767 or 64K.

2.4   Subroutine Call Instructions

A subroutine is a subprogram called by the main program a number of times to do a certain task. The subroutine is usually called by either BSR or JSR in the main program. BSR is used if the range is –128 to +127, and JSR is used beyond that. Its range is –32,768 to +32,767.

As shown in Figure 2.9, at BSR, execution is transferred from the main program to the subroutine. The subroutine must end with the Return from Subroutine (RTS) instruction. When the program executes RTS, it returns to the instruction following the location where it called the subroutine (next instruction), then it resumes its execution. How does the CPU know the return address? When the subroutine is called, the CPU automatically pushes the return address onto the stack. When RTS is executed, this return address is pulled off the stack and loaded to the program counter (PC).

Also, a subroutine can call another subroutine and the latter can call a third one and so on. This is called *nested subroutines*. When a subroutine is called, the contents of ACCA, ACCB, and the CCR flags might change. So these contents may be saved onto the stack before calling any subroutine. When the program returns from the subroutine, these values may be retrieved from the stack and used in the main program without losing the original values of these registers. Table 2.8 includes the subroutine calls and return instructions.

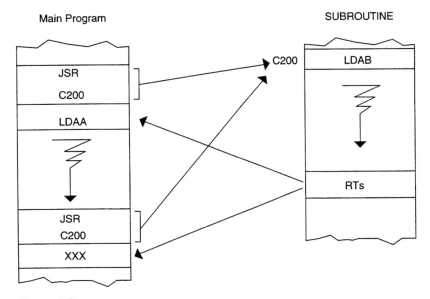

Figure 2.9

Table 2.8

Function	Mnemonic	REL	DIR	EXT	INDX	INDY	INH
Branch to Subroutine	BSR	X					
Jump to Subroutine	JSR		X	X	X	X	
Return from Subroutine	RTS						X

Copyright of Motorola. Used by permission.

CPU Control Instructions

Condition Code Register Instructions

Table 2.9 summarizes the CCR instructions. These instructions allow the programmer to set/clear carry C, interrupt I, and overflow V flags. The TPA instruction transfers (copies) the contents of CCR to ACCA. This is usually done if we want to save the contents of CCR before executing a subroutine. Also, TAP does the opposite and is used to retrieve the contents of CCR after executing the subroutine.

Stack Pointer and Index Register Instructions

Table 2.10 summarizes these instructions and their addressing modes. Most of these instructions are similar to another group of instructions that have already been explained. For example, ABX or ABY adds an 8-bit element to a 16-bit element, and the result is a 16-bit element.

Interrupt Handling Instructions

Table 2.11 lists the three instructions for this group.

We explain interrupts in detail in Chapter 5. Here we will briefly explain what happens when these instructions are executed. The software interrupt (SWI) instruction is similar to the JSR instruction except that the contents of all working registers are saved on the stack rather than just the return address.

Table 2.9

Function	Mnemonic	INH
Clear Carry Bit	CLC	X
Clear Interrupt Mask Bit	CLI	X
Clear Overflow Bit	CLV	X
Set Carry Bit	SEC	X
Set Interrupt Mask Bit	SEI	X
Set Overflow Bit	SEV	X
Transfer A to CCR	TAP	X
Transfer CCR to A	TPA	X

Copyright of Motorola. Used by permission.

Table 2.10

Function	Mnemonic	IMM	DIR	EXT	INDX	INDY	INH
Add Accumulator B to X	ABX						X
Add Accumulator B to Y	ABY						X
Compare X to Memory (16-Bit)	CPX	X	X	X	X	X	
Compare Y to Memory (16-Bit)	CPY	X	X	X	X	X	
Decrement Stack Pointer	DES						X
Decrement Index Register X	DEX						X
Decrement Index Register Y	DEY						X
Increment Stack Pointer	INS						X
Increment Index Register X	INX						X
Increment Index Register Y	INY						X
Load Index Register X	LDX	X	X	X	X	X	
Load Index Register Y	LDY	X	X	X	X	X	
Load Stack Pointer	LDS	X	X	X	X	X	
Pull X from Stack	PULX						X
Pull Y from Stack	PULY						X
Push X onto Stack	PSHX						X
Push Y onto Stack	PSHY						X
Store Index Register X	STX		X	X	X	X	
Store Index Register Y	STY		X	X	X	X	
Store Stack Pointer	STS		X	X	X	X	
Transfer SP to X	TSX						X
Transfer SP to Y	TSY						X
Transfer X to SP	TXS						X
Transfer Y to SP	TYX						X
Exchange D with X	XGDX						X
Exchange D with Y	XGDY						X

Table 2.11

Function	Mnemonic	INH
Return from Interrupt	RTI	X
Software Interrupt	SWI	X
Wait for Interrupt	WAI	X

Figure 2.10a shows that when SWI is executed, the following registers are pushed onto the stack in this order: PCL, PCH, YL, YH, XL, XH, ACCA, ACCB and CCR. Then the CPU executes what we call an Interrupt Service Routine (ISR), which is analogous to a subroutine. This ISR must end with Return from Interrupt (RTI), which is analogous to RTS in subroutines. When the CPU executes the RTI instruction, these registers on the stack are pulled off—as shown in Figure 2.10b—and the main program resumes.

The Wait for Interrupt (WAI) places the MCU in a reduced-power-consumption standby mode until some interrupt occurs.

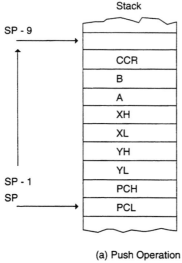
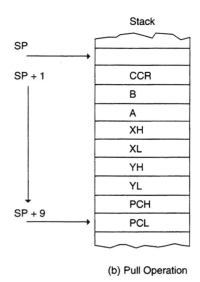

(a) Push Operation   (b) Pull Operation

Figure 2.10

2.5   Miscellaneous Instructions

These instructions are No Operation (NOP), STOP, and TEST, as listed in Table 2.12.

The NOP instruction adds a time delay of two clock cycles; thus it is used to add clock cycles in time-delay loops.

The STOP instruction causes the oscillator and all MCU clocks to freeze and is used to reduce power consumption. The TEST instruction is used during factory testing; it is not used by the programmer.

Exercises

1. Execute the following instructions by hand, then find the final contents of ACCA, SP and ACCB and the contents of locations $C330→$C333.

```
LDAA    $C330
LDAB    $C331
ABA
ADDA    $C332
MUL
```

Table 2.12

Function	Mnemonic	INH
No Operation (2-cycle delay)	NOP	X
Stop Clocks	STOP	X
Test	TEST	X

```
STD    $C331
LDS    #$A4
PSHA
PSHB
```

C330 53 2A BE FF XX XX

2. What happened to the contents of ACCA and ACCB when executing these instructions:

```
LDS    #$4A
LDAB   #$40
LDAA   #$50
PSHA
PSHB
PULA
PULB
```

   What are the contents of SP?

3. Write a sequence of instructions to clear bit 3 and set bit 4 of a data word.

4. There are two methods to clear C, I, V of CCR. Write two programs to do so and compare them.

5. What are the contents of ACCA and the status of the affected CCR flags when executing this program:

```
LDAA   #$F0
ANDA   $C050
ORAA   $C051
EORA   $C052
```

   C050   0F 80 87 XX XX XX

6. Write a program to do the following arithmetic operations using shift instructions:

   (a) 4A   (b) 4)̄70   (c) 3B
        $\times 4$ $\times 6$

7. Write a program to accomplish the following formula using *shift instructions:*

   (a) 7 (ACCA) + 5 (ACCB) $\xrightarrow{\text{store at}}$ $C060

   (b) 6 (ACCA) – 1/4 (ACCB) $\xrightarrow{\text{store at}}$ $C062

8. Find out which branch is going to occur in the following programs:

   (a)
```
                        LDX    #$C020
                        LDAA   $C040
                        CMPA   21,x
                           BHS    SS
                            │
                            │
                            │
        SS                  │
                            LDAB   xx
```

   C040   28 2A XX XX

(b)
```
                              LDY   #$C030
                              LDAB  0,Y
                              CMPB  1,Y
                              BGT   NN



NN                            XX XX
C030      B5 87 XX XX
```

(c)
```
                              LDAA  #$A4
                              STAA  $C055
                              BRCLR $C055,$40,NEXT



NEXT                          XX    XX



          END
```

3   Program Development

Some programmers try to construct their program by immediately attempting to code it using assembly or high-level languages. This produces a *spaghetti program*. Good programmers will follow program design approaches.

Creating accurate real-time software involves three phases:

1. Design phase
2. Coding phase
3. Testing and debugging phase

We are going to discuss the design and coding phases in this chapter and the testing and debugging phase in Chapter 11.

3.1.   Design Phase

Some program design techniques must be taken into consideration. These include top-down design, modular design, and structured programming.

Top-Down Design

In this approach, the programmer should determine the programming task called the *main task*. This main task is broken into a number of subtasks. Each subtask is also broken into further subtasks and so on. Figure 3.1 shows this process. The main task is divided into subtask 1 and subtask 2. Subtask 1 is also divided into subtask 11 and subtask 12. Subtask 11 is broken into subtask 111 subtask 112, . . . , subtask 11n. This is true also for the other subtasks.

The coding of this program proceeds from the bottom up—that is, it moves from the lowest subtask to the top.

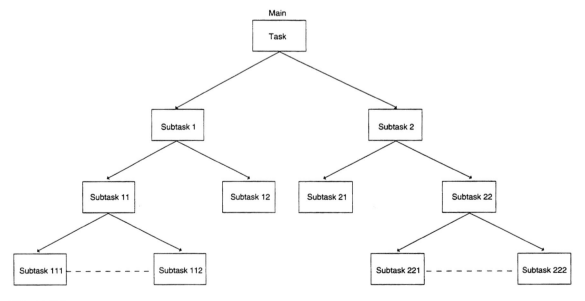

Figure 3.1

Modular Design

The most popular software design today is modular design, particularly in large, complex programs. The program is divided into many modules. Each module performs a certain task. For example, in engine control systems the program is divided into more than a hundred modules, such as fuel, spark, fuel-air ratio, crankshaft position, and throttle position modules. All the modules are assembled individually, then they are linked. As an example, code modules can be classified into these types of modules:

1. Port declaration
2. RAM declaration
3. ROM declaration
4. Initialization
5. Main loop
6. Algorithm
7. Arithmetic (math library)
8. Communication bus
9. I/O
10. Logic
11. Subroutines
13. Interrupts
14. Diagnostics

Two factors affect program modularity: coupling and strength. *Coupling* is a measure of how data are shared between modules. Tightly coupled modules share common data and are not desirable because it is hard to isolate the problem and debug the module. Loosely coupled modules have independent data, making it easy to isolate the problem.

Strength is a measure of performing a number of tasks. A strong module performs one task only. A weak module performs more than one task. The advantage of this approach is that each module can be coded, tested, and verified independently.

Structured Programming

A structured program has three elements: *sequence, looping,* and *decision.* The program should be in sequence. The sequence consists of a list of actions executed in order. The program also should be in a looping mechanism—that is, it should be executed a number of times. Finally, the program should have a decision after being executed.

Programmers have many tools to help them write good structured programs. It is convenient to write down an algorithm before coding in assembly language. *Pseudocode, flowcharts, data flow diagrams,* and *petri nets* are some tools that can be used for this purpose.

Pseudocode

Pseudocode represents a medium language between a high-level language and assembly language because it uses plain English. The general format of pseudocode looks like this:

```
Initialization:
            Variables, I/O initialization
If-Then_Else:
            if condition
               then action_1
               else action_2
            endif
Case:
            case 1:
            case 2:
            . . . . . .
            endcase
Repeat:
            repeat
            forever
               or
            repeat
            until condition
While:
            while condition do
            endwhile
```

We will use this method in some of the programming design examples in the next chapter.

Flowchart

A flowchart is a graphic representation showing the logical sequence that a program follows. Flowcharts are the easiest and oldest software system-modeling tools. They are useful for systems with less than 10,000 instructions. There is no way to describe the interaction between several processes. Some of the symbols used in programming flowcharts are shown in Figure 3.2.

Input/Output. The input/output block is used to represent operations involving the MCU reading data from an input device or writing data to an output device.

Processing. The processing block is used to represent any operations that the MCU performs (except I/O operations). These operations could be arithmetic, logic, shifting, and so on.

Decision. The decision block is used whenever the MCU is instructed to jump or branch to a subroutine.

Beginning and Termination. Beginning and termination are used to represent terminal points in a program or subroutine.

Connector. The flowchart can continue on a following page with an off-page connector symbol containing the same letter or number.

Structured Modeling

There are two steps for system development: analysis and design. Analysis is the process of defining the problem to be solved. Design involves using the *model* of system requirements built during analysis to create the best solution to the defined problem. Constructing a model of a thing before you build it allows you to see what the thing will look like when it is finished.

Modeling is also used in technical applications such as software engineering, hardware engineering, electrical power systems, and many more. Modeling can estimate time-scales, costs, required processing power, and so on.

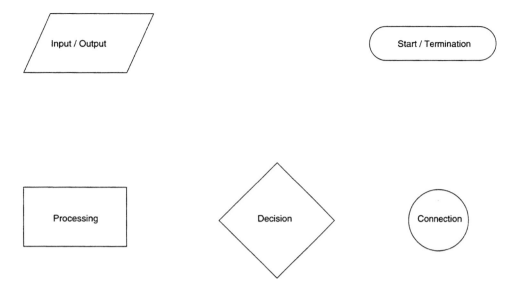

Figure 3.2

In the analysis stage, the requirements define the system from the customer's point of view. The system is represented by an *essential enviromental model.* This model shows the system from the outside and consists of:

1. The context diagram
2. The key event list

The context diagram consists of a single circle called the *process,* which represents the system, and a number of *terminators.* The terminators are rectangular and represent things with which the system has to interface. Figure 3.3 shows an example of a data context

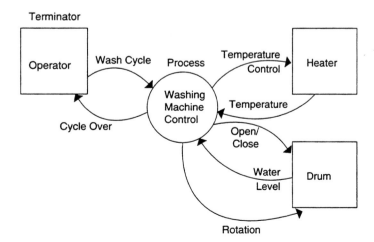

Figure 3.3

diagram. The context diagram is the top level of the data flow diagram (DFD). The arrows between terminators and the process represent either data flow or control flow. The diagram in Figure 3.3 shows the operation of a washing machine. It consists of three terminators (operator, heater, and drum), a process (the machine control), and data flows.

The subject of software design using data and control flow diagrams cannot be covered in this book because of space limitations. A brief idea of what these diagrams look like and how they are used in the design will have to suffice.

Another model—called an *essential behavioral model*—consists of:

1. DFDs
2. State transition diagrams
3. Entity relationship diagrams
4. Dictionary

Data Flow Diagrams

DFDs are used as a structured analysis tool for modeling software systems. Some of the symbols used in the construction of DFDs are shown in Figure 3.4.

DFDs are developed and decomposed to identify major components of each system and subsystem. Computer-aided software engineering (CASE) tools or Easycase tools are used to graph DFDs and write the data dictionary.

DeMarco's Rules. The following is a summary of rules given by DeMarco for the construction of DFDs:

1. Identify all input and output flows.
2. Work your way from inputs to outputs, backward from outputs to inputs.
3. Label all the interface data flows.
4. Label processes in terms of their inputs and outputs.
5. Ignore initialization and termination.
6. Omit details of trivial error paths.
7. Do not show control information.
8. Start over again.

Hatley and Pribhai's Extensions. Hatley and Pribhai have extended DeMarco's DFD to include the control flow diagram (CFD), control specifications, and data dictionary.

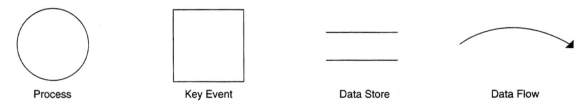

Process   Key Event   Data Store   Data Flow

Figure 3.4

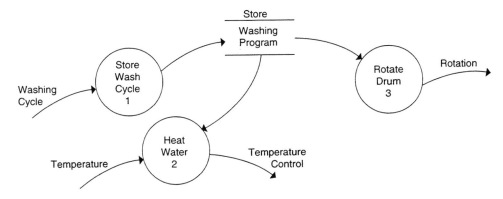

Figure 3.5

EXAMPLE 1
Figure 3.5 illustrates a DFD for the washing machine. It consists of several processes, data flows, and data stores, but there are no terminators. Data stores represent data storage for later use by the process. Data flows are usually solid lines. Each process can be decomposed into higher levels of subprocesses and is assigned the same name and number.

Control Flow Diagrams. CFDs are similar to DFDs except that the control flows are represented with dotted lines.

Petri Nets

Petri nets are used in multitasking or multiprocessing applications. A series of circles called *places* represent data stores or processes. Rectangular boxes represent transitions or operations. The processes and transitions are labeled with a data count and transition function. They are connected by unidirectional arcs, as shown in Figure 3.6.

The initial graph represents the initial data count in processes, labeled m_0. The system advances as transitions change (called *firing*). New markings are the result of firing of these transitions. A firing table is used to describe the status of places before and after firing.

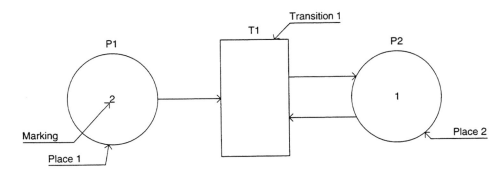

Figure 3.6

3.2 Coding Phase

Coding a program means translating a software design approach into an assembly language or a high-level language.

Coding Guidelines

The coding guidelines are:

1. Too few modules make a program difficult to maintain and reuse. Too many modules may not be efficient in terms of memory resource utilization. Good judgment must be ruled out.
2. Each module should have only one entry point and one exit point.
3. The code should be written in a clear manner and should contain useful comments.
4. The variables of each module should be classified as either *public* (global) or *private* (local). The public variables are those used by several modules. The private variables are those used by one module only.

Structured programming will improve programmer productivity and make it easier for the programmer to write and understand the code correctly.

3.3 Types of Programming Languages

There are three types of languages: machine, assembly, and high-level languages. Each has advantages and disadvantages, as explained below.

Machine Language

Machine language is the language of 0s and 1s, which is the only language the computer can understand. Other languages must be converted to machine language so the computer (MPU or MCU) can understand and execute the program. Each computer has its own machine language. For example, the opcode of an "Addition" instruction of Motorola's product is different from Intel's product. This means that the machine languages of different manufacturers are not compatible. The programmer uses the hex code to represent the opcode, operand addresses, and data since numbers can be represented in hex form much more easily than in binary form. This hex program is entered into the computer's memory through the hex keyboard. The keyboard's monitor program converts the hex code into binary code, then stores it to the memory.

Assembly Language

Assembly language is considered a medium language—that is, it is lower than the high-level languages and higher than machine language. It is the language of mnemonics. The programmer has to know the computer's internal architecture and its instruction set. Assembly language is still the most popular language used in real-time applications employing microprocessors and microcontrollers. Assembly language statements are written to form a program called a *source program*.

High-Level Language

A high-level language uses ordinary words. The programmer does not have to know the computer's internal architecture or its instruction set. High-level languages include Pascal, Cobol, Basic, and C. C is the most popular.

Comparison Between High-Level Languages and Assembly Language

1. The high-level languages use ordinary words, while assembly language uses mnemonics.
2. Both create the object file.
3. Assembly languages require knowledge of the microcomputer's internal structure, while high-level languages do not.
4. High-level languages are slower and use more memory than assembly language.
5. The greatest advantage of high-level languages over assembly language is that the same program written in a high-level language like C can be run on different microcontrollers using the right compiler. For example, if there is a program written on C, it can be run on the 68HC05 or 68HC11 or 68HC16 (by changing its header file) and each MCU using its own compiler. This saves a lot of time and money.

3.4   Assemblers and Compilers

Assembler

An assembler is a program that translates assembly language into machine language (object file) in Hex format (Figure 3.7).

Compiler

The compiler is a program that translates a high-level language into machine language (Figure 3.8.)

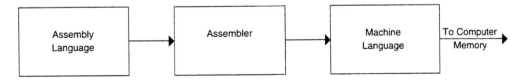

Figure 3.7

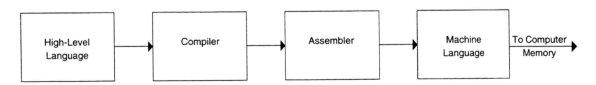

Figure 3.8

3.5   Assembling the Program

Source and Object File

A program written in assembly language is called the *source file* and usually has the extension .SRC or .ASM, depending on the assembler used. When "assembling" the source file (which is in mnemonics), two things take place:

1. Two files are generated: the *object file* with extension .OBJ and the *listing file* with extension .LST. The object file can be expressed in two ways: the hex format as in Intel's applications assemblers or the ASCII-HEX format as in Motorola's application assemblers. The listing file is a combination of the source file and the object file. We will consider listing files in detail later.

2. Assemblers also look for programmer errors. The programmer may make mistakes unintentionally in writing a code or may fail to declare variables and constants. In these cases, the assembler points at the line where the error occurred.

Listing File

A listing file (or program listing) is a result of assembling the source file, as mentioned above. Figure 3.9 shows a sample of a listing file. The listing file could have seven columns (fields):

1. Address field
2. Object code field
3. Line number field
4. Label field
5. Opcode field
6. Operand(s) field
7. Comments field

address	object	line#	label	opcode	operand	comments
		1				
		2				
C000		3		ORG	$C000	
		4				
C000	A656	5		LDAA	#$56	;accA has value of ;$56
C002	B757	6	BEG	STAA	$57	;save it in address ;$57
C004	4A	7		DECA		;is it zero yet?
C005	26FB	8		BNE	BEG	;no, branch
C007		9		END		;stop

Figure 3.9

Address Field

The address field corresponds to the location in memory for each byte. The starting address is determined by the "ORG" statement (directive). Addresses are listed as four digits (there may be more for other 16- and 32-bit MCUs) in hex and are automatically incremented by the number of bytes required by each instruction.

Object Field

Each instruction is translated into hex code. The hex code can be 1 to 4 bytes long depending on the instruction and addressing modes.

Line Number Field

Each line of code is numbered. Line numbers are in decimal form and the maximum limit is 65,536. These numbers are helpful in debugging the program. The line number field could also be in the first column, depending on the assembler.

Label Field

Mnemonics that apply to addresses are called *labels*. For example, in the last program listing, the label "BEG" is assigned to the address $C002. Every time we wish to refer to this address in the program we use the label "BEG."

Opcode Field

The operation code (opcode) is part of the instruction structure, as shown in Figure 3.10.

The opcode is usually 1 byte long and represents the command part of the instruction. Examples include ldaa, adda, staa, and so on.

Operand Field

As Figure 3.10 shows, the operand(s) is part of the instruction. It contains a value, an address, or a label. The operand could be 0, 1, 2 or 3 bytes. Inherent instructions do not need operands.

Comment Field

A programmer uses comments to explain each line. Good programmers provide good documentation of comments. These comments are very helpful for the next programmer to use a piece of code. Comments are usually preceded by an asterisk (*) or a semicolon (;), depending on the type of assembler.

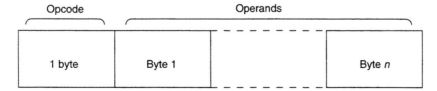

Figure 3.10

3.6 Assembler Directives

When writing a program, a programmer can use certain options that make it possible to reserve memory bytes for data, specify starting and ending addresses of the program, and select the format of the assembler output. These options are called *assembler directives*. The directives are divided into three categories:

1. Assembly control directives, which name the program, define its starting address, and tell the assembler when to end the program
2. Listing control directives, which specify the output format of the assembly
3. Data definition directives, which specify the type or data and where they will be stored in memory

These directives are all three-letter mnemonics except "page," which has four letters.

Table 3.1 summarizes some directives used for an assembler. Some of these directives are common between assemblers and others are different. Now we will provide further detail on some popular directives.

NAM (Name)

This directive assigns a name to a specific file and is usually the first statement of the program. In the example,

```
NAM FUEL
```

this particular file (module) has assigned a name called "FUEL".

EQU (Equate)

The EQU directive is used to assign a "permanent" address to a label. For example,

```
TEMP EQU $0020
```

assigns the address $0020 to the label "TEMP".

ORG (Origin)

This directive is used to define the starting address of the program or to define the starting address of different sections of the program. In the sample program in Figure 3.9, we used "ORG $C000" to tell the assembler that the starting address of the program is at $C000. In source files, we do not have to assign any addresses for each line of code. The assembler will do this job automatically starting from the specified starting address.

FCB, FDB, and FCC

The Form Constant Byte (FCB) and the Form Double Bytes, (FDB) directives are used to assign data (constant values). A 1-byte constant is defined by FCB. The directives must be followed by decimal, hex, octal, or binary numbers. For example, in

```
CONSTN  FCB $12
```

the hex number 12 is assigned to "CONSTN" during the program. FCB also can have more than one value in the operand field. Each value is separated by a comma, as in this example:

Table 3.1

Directive	Function
Assembly control	
NAM XXXXXXXX[a]	Program name (8 letters maximum)
ORG nnnn[b]	Origin (any number base)
END	Program end
Listing control	
PAGE	Top of Page
SPC n	Skip "n" lines
OPT NOO	No object tape
OPT O	The Assembler will generate an object tape (selected by default).
OPT M	The Assembler will write machine code to memory.
OPT NOM	No memory (selected by default).
OPT S	The Assembler will print the symbols at the end of Pass 2.
OPT NOS	No printing of symbols (selected by default).
OPT NOL	The Assembler will not print a listing of the assembler data.
OPT L	The listing of assembled data will be printed (selected by default).
OPT NOP	The Assembler will inhibit format paging of the assembly listing.
OPT P	The listing will be paged (selected by default).
OPT NOG	Causes only 1 line of data to be listed from the assembler directions FCC, FCB, and FDB.
OPT G	All data generated by the FCC, FCB, and FDB directions will be printed (selected by default).
Data definition storage allocation	
FCC "MESSAGE"	Character string data ⎱
FCB XX, YY, ZZ, etc.	One-byte data ⎰ Generates data
FDB XXXX, YYYY, ZZZZ, etc.	Double-byte data
RMB nnnn	Reserve n memory bytes ⎱
Symbol definition	⎰ No data generated
EQU nnnn	Assign permanent value

[a]Xs indicate alphanumeric characters.
[b]ns refer to any number base: 65420, $F100, @7756. We usually show a 4 ns since the number we use is often a 4-digit hex number.
From J. Greenfield and W. Wray, *Using Microprocessors and Microcomputers: The 6800 Family.* Copyright 1988 by Prentice Hall. Reprinted by permission of Prentice Hall, Upper Saddle River, New Jersey.

```
SHFTCONS   FCB $45,E1,4A,FF
```

A 2-byte constant is defined by FDB. For instance, in

```
CONST   FDB #400
```

the decimal number 400 is assigned to "CONST" during the program.

The Form Constant Character (FCC) is used to define ASCII characters in the program. FCC is followed by an ASCII character between apostrophes. An example is

```
TRNSMT    FCC    'A'
```

A is defined as TRNSMT during the program.

RMB

The Reserve Memory Byte (RMB) directive reserves location(s) where data is to be stored. For example,

```
TEMP    RMB 5
```

will reserve 5 bytes for the label "TEMP". The bytes are consecutive. TEMP will be the address of the first byte in the 5-byte memory (RAM).

END

The END directive indicates the end of the source code and must be included.

Note that these directives *are not* part of assembly language—that is, they are not part of the instruction set. They are symbols or mnemonics created for the assembler only. These directives make the job very easy for the programmer. For example, if we have this data assignment

```
CONST    FCB    $12
```

many instructions in the program could use this directive. If the programmer decided to change this value to $20, it is only necessary to change the directive value

```
CONST    FCB    $20
```

once and other occurrences of this value used by all instructions in the program will be changed automatically. Without using this directive in the program, the programmer has to change each of these constants throughout the program manually. The task would be unrealistic and a waste of time.

3.7   Advanced Assemblers

There are two types of assemblers:

1. *Coresident assemblers,* used by beginners. They have limited capabilities.
2. *Macro assemblers,* used by professional programmers. These assemblers have the capability of linking, macro operations, and relocation.

Linking the Program

In real-world programming, the program is divided into several modules. Each module consists of hundreds or even thousands of code lines. A longer program could have 100 or more of these modules. Usually there is a team of programmers working on one project. In cases like that, modules should be "linked" together. Linking combines all object files together to produce a hex file. It provides the following capabilities:

1. Specification of initialized and uninitialized memory.
2. Detection of any duplicated variables or constants.
3. Relocation of each module in the memory.
4. Generation of "absolute" and "map" files.

The absolute file has a special format called an *s-record*. This file is loaded to the ROM and RAM of the microcontroller to execute the program. We are going to explain s-records later. Some assemblers have this file with extension .ABS.

Another file generated after linking is called *Map file* with extension .MAP. The map file contains the new PSCT (program section) of each module, which is the starting address of each one. After relocation, this file also shows the BSCT (base section) and DSCT (data section) of each module. In addition, it contains the addresses of variables and constants.

The linker can also generate a symbol file that shows the final assigned addresses for each label and the value of each label.

S-Record Information

S-Record Contents

The s-records are the code generated, as we have noted, after linking the program. They are character strings of several fields that identify the record type, length, memory address, code/data, and checksum, as indicated in Figure 3.11.

The line fields of an s-record are:

Field	No. of Characters	Contents
Type	2	s-record type:S0 or S1 or S9
Record length	2	no. of character pairs in the record excluding type and length field
Address	4, 6, or 8	the 2-, 3-, or 4-byte address at which the data field is to be loaded to memory
Code/data	0–2n	from 0 to 2 bytes of executable code
Checksum	2	the 1s complement of the sum of the values represented by the pairs of characters, address, code/data

The checksum technique can detect errors but cannot correct them.

S-Record Types

S0 is the header record for each block of s-record. S1 is a record containing code/data and the 2-byte address at which the code/data is to reside. S9 is a termination record for a block of s-records.

Type	Record Length	Address	Code/Data	Check sum

Figure 3.11

EXAMPLE 2

Consider the following example:

```
S00600004844521B.
S1130000285E261F2012226A000424290008247C2A.
S9030000FC.
```

Let's explain each line.

First line. The S0 indicates a header record, 06 indicates the length of record (6 bytes) excluding the type and the record length, and 0000 indicates a four-character, 2-byte address. In addition, 484452 indicates ASCII for H, D, R letters (HDR) and 1B indicates checksum (the 1s complement of the S0 record).

Second line. The S1 indicates a code/data record. The 13 indicates length of record (13 bytes in hex). Moreover, 0000 indicates a four-character, 2-byte address. The numbers 28 to 7C are 16-character pairs of ASCII bytes of the actual program code/data. The 2A indicates the checksum of the S1 record.

Third line. The S9 indicates a termination record, the 03 indicates 3 bytes of length, and the 0000 indicates a 2-byte address field. FC indicates the checksum of the S9 record.

Editors

The editor allows a user to enter data into the system from a terminal and then corrects or alters the data as required. Editors commonly used with MPUs/MCUs employ the ASCII code to represent letters, numbers, and so on. Word processors with non-ASCII codes should not be used to write programs.

Types of Editors

1. Line-oriented
2. Screen-oriented
3. Language-oriented

In line-oriented editors, the user must specify the line number to make a change. This kind of editor is obsolete. In screen-oriented editors, the user can move the cursor around the screen to make changes, insertions, and deletions in the entered text. Screen-oriented editors are easy to use and immediately show the result of an editing operation. Language-oriented editors are designed to be used with a particular programming language. These editors check the program for errors in syntax as it is being entered.

Macros

A macro is a repeated sequence of instructions that requires *variable entries* at each insertion. The macro is defined by a name at the beginning that becomes the mnemonic by which the macro can be invoked. This name precedes the directive *macro*. Then the macro is ended by the *ENDM* directive. The programmer can use the macro statement as needed. The macro is similar to the subroutine, with one difference: the macro contains variable parameters (substitutable arguments) while a subroutine does not. A substitutable argument could be an instruction or any assembler directive.

EXAMPLE 3

If a program contains a repeated piece of code like this

```
LDAA    #CONST_1
ADDA    #CONST_2
SUBA    #CONST_3
   |
LDAA    #CONST_4
ADDA    #CONST_5
SUBA    #CONST_6
   |
LDAA    #CONST_N
ADDA    #CONST_M
SUBA    #CONST_L
```

we can use a macro to replace the repeated code.

```
MATH    MACRO           ; defines macro
        LDAA \F1         ; these are the instructions
        ADDA \F2         ; contains substitutable
        SUBA \F3         ; arguments
        ENDM            ; ends macro
```

where \F1, \F2, \F3 are the substitutable arguments. Then the substitutable arguments are defined as

```
MATH    #CONST_1, #CONST_2, #CONST_3
  .
  .
  .
MATH    #CONST_4, #CONST_5, #CONST_6
  .
  .
  .
MATH    #CONST_N, #CONST_M, #CONST_L
```

When the MATH calls the macro for the first time, CONST_1 replaces F1, CONST_2 replaces F2, and CONST_3 replaces F3 in the macro code. During the second call of the macro, CONST_4 replaces F1 and so on. The assembler recognizes these substitutable arguments by the presence of the backslash (\).

Exercises

1. Explain the main elements of a program design.

2. Explain the following terms in modern programming:

    coupling, strength, sequence, and looping

3. Define the following:

   machine language, assembly language, high-level language

4. Compare high-level and assembly languages.

5. Explain the advantages of assembling a program.

6. Explain the advantages of linking a program.

7. What is the difference between a macro and a subroutine?

8. What are the advantages of the assembler directives?

9. Use an appropriate combination of the design techniques discussed in this chapter to show the operation of
   a. a washing machine cycle
   b. a radar receiver

4 Programming Techniques

Now that we have become familiar with the MCU M68HC11 instruction set and the assembler directives, we are going to construct and execute useful programs using different techniques. These programs use the MCU's programming model and its on-chip RAM only. In the following chapters, we will take advantage of all the other resources of this chip. We would like to use pseudocode and flowcharts in some of these programs to explain the program sequence. These programs use assembly directives; if you have trouble with any of them, refer to Chapter 3 for the details. There will be some exercises after each section, and you are encouraged to solve all of them to gain programming experience. These programs will prepare you for later chapters.

4.1   Memory-Mapped I/O

In *memory-mapped I/O*, we treat I/O ports as if they were memory locations, as in the M68HC11 memory map. For the I/O data transfer, the same instructions are used to transfer data between the CPU and the I/Os. The I/O ports are assigned addresses from the memory address space. Figure 4.1 shows the memory map of MC68HC11E9.

The memory map of the 68HC11EVB evaluation system is depicted in Figure 4.2. The EVB is used as an evaluation/debugging tool. More details will be provided in the lab manual.

The internal RAM range is from $0000 to $00FF. The user can utilize RAM locations $0000 to $0035 (54 bytes) for the RAM variable declaration of the program. Locations $0036 to $004A are used for the stack. Locations $004B to $00C3 are employed for monitor variables. Locations $00C4 to $00FF are used for the vector interrupt jump table.

The *separate I/O* (isolated I/O) has a separate address space and its own instructions. Each technique has advantages and disadvantages. In memory-mapped I/O, the number of addresses assigned to the I/O ports is limited and memory reference instructions are used that have more flexibility, but this approach requires address decoding circuits. On the other hand, in the separate I/O, the instructions take less time to fetch and execute, and require no address decoding circuits, but they are limited in their address space and require their own instructions.

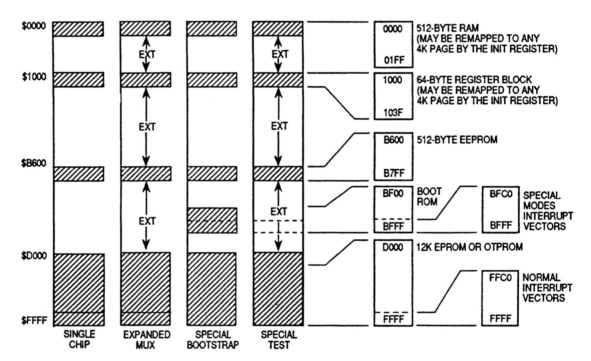

Figure 4.1
Copyright of Motorola. Used by permission.

Figure 4.2
Copyright of Motorola. Used by
permission.

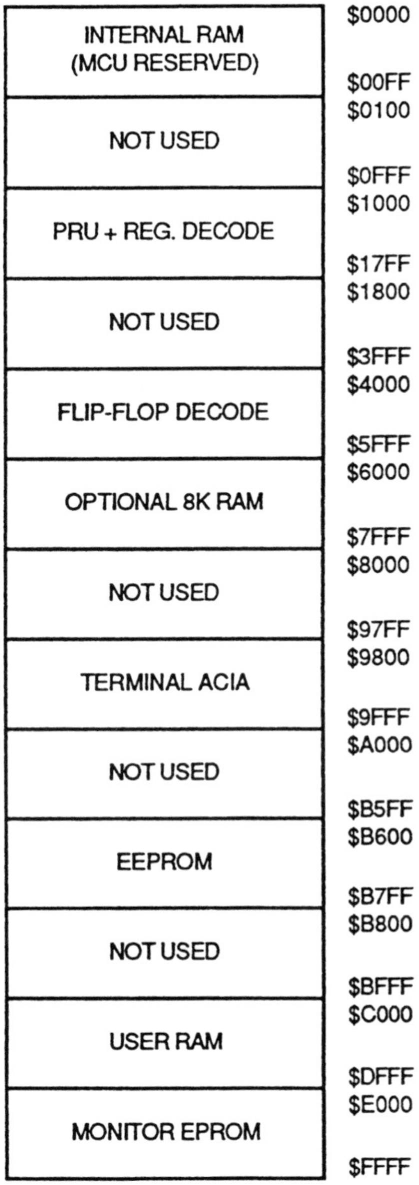

INTERNAL RAM (MCU RESERVED)	$0000 — $00FF
NOT USED	$0100 — $0FFF
PRU + REG. DECODE	$1000 — $17FF
NOT USED	$1800 — $3FFF
FLIP-FLOP DECODE	$4000 — $5FFF
OPTIONAL 8K RAM	$6000 — $7FFF
NOT USED	$8000 — $97FF
TERMINAL ACIA	$9800 — $9FFF
NOT USED	$A000 — $B5FF
EEPROM	$B600 — $B7FF
NOT USED	$B800 — $BFFF
USER RAM	$C000 — $DFFF
MONITOR EPROM	$E000 — $FFFF

4.2   Programming Examples

Program 1:   Clearing a Block of Memory (an Array)

This program explains how to clear RAM locations called a *block*. Each block has its memory addresses (called also *locations*) and its data (called *contents*). We call this block of memory an *array* and the contents *elements*.

Figure 4.3

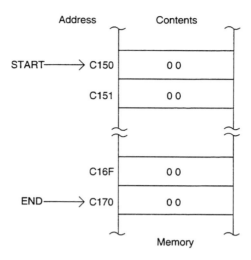

In this example, a block starts at address $C150 and ends at $C170. These addresses have random contents. By clearing these locations, we mean that all the contents become zeros (Figure 4.3).

Algorithm Description

1. Assign pointer(s) wherever applicable .
2. Assign counter(s) if necessary.
3. Write code to execute the program for one step only.
4. Create loop(s) to go through all the steps until the program ends.

Design Phase

Pseudocode

Let's first use pseudocode to help in writing the assembly language code using the above rules. The pseudocode is as follows:

Initialization:

```
            Equate START       ;top address
            Equate END         ;bottom address
    End
    Program:
            Assign pointer to point to the top of the block
            Clear this location
            Increment pointer by 1 to point to next location
            Check if this is the last address to be cleared
            If not, loop again to clear the next location
    End
```

Second, we use the flowchart in Figure 4.4 to implement the above rules.

Figure 4.4

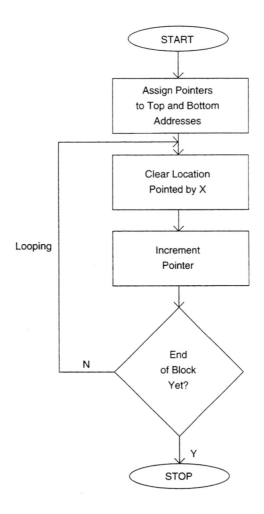

Coding Phase

For the assembly language, we use equate directives to assign labels to the top and bottom addresses of the block. This is called *initialization* or *declaration*.

```
START   EQU $C150               ;top address of the block
END     EQU $C170               ;bottom address
```

For rule 1, to assign a pointer to START location, we write

```
            LDX   #START
```

This means that the X register contents equal $C150. If we want to move the pointer from the $C150 location to the $C151 location, we increment the X register as follows:

```
            INX
```

Its contents then become $C151. In other words, it points to the next location, $C151. For rule 2, we are not going to assign any counter in this example, but we will in the next. According to rule 3 we have to clear one location (the first address in this case). Now the contents of $C150 equal 0. Next, using rule 4 we have to clear the rest of the locations by creating a loop that goes through them.

Every time we go to the next location, we have to check if this is the end of the block. If it is, we stop executing the program. If it is not, we keep incrementing the pointer and clearing the corresponding location.

This process is called *looping*. After we reach the END location, the program does not loop again but stops. Usually we use index registers X and Y as pointers. Generally, the MCU's index registers are used as pointers. Let's call this program *clear.asm*. *Clear* is the filename and *.asm* is its extension.

Clear.asm

```
START    EQU    $C150
END      EQU    $C170

                    ORG   $C000

         LDX  #START        ;x points to the beginning of array
AGN      CLR  0,X           ;clear location pointed by x        ·
         INX               ;go to next location
         CPX  #END          ;end of block yet?
         BNE  AGN           ;no, branch
         END               ;yes, stop
```

As we mentioned about directives in the last chapter, AGN is called a *label*. It marks the line containing CLR 0,X, so if we want to branch to this line we have to refer to it by *AGN*. We created one loop by writing the instruction BNE, which makes the program branch backward to the AGN label and continue checking for the END address of the block until it reaches it. Then the program ends.

Notice that the location $C170 is not going to be affected—that is, is not going to be cleared. Why not? Because when the X register reaches this location, the program stops without going through the clearing instruction.

So after executing the program, we find that only locations $C150 to $C16F are actually cleared. This is an important point. If we want to include the $C170 location, we can change the instruction BNE AGN to BLS AGN. Try this instruction in the above program.

In later programs, we will use the same techniques, so we will not have to repeat the explanation unless something new comes up.

How to Debug a Program

There are many tools to debug a program written in an assembly language. These tools include evaluation boards, evaluation modules, emulators, logic analyzers, and simulation packages, as explained in Chapter 11. All these tools use at least two popular methods: breakpoints and single-step methods.

We can assign as many *breakpoints* as desired at particular code lines in the program to examine the contents of a register or a memory location. For example, in the clear.asm program just executed, we can assign a breakpoint at the instruction INX to see how it increments. By executing the program for the first loop and checking the contents of register X, we find that it has the contents $C150. Then, by executing the program for the next loops, register X should have contents of $C151, $C152, and so on.

We can assign another breakpoint at the instruction CLR 0,X to see how the locations $C150 to $C170 are being cleared at one location per loop.

Another method used for debugging is called *single-step* execution. Using this feature, we can execute a block of code, line by line, and examine the contents of registers or memory locations. The disadvantage of this method is that we cannot debug real-time running timers or I/O driven interrupts.

Exercise

In the last example we cleared the block from top to bottom (in the forward direction). Modify the program to clear it, proceeding from bottom to top. The locations to be cleared are $C170 to $C185. Use pseudocode or a flowchart before you write the assembly code.

Program 2: Clearing a Block of Memory Using a "Counter"

Let's use the previous example to clear a block of memory but using a different technique.

In this example, we are going to use a counter according to rule 2 from the programming rules. The counter contains the number of locations to be cleared. The number of locations in hex = 70 – 50 + 1 = 21 H (hex) = 33 D (decimal). We do not have to assign the label END.

Let's try to construct the program using pseudocode and the flowchart before coding.

Design Phase

Pseudocode

```
Initialization:
          Equate START        ;top address
          Equate COUNTER       ;number of locations to be cleared
End
Program:
          Assign pointer at the top address
          Assign a counter to keep tracking cleared locations
          Clear location pointed by X
          Increment pointer by 1
          Decrement counter by 1
          Check if the counter is 0
          If not, loop again to clear the next location
End
```

Flowchart

In the flowchart in Figure 4.5, we have assigned a counter to keep tracking the number of locations being cleared. Also, every time we move forward by incrementing the pointer by

1, we decrement the counter by 1. We check the counter contents every time we decrement it because when it reaches 0, we have reached the last location and the program should stop.

Coding Phase

The following is assembly code:

Clear1.asm

```
START   EQU  $C150
COUNTER EQU  $21
```

Figure 4.5

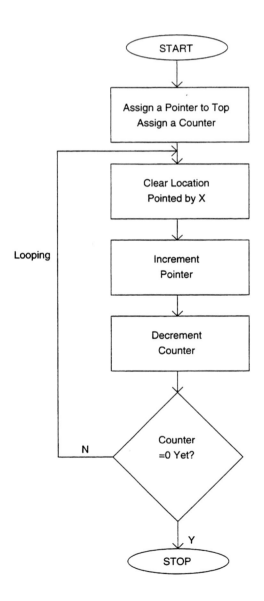

```
                              ORG  $C000

              LDAA  #COUNTER
              LDX   #START      ;X point to the beginning
      AGN     CLR   0,X         ;clear location pointed by X
              INX               ;go next location
              DECA              ;decrement counter
              BNE   AGN         ;if not zero, branch
              END               ;stop
```

Note that BNE is checking only the immediately previous instruction. In this case it checks the contents of ACCA and not the X register. So when the contents of ACCA reach 0 after clearing 32 locations, the program stops.

To calculate the counter value, we subtract the last location from the first location plus 1.

Exercise

In real-world applications, a certain block of RAM is cleared in the initialization routine before executing the main program loop or after coming out of reset. This block of RAM is used as a scratch RAM for math routines or for temporary storage. Modify the program clearl.asm so that index register X points to "00" instead of to START. Clear the same locations, $C150 to $C170. Also show how you debug such a program using breakpoints or single-step techniques.

Program 3: Moving a Block of Memory from One Location to Another
Algorithm Description

In this program, we want to move the contents of block 1 ($C100 to $C121) to block 2 ($C150 to $C171). This means that the contents of $C100 (44H) will move to the contents of $C150 and overwrite them and so on. After executing the program, the contents of the two blocks should be same.

Remember, when we write any new data over existing data, the latter will be lost and replaced by the new data. Using the previous programming rules, the program looks like this:

Move.asm

```
START1  EQU $C100
END     EQU $C122
START2  EQU $C150


                    ORG  $C000

            LDX   #START1    ;x points to 1st byte in block 1
            LDY   #START2    ;y points to 1st byte in block 2
    AGAIN   LDAA  0,X        ;get data from block1
            STAA  0,Y        ;store it in block2
            INX              ;increment 1st pointer
            INY              ;increment 2nd pointer
            CPX   #END       ;reached bottom of the block yet?
```

```
                     BNE   AGAIN          ;no, branch
                     END                  ;yes, stop
```

Execute the program, then check the contents of locations $C150 to $C171. They should be the same as the contents of $C100 to $C121. Notice also that we did not assign a pointer to the end of block 2. Why? Because when the pointer reaches the END label ($C122), it stops and the last location ($C121) in block 1 has moved to $C171 in block 2.

Exercise

Automotive applications are of the most suitable applications for moving a block of memory. A block of RAM called Keep-Alive Memory (KAM) is kept alive by connecting it to the battery. This memory keeps many constants that are used by the engine's or the transmission's control program. If the battery is disconnected, these constants are lost. When the battery is reconnected, initial constants stored in ROM called *learned constants* are copied to the KAM just mentioned. Modify the "move" program and use a counter instead of assigning the END label. What are the contents of the X register and ACCA at the end?

Program 4:   Searching for an Element in an Array

Assume we are going to search for the element FE in locations $C150 to $C175. If we find that element, we will save it in location $C180.

Algorithm Description

The idea here is to compare the contents of this array of elements with the value FE. If it is found, we save it in a location called *FOUND*. Here is the program:

```
C150   XX XX XX XX XX XX FE XX XX ......
C160   XX XX XX .......
C170   XX XX XX ........
```

Method 1

Search.asm

```
START  EQU $C150
END    EQU $C176
FOUND  EQU $C180


                    ORG $C100

                    CLR   FOUND
                    LDX   #START-1
REPEAT              INX                  ;point to next location
                    CPX   #END           ;end of array yet?
                    BEQ   EXIT           ;yes,stop
```

```
          LDAA 0,X              ;no,get the contents of this
                                ;location
          CMPA #$FE             ;contents match FE?
          BNE  REPEAT           ;no, search again
          STAA FOUND            ;yes, save element,otherwise 0
EXIT      NOP                   ;end of array/search, stop
```

Initially, X register is loaded with location (start-1) by location $C14f. Can we do that? Yes, we can write any equate and add or subtract any number to or from it to point to the proper location. We write it in such a way that when we reach the next instruction INX, we will start searching from address $C150. After executing the program, check location $C180. If you find #$00, this means there was no such element FE. Otherwise you will find element FE.

Method 2

Search1.asm

```
START   EQU $C150
END     EQU $C176
FOUND   EQU $C180


                ORG $C100


          LDX   #START         ;point to the begining of array
REPEAT    LDAA 0,X             ;get the contents of this location
          CMPA #$FE            ;contents match FE?
          BEQ  EXIT            ;yes, save it
          INX                  ;no,go to next location
          CPX  #END            ;end of array yet?
          BNE  REPEAT          ;no,keep searching
          CLRA
EXIT      STAA FOUND           ;save element,otherwise 0
          NOP
```

Compare the two methods.

Exercises

1. Write a routine to find the smallest signed byte in a block of data. The block address range is $C120 to $C160.

2. Repeat the previous exercise to find the largest unsigned byte in the same range.

Program 5:   Counting the Number of Elements in a Block of Memory

We would like to count the number of elements, each (44H) occurring in a block ranging from $C160 to $C180. The number of occurrences should be saved in location $C190. In the real world, the block range could be much larger than the one we use here, but the principle is the same.

Algorithm Description

The idea here is to compare the contents each time with the element (44H). If found, then add 1 to the counter (CTR) location.

Here is the program:

```
C160    XX XX XX XX 44 XX XX XX 44 XX XX ...
C170    XX 44 XX...
C180    44 XX ....
```

count.asm

```
START EQU $C160
END   EQU $C181
CTR   EQU $C190

            ORG $C000

        CLR  CTR
        LDX  #START          ;point to the start of the block
AGN     LDAA 0,X
        CMPA #$44            ;element found yet?
        BNE  SKIP            ;no, branch
        INC  CTR             ;yes, increment counter
SKIP    INX
        CPX  #END            ;end of block yet?
        BNE  AGN             ;no, try again
        NOP                 ;yes, stop
        END
```

Notice that we had to clear location $C190 (counter) at the beginning.

Program 6:   Isolating the Positive from the Negative Elements in a Block of Memory with *Signed* Numbers (8 bits each)

The original block contains positive and negative signed numbers. As you will recall, if the 8-bit signed number has a 1 in its MSB (sign bit), it is considered a negative number. If it has a 0, it is considered a positive number.

Now can you guess what the critical number is that differentiates between negative and positive signed numbers? You might think that it is 00H. This is true if the numbers are unsigned, but in signed numbers the critical value 7FH = 01111111 is the maximum positive number in the range of 256.

As shown in Figure 4.6, the numbers ranging from 00 to 7F H are positive numbers, and the numbers ranging from 80 to FF are negative numbers. The total numbers are 256.

Algorithm Description

In this program, we compare each element with 7F H. Any number equal to or lower than this value will be positive; any number higher than this value will be negative. The original block ($C200 to $C210) has the positive and negative numbers mixed together. We would

Figure 4.6

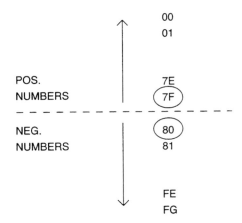

like to isolate the positive numbers and save them in block 1 with starting address $C250 and the negative numbers in block 2 with starting address $C260.

```
C200    6C 00 E2 91 7F 80 21 FF 3A 4B 5C C5 B4 11 00 70
C210    XX XX ....
```

Here is the program:

Pos_Neg.asm

```
START   EQU   $C200
END     EQU   $C210
POS     EQU   $C250
NEG     EQU   $C260
TEMP0   EQU   $C275
TEMP1   EQU   $C276


               ORG $C100

          LDX   #NEG
          STX   TEMP0      ;save neg block start address
          LDX   #START     ;x points to main block
          LDY   #POS       ;y points to pos block

BEGIN     CPX   #END       ;end of original block yet?
          BEQ   EXIT       ;yes, exit
          LDAA  0,X        ;no, get a byte of original
          INX              ;go to next
          STX   TEMP1      ;save next pointer of the original
                           ;block
          CMPA  #$7F       ;is this byte pos?
          BLS   POSITIVE   ;yes, branch to pos block

NEGATIVE  LDX   TEMP0      ;no, negative
          STAA  0,X        ;save neg byte in neg block
```

```
              INX                ;go to next
              STX   TEMP0        ;save the new pointer
              LDX   TEMP1
              BRA   BEGIN        ;repeat

POSITIVE      STAA 0,Y           ;save pos byte
              INY                ;go to next
              BRA   BEGIN        ;repeat

EXIT          NOP
              END
```

Run this program then check the locations in blocks 1 and 2.

Exercise

Write a subroutine comparing two equal-length blocks of data to see if the data are identical. If they are identical, indicate that by setting the carry flag or by storing the notation *OK*. If the data are not identical, indicate that by clearing the flag or by storing the word *diff*.

Program 7:   Adding All Elements in a Block of Memory

We would like to add all the contents of the locations starting from $C100 to $C120. The result of addition could be 2 bytes long. So we want to save the results in locations $C121 and $C122. The sum of the contents of a block of RAM may be used to calculate the checksum. The checksum is the 1s complement of the sum of these contents.

ADDB1K.asm

```
START   EQU   $C100
END     EQU   $C120
SUM     EQU   $C121


              ;same equates

                    ORG $C000

              LDX   #START       ;x points to the start
              LDD   #0000        ;clear a,b
ADDEM         ADDB  0,X          ;add byte to ACCB
              ADCA  #00          ;add 1 to ACCA if c=1
              INX                ;go to next location
              CPX   #END         ;end of block yet?
              BLS   ADDEM        ;no, repeat
              STD   SUM          ;yes, store results
              END
```

Exercises

1. Modify the above program so you can add the contents of every other location. For example, you will add the contents of $C100, $C102, $C104, and so on.
2. Design an algorithm to sort elements in an array using the *bubble sorting* method. *Hint:* We want to write a program to sort or rearrange the elements in a block of memory in ascending order. As shown in Figure 4.7, we have random numbers in locations $C160 to $C163 and we wish to sort them in ascending order.

Algorithm Description for Exercise 2

The technique here uses the bubble sorting method. We compare the first location contents with the rest of the locations in the block. If we find that the element being used for comparison purposes is higher than the one compared to, we swap the two elements. Otherwise we leave them alone.

To begin, we compare the element in the first location (20H) with the element in the second location (15H). Since 20H is higher than 15H, we swap the two elements.

Next, we compare the element in the first location $C100 (15H) to the element in the third location (30H). Since it is smaller, we leave them alone. We keep doing this until we reach the end of the block. Then we take the element in the second location (20H) and compare it with the other locations' elements using the same technique. After that, we compare the element in the third location and compare it with the rest and so on.

Exercise

Rewrite the above program to sort these elements in descending order.

How to Write a Logic Code

In real-world applications, many conditions have to be met before certain decisions are made. These conditions are written in a logical form such as AND or OR. The logic code involves variables and status flags. The logic form looks like this:

If < logic code > **Then** action1 **Else** action2

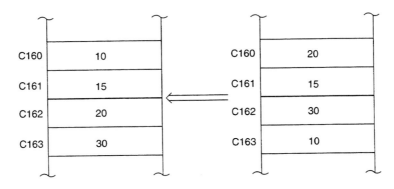

Figure 4.7

Program 8:   ORed Terms

Assume we have 4 RAM bytes containing the variables G, H, K, L. These variables have to meet certain conditions. If some or all of these conditions are true, the decision is made; otherwise nothing happens.

Consider this logic:

```
          1st logic               2nd logic
      term1     term2         term3     term4
```
If (G > 2 **OR** H < 3) **AND** (K < 15 **OR** L > 20) **Then** turn on crankshaft (action 1).
Else
          exit (action 2).

This logic code says: If (term1 OR term2) AND (term3 OR term4) are true, turn on crankshaft; otherwise exit. Term1 is ORed with term2 in the first logic. Term3 is ORed with term4 in the second logic and both logic are ANDed together. In the first logic, one of the two variables has to be true. So if G > 2, this term becomes true (logic 1) and we do not have to check term2. Now the first logic is true. If G <= 2, term1 is false (logic 0) and we have to check term2 as to whether it is true or false. This is the meaning of OR logic.

If H < 3, the first logic is true, but if H >= 3, this logic is false. If the first logic is false (0 OR 0), then ANDing this term with any logic makes the result logic is 0 and no action is taken. The different logic combinations are:

If (true **OR** xx) **AND** (true **OR** xx) **Then** action1
If (false **OR** true) **AND** (true **OR** xx) **Then** action1
If (true **OR** xx) **AND** (false **OR** true) **Then** action1
If (false **OR** true) **AND** (false **OR** true) **Then** action1
If (false **OR** false) **AND** (xx **OR** xx) **Then** action2
If (true **OR** xx) **AND** (false **OR** false) **Then** action2

where xx = do not care case and could be true or false.

The above boolean expression can be represented by ladder logic. Ladder logic involves a sequence of ladder rungs. Each rung consists of a set of input conditions and outputs at the end of the rung.

The input conditions are boolean logic, as explained above, but in ladder form. For example, the above logic can be represented using ladder logic as shown in Figure 4.8.

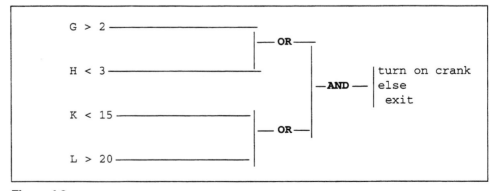

Figure 4.8

Let's write a program to demonstrate this logic:

```
              LDAA  G          ; G is a variable
              CMPA  #2         ; G>2 ?
              BHI   SECOND     ; yes, go to 2nd logic
              LDAA  H          ; no, check term 2
              CMPA  #3         ; H<3 ?
              BHS   EXIT       ; no, exit

SECOND        LDAA  K          ; yes, check term 3
              CMPA  #$15       ; K<15 ?
              BLO   ACTION     ; yes, take action
              LDAA  L          ; no, check term 4
              CMPA  #$20       ; L>20?
              BLS   EXIT       ; no, exit
ACTION                         ; yes,turn on crank shaft
EXIT          RTS              ; no action
```

Program 9: **AND Terms**

We will use the same technique as mentioned before using ANDed terms instead of ORed terms. Consider the following logic:

IF $(G > 4$ **AND** $H < 7)$ **OR** $(K < 15$ **AND** $L > 22)$ **Then** turn on gas pump
ELSE
        exit.

where G, H,K, and L are variables.

In the last example, to be true, one term at least in each logic had to be true for the resulting logic to become true. As you can see, in AND logic at least one logic (both terms) has to be true for the result logic (output) to become true.

Using the previous technique to explain this logic, if $G > 4$ **AND** $H < 7$, then the first logic is true. If term1 or term2 is false, the first logic is false. Assuming that the first logic is true, there is no need to check term3 and term4 of the second logic, and the whole statement becomes true and the action is made. But if the first logic is false, in the next step we have to examine term3 and term4. These terms must be true for the statement to become true. If one term in the second logic is false, the total statement is false and no action is taken. The different combinations are:

If (true) **OR** (xx) **Then** action1
If (false) **OR** (true) **Then** action1
If (false) **OR** (false) **Then** action2

The above logic can be represented by ladder logic, as Figure 4.9 shows.

Let's write a program to accomplish this:

```
              LDAA  G
```

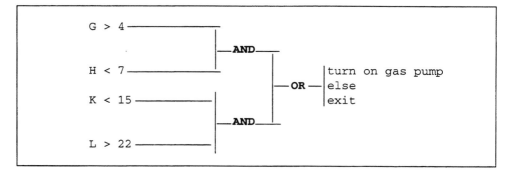

Figure 4.9

```
               CMPA #4        ; G > 4?
               BLS  SECOND    ; no,branch to 2nd logic
               LDAA H         ; yes,check term 2
               CMPA #7        ; H < 7 ?
               BLO  ACTION    ; yes,tab action

SECOND         LDAA K         ; no,check the 2nd logic
               CMPA #$15      ; K < 15?
               BHS  EXIT      ; no, exit
               LDAA L         ; yes, check term 4
               CMPA #$22      ; L < 22 ?
               BLS  EXIT      ; no,exit
ACTION         |             ; yes,turn on pump
EXIT           RTS            ; exit
```

Exercise

Write pseudocode and the assembly program to implement the following logic:

   a. **If**(c>1 **OR** d<3 **OR** e<5) **AND** (f<2 **OR** g<5 **OR** h>3) **Then** action **else** exit.
   b. **If**(e>3 **AND** f>=6 **AND** 9<=3) **OR** (h>=0 **AND** i<3 **AND** j>5) **Then** action1 **else** action2.
   c. **If**{(c>=4 **OR** d<5) **AND** (f>=0 **OR** g<1 **OR** h>13)} **OR** j>7 **Then** action **else** exit.
   d. **If**{(r>66 **AND** s<25 **AND** t>=20) **OR** (u>0 **AND** v<=13)} **AND** j<7 **Then** action1 **else**
                                                                      action2.
   e. Figure 4.10 demonstrates the logic.

Program 10: Status Flags in Logic Code

Status flags are used most often with the variables to form a logic code. A status flag is a bit in a status byte that is set or reset to meet certain condition(s). Let's modify the logic in the program in Figure 4.8 as follows:

If (q_flg **OR** G>2 **OR** H<3) **AND** (K<15 **OR** L>20) **Then** turn on crankshaft
ELSE
        exit.

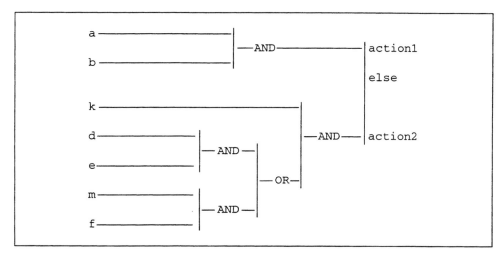

Figure 4.10

where q_flg = a status flag set to 1—that is, q_flg = 1. We use the same technique that we explained earlier. To examine a status flag, we use the instructions BRCLR or BRSET. Now the program is modified as follows:

```
              |

BRSET Sbyte, $q_flg,NEXT ; if QF is set, branch
LDAA  G
CMPA  #2                 ; is G > 2 ?
BHI   NEXT               ; yes, branch
              |          ; no, continue

NEXT          |
```

The same technique is used with the ANDed term example.

Exercises

1. Draw a flowchart and write a program in assembly language to implement this logic:

   If(q̄_flg OR G >6 OR H<5) AND (K <10 OR L>19) Then action1
   Else action2.

   where q̄_flg = cleared status flag in a status byte or q_flg = 0

2. Write an assembly language program to implement this logic, as shown in Figure 4.11.

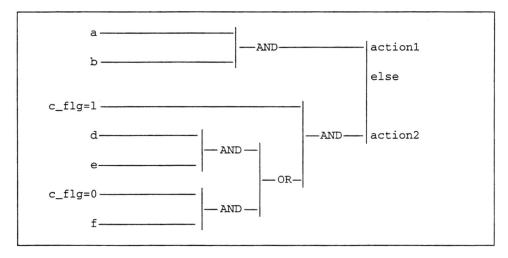

Figure 4.11

Program 11: Adding a 16-bit to a 16-bit Number (Double Precision)

Algorithm Description

To add a 16-bit element to another 16-bit element, the result takes a maximum of 3 bytes (Figure 4.12). The most significant byte of the result could be 0 or 1 only. In adding numbers higher than 8 or 16 bits long, we could also use ADC to add the carry to the higher bytes.

Here is the program:

AD16to16.ASM

```
                        ORG  $C200

MHI     RMB    1                ;hi-byte of M element
MLO     RMB    1                ;lo-byte of M element
NHI     RMB    1                ;hi-byte of N element
NLO     RMB    1                ;lo-byte of N element
ADHII   RMB    1                ;MSB if the sum
ADDHI   RMB    2                ;hi-byte of the sum

                        ORG   $C100

        CLR    ADHHII
        LDD    MHI
        ADDD   NHI              ;add the 2nd element
        STD    ADDHI            ;store the in result low byte
        BCC    EXIT             ;if result <=FFFFH, branch
        INC    ADHHII           ;if not, add 1 to MSB of result
EXIT    NOP
        END
```

Figure 4.12

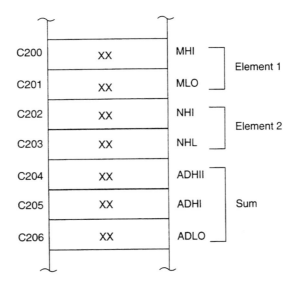

Program 12: Adding Two Elements When Each Is 32-bits Long
Algorithm Description

In this problem, we want to add two elements, each of which is a long word (4 bytes) using math buffers. As shown below, we are going to add the contents of math buffers M0, M1, M2, M3 to the contents of locations pointed to by the X index register:—X, (X+1), (X+2), and (X+3)—and store the results back to these math buffers.

$$
\begin{array}{r}
\text{M3} \quad \text{M2} \quad \text{M1} \quad \text{M0} \\
+ \quad \text{X} \quad . \quad \text{X+1} \quad \text{X+2} \quad \text{X+3} \\
\hline
\text{M3} \quad \text{M2} \quad \text{M1} \quad \text{M0}
\end{array}
$$

Here is the program:

```
                ORG    $C100
M3    RMB    1
M2    RMB    1
M1    RMB    1
M0    RMB    1
LL    RMB    4


                ORG    $C200

        LDX    #LL      ;X points to MSB of 2nd element ($C104)
        LDD    M1       ;M1,M0 loaded to accD
        ADDD   2,X      ;add bytes at x+2,x+3
        STD    M1       ;store results back
        LDD    M3       ;M3,M2 loaded to accD
        ADCB   1,X      ;add byte at x+1 to ACCB
        ADCA   0,X      ;add byte at x to ACCA
        STD    M3       ;store results back
```

```
        RTS
        END
```

We wrote this code as a subroutine because you might need it frequently to do arithmetic calculations.

Note: We used ADCA and ADCB instructions starting at the second byte from the right. We also ignored the C flag.

Exercises

1. Write a subroutine to subtract two elements—each 32 bits long—using the above method.

2. Design an algorithm to perform a square of an integer n by summing the first $|n|$ odd numbers. For example,

$$3^2 = 1 + 3 + 5$$
$$5^2 = 1 + 3 + 5 + 7 + 9$$

Program 13: Multiplying a 16-bit by a 16-bit Number with a 32-bit Product

The following algorithm is used to multiply two unsigned elements, each 16 bits. The multiplicand has mpdl, mpdh for low and high bytes respectively and the multiplier has mlyl, mlyh for low and high bytes respectively. The product is stored in res1, res2, res3, and res4, where res1 is the MSB and res4 is the LSB of the 4-byte result.

Algorithm Description

The technique used here is the same as in decimal multiplication. Every time we multiply two numbers, we have to shift their product to the left and add to the previous results. For example, let's multiply two decimal numbers:

```
            67
          × 34
          ─────
            28   ◄──────── shift
            24   ◄──────── add
          ─────
           268   ◄──────── shift
            21   ◄──────── add
          ─────
           478   ◄──────── shift
            18   ◄──────── add
          ─────
          2278   ◄──────── results
```

The following algorithm will implement this technique in multiplying unsigned 16-bit by 16-bit elements as follows:

$$
\begin{array}{c}
\text{MPDH} \quad \text{MPDL} \\
\times \text{MPYH} \quad \text{MPYL} \\
\hline
\text{RES1 RES2 RES3 RES4}
\end{array}
$$

Here is the routine:

```
                              ORG  $C100

RES1 RMB 1                             ;product MSB
RES2 RMB 1
RES3 RMB 1
RES4 RMB 1                             ;product LSB
MLYH RMB 1                             ;multiplier MSB
MLYL RMB 1                             ;multiplier LSB
MPDH RMB 1                             ;multiplicand MSB
MPDL RMB 1                             ;multiplicand LSB

                              ORG  $C000

              LDD   #00               ;clear product result bytes
              STD   RES1
              STD   RES3
              LDAA  MLYL              ;multiplier LSB loaded to ACCA
              LDAB  MPDL              ;multiplicand LSB loaded to ACCB
              MUL                     ;multiply them
              STD   RES3              ;results to RES3,RES4
              LDAA  MLYL              ;multiplier LSB loaded to ACCA
              LDAB  MPDH              ;multiplicand MSB loaded to ACCB
              MUL                     ;multiply
              ADDD  RES2              ;add RES2,RES3
              STD   RES2              ;store back
              BCC   NEXT              ;branch if c=0
              INC   RES1              ;otherwise increment RES1

NEXT          LDAA  MLYH              ;multiplier MSB loaded to ACCA
              LDAB  MPDL              ;multiplicand LSB loaded to ACCB
              MUL                     ;multiply
              ADDD  RES2              ;add RES2,RES3
              STD   RES2              ;store back
              BCC   NEXT_1            ;branch if c=0
              INC   RES1              ;otherwise increment RES1

NEXT_1        LDAA  MLYH              ;multiplier MSB loaded to ACCA
              LDAB  MPDH              ;multiplicand MSB loaded to ACCB
              MUL                     ;multiply
              ADDD  RES1              ;add RES1,RES2
              STD   RES1              ;store back
              END
```

Program 14: Dividing a 16-bit by a 16-bit Number with a 32-bit Quotient

We have seen integer and fraction division before. The following routine is used to calculate both the integer part and the fraction part of a quotient.

Algorithm Description

Dividend / Divisor = Quotient (integer) + Remainder (fraction)

The 16-bit dividend is stored in DVDNH, DVDNL and the 16-bit divisor is stored in DVSRH, DVSRL. The 16-bit integer is stored in Q1, Q2 and the 16-bit remainder in ACCD is divided again, with the fraction result stored in Q3 and Q4.

```
                    ORG $0000

Q1    RMB 1                 ;quotient MSB
Q2    RMB 1
Q3    RMB 1
Q4    RMB 1                 ;remainder LSB
DVDNH RMB 1                 ;dividend MSB
DVDNL RMB 1                 ;dividend LSB
DVSRH RMB 1                 ;divisor MSB
DVSRL RMB 1                 ;divisor LSB

                    ORG $C000

        LDD DVDNH           ;dividend loaded to ACCD
        LDX DVSRH           ;divisor loaded to reg X
        IDIV                ;integer divide
        STX Q1              ;store quotient in Q1,Q2
        LDX DVSRH           ;divisor loaded to reg X again
        FDIV                ;fraction divide
        STX Q3              ;store remainder in Q3,Q4
        END
```

For example, if we divide $8844 by $2212, the integer is $0003 and the fraction $FFE1 is a binary-weighted fraction. The equivalent decimal value can be calculated as follows:

$$b_{15} \times 2^{-1} + b_{14} \times 2^{-2} + \ldots + b_0 \times 2^{-16}$$

Exercises

1. Modify Program 13 if both multiplicand and multiplier are signed numbers.

2. Design an algorithm to perform a sum of cubes of first-order n positive integers. For example,

$$1^3 + 2^3 + 3^3 + \ldots + n^3 = [n(n+1)/2]^2 \quad 1 \leq n \leq 22$$

3. Write a routine to multiply 32-bit by 32-bit numbers.

4. Design an algorithm to perform integer exponentation. Use 2 bytes, one for BASE and the other for EXPONENT. For example,

$$2^2 = 2 \times 2$$
$$2^3 = 2 \times 2 \times 2$$
$$2^n = 2 \times 2 \times \ldots n \text{ times and } n < 5.$$

5. Write a routine to solve the following equation:

$$L = S + \frac{R-Q}{M-N}$$

where the elements are 16-bit signed each.

6. Find the scale of RAM1 and RAM2 after executing the following program:

```
RPM    RMB 2          ;RPM value, scale 2^0
THRTL  RMB 1          ;throttle value, scale 2^3
CONST  RMB 2          ;constant value, scale 2^-2
```

```
a.  LDD   RPM
    LDAB  THRTL
    MUL
    LSLD
    LSLD
    STD RAM1
    END
```

```
b.  LDD RPM
    LSLD
    LSLD
    LSLD
    LDX   CONST
    IDIV
    STX   RAM2
    XGDX
    LSLD
    END
```

5 On-Chip I/O Resources

5.1   Crystal Oscillator and E-Clock

A quartz crystal exhibits the property that when mechanical stress is applied across the faces of the crystal, a difference of voltage develops across these faces. This property is called the *piezoelectric effect*. Similarly, if AC voltage is applied across the faces of the crystal, mechanical vibrations are generated that have a natural resonant frequency dependent on the crystal. The equivalent electrical resonant circuit of a crystal is shown in Figure 5.1. Cm represents the capacitance due to mechanical mounting of the crystal. The equivalent crystal quality factor (Q) is high, typically 20,000. The crystal as represented by the equivalent electrical circuit in Figure 5.1 can have two resonant frequencies. One resonant is the *series-resonant* frequency, where impedance is very low. The other resonant is the *parallel-resonant*, where impedance is very high.

You can tell if a crystal oscillator is a series- or parallel-resonant by the way the crystal itself is connected to the circuit. If the crystal is connected in series with the circuit, the oscillator is a series-resonant, but if it is connected in parallel, the oscillator is a parallel resonant. In the M68HC11, the crystal oscillator is a pierce parallel-resonant type. It consists of a two-input NAND gate.

Figure 5.1

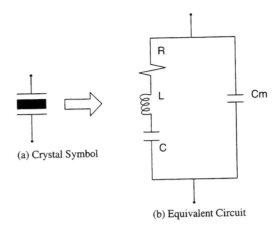

(a) Crystal Symbol

(b) Equivalent Circuit

One of the inputs to this gate is driven by an internal signal that disables the oscillator when the STOP instruction is executed. The other input is the EXTAL input pin of the MCU. The output of this NAND gate is the XTAL output pin of the MCU.

The frequency applied to these pins is four-times higher than the desired E-clock frequency. The E-clock (system clock) is the bus frequency clock output, which is used as a basic timing reference signal—that is,

$$E = \frac{\text{crystal freq.}}{4}$$

5.2 Parallel I/O Ports

Almost all MCUs have on-chip I/O parallel ports. These ports are used to interface memory and outside I/O devices to the MCUs. The design of I/O interfaces differs from the design of a memory interface. Some of the differences between memory and I/O devices are as follows:

1. When the memory cannot read or write fast enough, the CPU waits by inserting wait states, but its speed is comparable to that of the CPU. In the case of I/O devices, the rate at which an I/O device can send or receive data may be slower than that of the CPU by many orders of magnitude, and insertion of wait states is not practical. Instead, the I/O device provides the CPU with an indication of whether the I/O device is ready to send or receive information using *status bits* (called *flags*) in a *status word*. The status flags are read by the CPU data bus.

2. Memory never initiates any data transfer by itself; the CPU usually does that. But an I/O device may take the initiative for a data transfer.

3. Memory performs only read and write operations, while an I/O device may perform many other operations, such as turning on, off, moving, or any kind of activation.

The I/O ports have special capabilities, including buffering, latching, and limited current driving. Bus buffers are appropriate for the buffering of address lines because of their unidirectional nature. Data buffers are buffered with *transceivers* (transmitters / receivers), which permit two-way communication (Figure 5.2). The direction of the data (D) depends

Figure 5.2

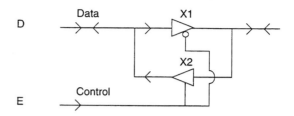

on the polarity of the enable (E) signal. If E is positive, X2 is enabled and X1 is disabled and the data direction is from right to left. If E is negative, X1 is enabled and X2 is disabled and the data direction is from left to right. Latches are used for address/data demultiplexing. They are used when I/O ports are used as outputs.

Either the internal RAM (512 bytes) or the internal registers (64 registers) can be remapped to any 4K boundary by software. In the M68HC11E9, internal RAM resides from $0000 to $01FF and the internal registers reside from $1000 to $103F, as shown in the memory map. The address $1000 is called the *base address* and is usually loaded to index register X at the beginning of the program. Then we use the indexed mode to access any register. For example, the address of port B is $1004. To output any data out of port B, we write:

```
LDX     #$1000      ;base address for registers
STAA    PORTB,X     ;write to port B
```

where port B is declared to equal $04.

5.3   I/O Ports in the MC68HC11E9

Port A

As mentioned in Chapter 1, Port A has three input-only pins (PA0 to PA2), three output-only pins (PA4 to PA6), and two bidirectional pins (PA3 and PA7). (See Figure 5.3.) PA7 is used for general-purpose I/O interfacing. After reset, it becomes input and can be programmed as a general-purpose output line if DDRA7 is set. Port A has a double role. The PA0 to PA2 lines serve as input capture (IC3 to IC1) for the programmable timer, as we will discuss in

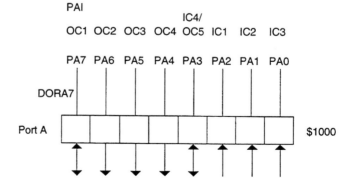

Figure 5.3

detail later. The PA4 to PA6 lines serve as output compare (OC4 to OC2). PA3 serves as IC4 or OC5 in the 68HC11E9. Bit 3 of the TMSK1 register determines this selection. PA7 can serve as the input to an 8-bit pulse accumulator facility for counting pulses if this accumulator is enabled or can be programmed as special output compare OC1.

Port B

Port B is a fixed output port and lines PB0 to PB7 are shown in Figure 5.4 in the single mode. In the expanded mode, port B lines are replaced by the upper half of the address bus A8 to A15.

Port C

Port C is a bidirectional port in single-chip mode. It can be programmed either as an input or an output port using data direction register C (DDRC). The DDR bits determine the direction of the I/O line. If a DDR bit is reset, the corresponding I/O line becomes an input to the port. If a DDR bit is set, the corresponding I/O line becomes an output to this port.

If the chip operates in expanded mode, port C lines become the multiplexed address and data lines of the lower half of the address bus, as shown in Figure 5.5. The low 8 address lines (total of 16 address lines) are multiplexed with data lines D0 to D7. To program port C as an input, DDRC must be cleared as follows:

```
LDAA #00
STAA DDRC     ;port C is input
```

and to program port C as an output, DDRC must be set as follows:

```
LDAA  #$FF
STAA  DDRC    ;port C is output
```

Ports B and C are also used in the single-strobed handshaking mode, as we will see in the next section.

Port D

PD0 to PD5 are bidirectional general-purpose I/O lines in expanded mode. PD6 and PD7 are used as AS and R/$\overline{\text{W}}$ lines. In single-chip mode, it is used as a serial communication port for SCI and SPI applications using six lines only. PD6 and PD7 are used in handshake only as STRA and STRB lines respectively.

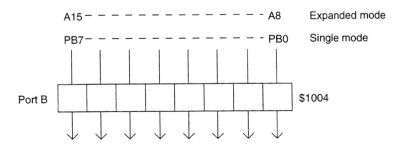

Figure 5.4

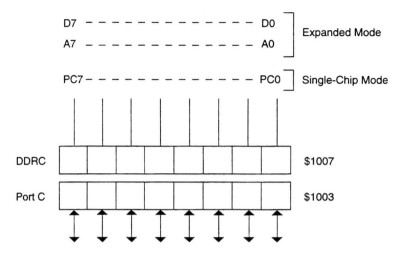

Figure 5.5

Port E

PE0 to PE7 are fixed input lines. They are used as general I/O lines or as ADC lines. If the ADC feature is enabled, these lines are called AN0 to AN7 (eight channels). Recall that if the MCU is a 48-pin DIP type, there are only four lines (AN0 to AN3) or four channels in ADC mode. If the chip is a 52-pin PLCC type, port E has eight lines.

Program 1: **Simple I/O Transfer**

Write a program to read data from port C and output it to port B.

```
ORG  $C000
LDX  #$1000      ; base address
LDAA #00
STAA DDRC,X      ; port C is input (if not default)
LDAA PORTC,X
STAA PORTB,X
END
```

where DDRC has an offset of $09, port C of $08, and port B of $04. Thus the DDRC address is 09 + 1000 = $1009 and so on for the rest.

5.4   Handshaking

In single-chip mode, the MCU 68HC11 can communicate with outside devices (peripherals) using handshaking techniques. There are two types of handshakes:

1. Simple strobed I/O handshake.

2. Full handshake I/O, which is divided into two parts:
   a. Interlocked (input or output)
   b. Pulsed (input or output)

Simple Strobed I/O Handshake

In this handshake, port B, port C, DDRC, port C latch (port CL), parallel I/O control register (PIOC), STRA, and STRB are involved. Ports B and C communicate with the peripheral as shown in Figure 5.6, where port B is used as an output and port C as an input port. Figure 5.7 illustrates PIOC bit functions.

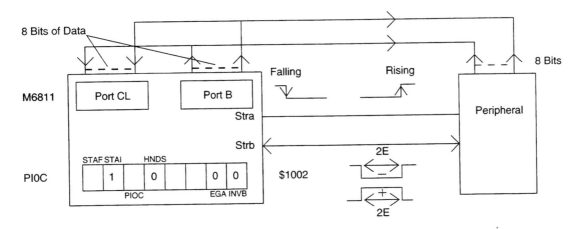

Figure 5.6

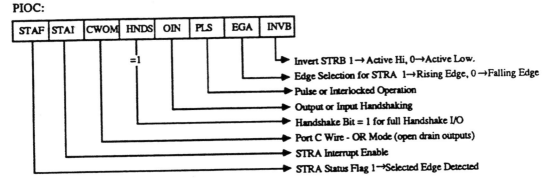

REGISTERS:
- DDRC — Select desired PORT C pins as inputs or outputs.
- PORTCL — Latch port C input data on STRA edge
    (used to read latched data)
- PORTC — Input pin values (not latched) on Port C pins
    (used to read current data)
- PIOC — Parallel I/O Control Register

PINS:
- PORT C – 0-7 (inputs or outputs)
- PORT D – 6, 7 (STRA, STRB)

Figure 5.7

The following explains how this simple-strobed handshake works:

1. The PIOC register must be configured first:
   a. HNDS bit B4 = 0.
   b. STRA is programmed to be either a rising or a falling edge (edge-sensitive) using B1.
   c. STRB is programmed to be either active low or hi (level sensitive) using B0.
   d. If you use the interrupt technique, then B6 (STAI) must be enabled—that is, it must equal 1. But if you use the polling technique, then B6 = 0; use the status flag B7 STAF.

2. The MCU tells the peripheral that it has data to be sent. This occurs by writing to port B, which generates a pulse (STRB). The polarity of this pulse depends on B0 of PIOC.

3. The MCU waits until it receives a pulse from the peripheral that it is ready to send new data. As mentioned, the MCU is programmed to detect either a rising or a falling edge. Once the MCU receives data, these data will be transferred from port C to port CL. The status flag STAF is set, and if STAI is set, the IRQ interrupt is requested.

4. This process is repeated over and over.

Note: Port C in this case is input by default (after reset). Also, STRB width is 2E clock cycles.

Exercise

Write a program for a simple strobed I/O handshake in which the input data are echoed to the output using the polling technique.

Full Handshake

In this mode port B is not involved anywhere. Port C is programmed as an input or an output port. As we mentioned, there are two cases:

1. Interlocked full handshake (input and output)
2. Pulsed full handshake (input and output)

The HNDS bit in this case is set to 1. If it is in an interlock mode, PLS bit = 0; if it is in a pulsed mode, PLS bit = 1. The difference between the interlocked and pulsed handshake is that in the interlocked, STRB will end its width only when a STRA is initiated, while in the pulsed handshake, the STRB width is equal to 2E. Figure 5.8 shows an input handshake for input interlocked mode. The same process applies to the output handshake (interlocked and pulsed), except that port C is an output.

Exercise

A high-speed paper tape reader is connected to the M68HC11 as indicated in Figure 5.9. The sprocket hole sends a positive signal through an electronic interface circuit on the STRA line. When the MCU detects the strobe signal, it sends an active high pulse to a stepper motor to move to the next character. What handshake mode does this circuit use? Write a program to illustrate that process.

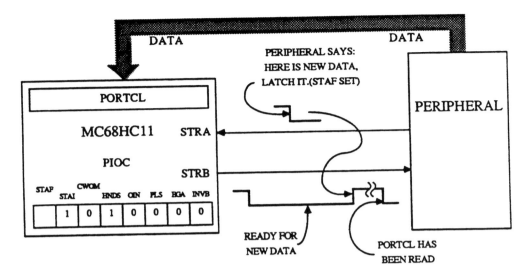

Figure 5.8
Copyright of Motorola. Used by permission.

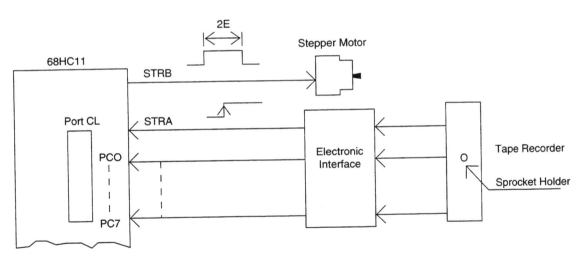

Figure 5.9

5.5 Reset and Interrupts

In the M68HC11, there are a total of 21 interrupt sources. Six are nonmaskable interrupts, and the other 15 are maskable interrupts. The word *maskable* means that it can be prevented or inhibited. The 6 nonmaskable interrupts sources are the following:

1. Reset
2. Clock monitor
3. COP watchdog

4. Illegal opcode

5. External interrupt request (XIRQ)

6. Software interrupt (SWI)

The 15 maskable interrupt sources are:

1. Interrupt request (IRQ)

2. RTI interrupt

3. Timer input capture 1

4. Timer input capture 2

5. Timer input capture 3

6. Timer output compare 1

7. Timer output compare 2

8. Timer output compare 3

9. Timer output compare 4

10. Timer input capture 4/output compare 5

11. Timer overflow

12. Pulse accumulated overflow

13. Pulse accumulated input edge

14. SPI Transfer Complete

15. SCI Serial System

Maskable Interrupts

Unlike resets, which are asynchronous, stopping the flow of the program in mid-instruction and forcing it to restart, interrupts are synchronous. Interrupts are considered pending until the current instruction is complete. These interrupts affect the I-bit in the CCR register and are in the above sequence by default priority. The maskable interrupts can be elevated to the highest priority (7) by writing to the High Priority (HPRIO) register bits B0 to B3. This priority can be changed only when the I bit = 1. Table 5.1 summarizes selected values to change the highest priority.

IRQ is the highest priority by default, as noted earlier. If we wish to change this default to let the timer input capture 1, for example, be the highest-priority interrupt, then we have to write $08 to the HPRIO register. The default configuration for the IRQ pin in the M68HC11 is a low-level-sensitive wired-OR network. IRQ can be programmed as a low-going edge-sensitive, via IRQE bit in the OPTION register.

Nonmaskable Interrupts

The nonmaskable interrupts have a fixed-priority interrupt relationship. SWI has the highest priority over the others.

Resets

The reset pin on the M68HC11 is a *bidirectional* line. It includes *external* and *internal resets*. In external resets, the reset line is an input to the MCU, but in internal resets, this line is an output.

Table 5.1

PSEL3	PSEL2	PSEL1	PSEL0	Interrupt Source Promoted
0	0	0	0	Timer Overflow
0	0	0	1	Pulse Accumulator Overflow
0	0	1	0	Pulse Accumulator Input Edge
0	0	1	1	SPI Transfer Complete
0	1	0	0	SCI Serial System
0	1	0	1	Reserved (Default to $\overline{IRQ}$)
0	1	1	0	$\overline{IRQ}$ (External Pin or Parallel I/O)
0	1	1	1	Real-Time Interrupt
1	0	0	0	Timer Input Capture 1
1	0	0	1	Timer Input Capture 2
1	0	1	0	Timer Input Capture 3
1	0	1	1	Timer Output Compare 1
1	1	0	0	Timer Output Compare 2
1	1	0	1	Timer Output Compare 3
1	1	1	0	Timer Output Compare 4
1	1	1	1	Timer Output Compare 5

External Reset

The reset pin is pulled down by a hardware switch for more than 6E cycles. Then the MCU executes the reset interrupt service routine at addresses $FFFE and $FFFF. All resets (external and internal) cause the CPU's internal registers and on-chip peripherals to be initialized. The external reset circuit is shown in Figure 5.10.

Internal Reset

Power-On Reset. At cold start, when the MCU is powered on, an R-C timing circuit holds the reset line low for 4064 E-clock cycles to allow the oscillator to stabilize. Then the MCU executes the reset interrupt service routine at addresses $FFFE and $FFFF.

Figure 5.10

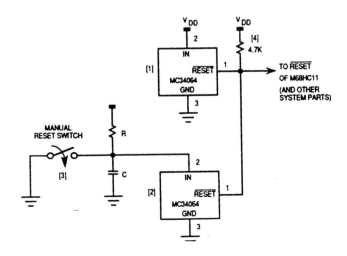

Clock Monitor. The clock monitor function is enabled/disabled by the CME control bit in the OPTION register. When the clock monitor is enabled and the MCU clock drops below a frequency of 10 KHz or if it stops, a system reset is generated and the reset line is driven low.

Computer Operating Properly

The Computer Operating Properly (COP) watchdog timer is intended to detect software processing errors. When the COP is being used, software is responsible for keeping a free-running watchdog timer from timing out. If the watchdog timer times out, it is an indication that software is no longer being executed in the intended sequence. Thus a system reset is generated.

To enable the COP system, the NOCOP bit must be programmed to zero, then the COP time-out period is selected as well as values of CR0 and CR1 of the option register. Table 5.2 summarizes the watchdog rates.

To avoid having the watchdog timer resetting the CPU, the system software must write $55 to the COPRST to assume the COP timer-clearing mechanism and then write $AA to clear the COP timer as follows:

```
LDAA   #$55
STAA   COPRST
COMA
STAA   COPRST
```

This code must be executed with the time between execution of the sequence *never* exceeding the selected time-out period (from Table 5.2). As the 68HC11 comes out of reset, the CPU does the following automatically:

1. It sets the X, I, and S bits in CCR so that interrupts and the STOP mode are initially disabled.

2. The INIT register is initialized to 01—that is, RAM is at $000 to $00FF and registers are at $1000 to $103F.

3. All bidirectional I/O lines are configured as inputs.

4. The CPU fetches the reset vector from $FFFE and $FFFF.

Table 5.2

			Crystal Frequency		
			2^{23} Hz	8 MHz	4 MHz
CR1	CR0	E ÷ 2^{15} Divided By	Nominal Time-Out		
0	0	1	15.625 mS	16.384 mS	32.768 mS
0	1	4	62.5 mS	65.536 mS	131.07 mS
1	0	16	250 mS	262.14 mS	524.29 mS
1	1	64	1 sec	1.049 sec	2.1 sec
			2.1 MHz	2 MHz	1 MHz
			Bus Frequency (E Clock)		

Interrupt Vector

Each interrupt source (maskable and nonmaskable) has a 2-byte PROM address, which is loaded into the program counter. This 2-byte interrupt vector (interrupt address) could contain the real address of the interrupt service routine (ISR). Table 5.3 shows the interrupt vectors for all possible interrupts. If the I or X bit is 0, the corresponding interrupt is enabled, but if any I or X is 1, the corresponding interrupt is disabled. The following interrupt sequences occur in case of an interrupt:

1. The CPU completes the instruction it is currently executing.
2. It stacks all CPU registers as Figure 5.11 indicates.
3. It sets the I/X bits to disable further interrupts.
4. It fetches the proper interrupt vector and executes the ISR.

In the EVB system, the contents of a vector address point to the beginning of a 3-byte location in RAM for each interrupt. The RAM locations are used to contain a jump

Table 5.3

Vector Address	Interrupt Source	CC Register Mask
FFCO,C1	RESERVED	
	$16 BYTES	
FFD4,D5	RESERVED	
FFD6,D7	SCI SERIAL SYSTEM	I
FFD8,D9	SPI SERIAL TRANSFER COMPLETE	I
FFDA,DB	PULSE ACCUMULATOR INPUT EDGE	I
FFDC,DD	PULSE ACCUMULATOR OVERFLOW	I
FFDE,DF	TIMER OVEFLOW	I
FFE0,E1	TIMER OUTPUT COMPARE 5	I
FFE2,E3	TIMER OUTPUT COMPARE 4	I
FFE4,E5	TIMER OUTPUT COMPARE 3	I
FFE6,E7	TIMER OUTPUT COMPARE 2	I
FFE8,E9	TIMER OUTPUT COMPARE 1	I
FFEA,EB	TIMER INPUT CAPTURE 3	I
FFEC,ED	TIMER INPUT CAPTURE 2	I
FFEE,EF	TIMER INPUT CAPTURE 1	I
FFF0,F1	PERIODIC INTERRUPT	I
FFF2,F3	$\overline{\text{IRQ}}$	I
FFF4,F5	$\overline{\text{XIRQ}}$	X
FFF6,F7	SWI	NONE
FFF8,F9	ILLEGAL OPCODE TRAP	NONE
FFFA,FB	COP FAILURE	NONE
FFFC,FD	CLOCK MONITOR FAILURE	NONE
FFFE,FF	$\overline{\text{RESET}}$	NONE

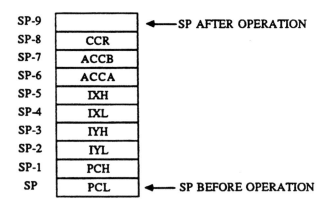

Figure 5.11
Copyright of Motorola. Used by permission.

instruction and an absolute address where the user interrupt service routine starts. These absolute addresses are called *pseudovectors*. Table 5.4 summarizes the interrupt vector jump table in RAM for the EVB.

Illegal Opcode Trap

When an illegal opcode is detected, an interrupt is requested to the illegal opcode vector. The illegal opcode vector should never be left uninitialized.

Table 5.4

Interrupt Vector	Field
Serial Communications Interface (SCI)	$00C4–$00C6
Serial Peripheral Interface (SPI)	$00C7–$00C9
Pulse Accumulator Input Edge	$00CA–$00CC
Pulse Accumulator Overflow	$00CD–$00CF
Timer Overflow	$00D0–$00D2
Timer Output Compare 5	$00D3–$00D5
Timer Output Compare 4	$00D6–$00D8
Timer Output Compare 3	$00D9–$00DB
Timer Output Compare 2	$00DC–$00DE
Timer Output Compare 1	$00DF–$00E1
Timer Input Capture 3	$00E2–$00E4
Timer Input Capture 2	$00E5–$00E7
Timer Input Capture 1	$00E8–$00EA
Real Time Interrupt	$00EB–$00ED
IRQ	$00EE–$00F0
XIRQ	$00F1–$00F3
Software Interrupt (SWI)	$00F4–$00F6
Illegal Opcode	$00F7–$00F9
Computer Operating Properly (COP)	$00FA–$00FC
Clock Monitor	$00FD–$00FF

Software Interrupt

The SWI instruction is executed in a manner similar to other maskable interrupts in that it sets the I bit, and the CPU registers are stacked, but it is not inhibited by interrupt mask bits I and X. The SWI instruction is commonly used in debug monitors to transfer control from a user program to the debug monitor. It is used to implement breakpoints and single-stepping.

External Interrupt Request

The XIRQ interrupt is the only hardware interrupt request (besides the hardware reset, whose only purpose is to reset the CPU). It is low-level sensitive. As mentioned earlier, the X bit of the CCR is set at reset time, disabling the hardware interrupt. The X bit is cleared to enable it by using a few instructions, as follows:

```
TPA                 ;transfer CCR to ACCA
ANDA #$BF           ;clear X bit in ACCA
TAP                 ;transfer ACCA to CCR
```

There is no single instruction to clear the X bit, as in the case of the I bit. Once the X bit has been cleared, the software cannot set it again. Thus, it is nonmaskable. After this interrupt occurs, both the X and I bits are set automatically. An RTI instruction in the ISR restores the X and I bits to their preinterrupt states.

One application using the features of XIRQ is an external power-sensing circuit to warn the MCU that the supply voltage has dropped below a certain level.

Return from Interrupt

When an interrupt has been serviced as needed in the ISR, the RTI instruction terminates the interrupt processing and returns to the main program. When the RTI is executed, all registers pushed into the stack are pulled out.

Exercises

1. What are the differences between maskable and nonmaskable interrupts? Give examples.
2. Explain the differences between IRQ and XIRQ.
3. Explain the function of the clock monitor and the COP watchdog timer, and describe how they reset the CPU.
4. How would you change the high-priority order of a certain maskable interrupt?
5. What is the difference between synchronous and asynchronous interrupts?

I/O Techniques

When interfacing the peripheral devices, three techniques are used to transfer data between these devices and the I/O ports.

Direct I/O

In the direct I/O, the CPU unconditionally reads one I/O port and writes to another without a need for prior status checking. A good example is the switches/LEDs interface program discussed in Chapter 6. The CPU reads the status of the switches and correspondingly turns on or off the LEDs. This technique is not common but is used in special cases.

Polled I/O

In this technique, the CPU first checks the status of the I/O device using status bits (flags). If the device is not ready, the CPU may jump to other tasks. After a timed interval, it comes back to check the status again. If the device is ready, the CPU services this device (transfers data). This is called *polling.* The disadvantage of polling is that the CPU time is wasted for status checks.

Interrupt-Driven I/O

In interrupt-driven I/O, data transfers are initiated by the I/O device, which uses the interrupt mechanism to notify the CPU of its readiness. The CPU suspends the current processing to service the interrupt. The ISR handles the required I/O processing. This technique is the most efficient because it eliminates the burden associated with status scanning.

5.6   Serial Communication Interface

There are two serial I/O systems used in the M68HC11 MCU: the Serial Communication Interface (SCI) and the Serial Peripheral Interface (SPI). The SCI is asynchronous and is used to communicate between the MCU and any other remote devices, such as other MCUs or a CRT. The SPI is synchronous and is used to communicate between the MCU and any devices on board or close by. Both SCI and SPI use port D lines.

SCI Features

Some SCI features are:

1. Two-wire serial interface
2. Standard NRZ (Non-Return-Zero) mark/space format
3. Full-duplex operation
4. Programmable word length (8 or 9 bits)
5. Interrupt-driven communication
6. Software-programmable internal baud rate generator

SCI Receiver Features

The SCI receiver has the following features:

1. Idle line detection
2. Framing error, noise, and overrun detection
3. Receiver data register full detection
4. Receiver wake-up capability

SCI Transmitter Features

The SCI transmitter has these capabilities:

1. Transmit data register empty detection
2. Transmit complete detection
3. Break transmission
4. Idle line generation

Character Format

In asynchronous transmission, data are sent in serial frames as depicted in Figure 5.12. No clock is involved in synchronizing the data, which is why it is called *asynchronous*. Each frame is defined by a *start bit* at the beginning and a *stop bit* at the end. It contains 8 or 9 bits of data called a *character*. The total bits including start and stop bits form a *frame*. In serial communications, high logic (1) is called a *mark* and low logic (0) is called a *space*. A start bit is always a space, while a stop bit is a mark. The parity bit is used to detect any fault occurring during transmission. It could be *even* or *odd* parity. Even parity means the number of marks in a frame including the parity bit should be even. In odd parity, the number of marks is odd.

SCI Transmitter

The software writes to the SCI data register buffer SCDT, then the data are transferred to the transmitter shift register. Start and stop bits are added to the data in this register through the transmitter control register. Also, the ninth bit (T8) could be added to make the character 9 bits long instead of 8. This requires programming the SCCR1 register. The data are shifted serially to the outside world through pin T_xD (PD1) in the serial format displayed in Figure 5.13. Notice that the data come in parallel form from the CPU (8 lines) and are transferred to serial form through the shift register to the outside world. The LSB of the data is shifted out first and the MSB last.

SCI Receiver

The software enables the shift register to receive data from the outside world in serial form through the R_xD pin (PD0). After detecting the stop bit (end of the frame), the data are transferred to the SCDR register without start and stop bits in parallel form to the CPU.

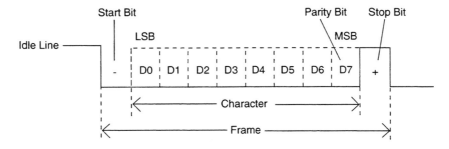

Figure 5.12

Parallel Data from CPU Data Bus

SCDT | D7 | | | | | | | D0

Shift Register | 1 | D7 | | | | | | | D0 | 0 | → Serial Data □ $T_x D$

STOP   START

Figure 5.13

Double Buffering

The SCDR data register has two functions: it is the transmit data register when written to, and the receive data register when read from. Both the transmitter and the receiver are double buffered.

SCI Registers

There are five registers in the SCI system; we will study each.

Baud Rate Control Register (Baud)

The baud register is used to select the baud rate for the SCI system (Figure 5.14). The baud rate is defined as the number of transmitted bits per second (bps). The data rate is the number of characters per second (see Figure 5.12). Both the transmitter and the receiver should have the same baud rate. The SCP0 and SCP1 bits are used as a prescales for the SCR0 to SCR2 bits to provide the multiple baud rate combination for a certain crystal frequency. For the baud rate prescale tables, see Tables 5.5 and 5.6. Bits TCLR and RCKB are used in factory testing only.

$$\text{Selected baud rate} = \frac{\text{Prescale baud rate}}{2^n}$$

where n = number of bits in SCR0 to SCR2

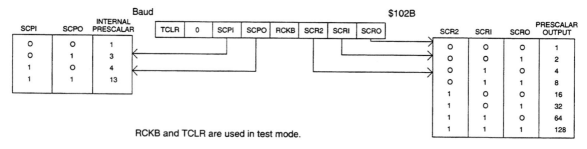

Figure 5.14

Table 5.5

			Crystal Frequency				
			2^{23} Hz	8 MHz	4.9152 MHz	4 MHz	3.6864 MHz
SCP1	SCP0	Division Factor	Highest Baud Rate				
0	0	1	131.072K Baud	125.000K Baud	76.80K Baud	62.50 Baud	57.60K Baud
0	1	3	43.691K Baud	41.667K Baud	25.60K Baud	20.883K Baud	19.20K Baud
1	0	4	32.768K Baud	31.250K Baud	19.20K Baud	15.625K Baud	14.40K Baud
1	1	13	10.082K Baud	9600 Baud	5.908K Baud	4800 Baud	4431K Baud
			2.1 MHz	2 MHz	1.2288 MHz	1 MHz	921.6 KHz
			Bus Frequency (E clock)				

Copyright of Motorola. Used by permission.

Table 5.6

				Highest Baud Rate				
				131.072K Baud	32.768K Baud	76.80K Baud	19.20K Baud	9600 Baud
SCR2	SCR1	SCR0	Division Factor	SCI Baud Rate				
0	0	0	1	131.072K Baud	32.768K Baud	76.80K Baud	19.20K Baud	9600 Baud
0	0	1	2	65.536K Baud	16.384K Baud	38.40K Baud	9600 Baud	4800 Baud
0	1	0	4	32.768K Baud	8192 Baud	19.20K Baud	4800 Baud	2400 Baud
0	1	1	8	16.384K Baud	4096 Baud	9600 Baud	2400 Baud	1200 Baud
1	0	0	16	8192 Baud	2048 Baud	4800 Baud	1200 Baud	600 Baud
1	0	1	32	4096 Baud	1024 Baud	2400 Baud	600 Baud	300 Baud
1	1	0	64	2048 Baud	512 Baud	1200 Baud	300 Baud	150 Baud
1	1	1	128	1024 Baud	256 Baud	600 Baud	150 Baud	75 Baud

Copyright of Motorola. Used by permission.

EXAMPLE 1

A crystal frequency is equal to 8 MHz. Find the status of the baud register bits to select a baud rate of 4800 bps.

Solution

From Table 5.5, SCP0 = SCP1 = 1 to get a preschool of 9600. Then from Table 5.6, under the 9600 baud column, the status of SCR0 is 1 and SCR1 and SCR2 are 0 (the division factor is 2), yielding a baud of 4800. The Baud register has 31 hex.

SCI Control Register 1 (SCCR1)

This register is shown in Figure 5.15. The WAKE bit is used to select one of two receiver wake-up methods: idle line or address mark. If the M bit is reset, it selects 8 bits of data; if it is set, it selects 9 bits of data. If M is set, bit T8 is loaded with the 9th bit during transmission. Also, bit R8 is loaded with the 9th bit during receiving. The 9th bit may be used for parity generation or detection. It may be also used to distinguish between ordinary data and an address in the network level of a protocol.

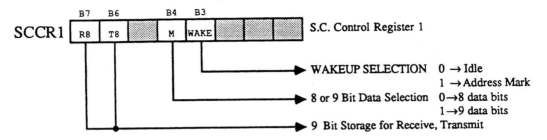

Figure 5.15
Copyright of Motorola. Used by permission.

SCI Control Register 2 (SCCR2)

The bit function of this register is indicated in Figure 5.16. The transmitter and receiver can be enabled through the TE and RE bits respectively. Transmitter interrupt and receiver interrupt can also be enabled through TIE and RIE bits respectively. To send break characters, the SBK is set. In this case, the T_xD line goes to zero. The character length for all characters is influenced by the M bit in SCCR1, as mentioned earlier. Break characters have to be Start or Stop bits.

If the idle line interrupt is enabled, the ILIE bit should be set. When the transmitter is not sending any characters, it is idling and T_xD is set to 1. The first time the transmitter is enabled, the idle character (10 or 11 bits) is transmitted and acts as a preamble. This ensures that any receiver connected to the transmitter will be synchronized. If the transmission is complete, interrupt could be enabled if TCIE is set. Also, if RIE is set, the interrupt is enabled when receiving is done.

SCI Status Register (SCSR)

The register bits are shown in Figure 5.17. These seven bits reflect the transmitter and receiver status.

- B_1 is set if there is a framing error in receiving data—that is, it should be a stop bit, which is high and comes at the end of the frame.
- B_2 is set if the receiver detects a short positive or negative pulse longer than one-sixteenth of a bit time and shorter than the bit time.

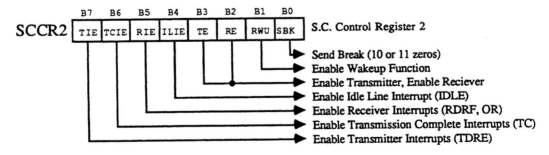

Figure 5.16
Copyright of Motorola. Used by permission.

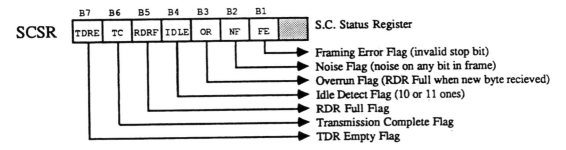

Figure 5.17
Copyright of Motorola. Used by permission.

- B_3 is set when the receiver overruns—that is, data have to be moved from the input shift register before the previously input data are read from the data register (SCDR).
- B_4 is set when it is idling—that is, the receiver input line remains high for longer than a full frame (10 or 11 bits) unless the SCI is in sleep mode.
- B_5 is set if the receiver data register is full, because a frame has been received.
- B_6 is set if the transmission is complete—that is, at the end of sending a frame.
- B_7 is set if the transmitter data register is empty—that is, when data are moved from the data register to the shift register.

SCI Data Register

As mentioned previously, the SCDR is actually two separate registers. When SCDR is read, the read-only RDR is accessed; when SCDR is written, the write-only TDR is accessed. Both registers have the same address.

EXAMPLE 2
Use the information in Example 1 to calculate:

a. Bit time

b. Data rate

c. Number of characters transmitted per second. Assume 8 bits of data.

Solution

a. Bit time $= \dfrac{1}{\text{baud rate}} = \dfrac{1}{4800 \text{ bps}} = 0.20833 \text{ mS} = 208.33 \text{ }\mu\text{S}$

b. Data rate = number of data bits only transmitted per second
   Since number of data bits are 8 out of total 10 bits, then

$$\text{data rate} = \frac{8}{10} \times \text{baud rate}$$

$$= \frac{8}{10} \times 4800 = 3840 \text{ bps}$$

c. Number of transmitted characters/sec $= \dfrac{1 \text{ sec}}{8 \times 208.33 \text{ }\mu\text{S}} = 600 \text{ characters}$

Exercise

In Example 2, if the data length is 9 bits, calculate:

a. Data rate

b. Character rate

c. Frame rate

Program 2: **Transmitting and Receiving Using SCI**

Write two routines: The first routine is to dump a block of data from memory onto a cassette tape. The second routine is to load a block of data onto a memory from a cassette tape. The baud rate in both cases is 9600.

First Routine. The program is to transmit data from a look-up table (memory) starting at $C100 and ending at $C135 to a cassette tape connected to T_xD and R_xD of the M68HC11 SCI system.

First, we have to define all register addresses, variables, and constants at the beginning of each program.

```
BAUD    EQU $2B
SCCR1   EQU $2C
SCCR2   EQU $2D
SCSR    EQU $2E
SCDR    EQU $2F
BASE    EQU $1000
BEGIN   EQU $C100
END     EQU $C135
```

We assume that our crystal frequency is 8 MHz. By examining Tables 5.5 and 5.6, we can select the baud rate.

```
                ORG    $C000


        ;   Initialization routine

        LDX    #BASE
        LDAA   #$30
        STAA   BAUD,X     ; 9600 baud is selected
        LDAA   #00
        STAA   SCCR1,X    ; 8 bits of data
        LDAA   #$0C
        STAA   SCCR2,X    ; Tx & Rx are enabled

        ;   Transmit routine

        LDY    #BEGIN     ; point to the beginning of table
NEXT    LDAA   0,Y

        STAA   SCDR,X
SEND    BRCLR  SCSR,X,$80,SEND
        INY               ; to next byte
```

```
            CPY    #END       ; end of table yet?
            BNE    NEXT       ; no, branch
   HERE     BRA    HERE       ; yes, stop
            END
```

In the initialization routine, we selected a baud rate of 9600 by storing $30 in the baud register, then we selected 8 bits of data length through the SCCR1 register. Finally, we enabled both transmission and receiver through the SCCR2 register. In the transmission routine, we load the first byte of the look-up table to SCDR. Once the SCDR is loaded with data from the CPU, this data will be transmitted automatically through pin $T_X D$. We check bit TDRE in register SCSR to be sure that the SCDR is empty and data are transmitted out. The TDRE bit is automatically cleared when the status register is read and a new character is written.

Then, we send the second byte of the data table and check if an end of table has been reached. If not, we increment the pointer to the next byte and send this byte out. We repeat this process until we reach end of the table, then the program stops.

Second Routine. This routine has the opposite function of the first one. This routine receives data (instead of transmitting it) from the cassette tape and stores it in memory.

```
   ; use the same equates and initialization routines, then proceed
   ; to the following code:

                ORG $C000

            LDY    #BEGIN
   RECEIVE  BRCLR  SCSR,X,$20,RECEIVE
            LDAA   SCDR,X     ; get the received byte
            STAA   0,Y        ; store it in data table
            INY               ; go to next byte
            CPY    #END       ; is end of table yet?
            BNE    RECEIVE    ; no, branch
   HERE     BRA    HERE       ; yes, stop
            END
```

In this receive routine, we have to wait until SCDR is full and then we take this byte of information and store it in the data table. We keep doing this until we reach the end of the table, then the program stops.

Program 3:   SCI Gets an Interrupt When Receiving Data

The SCI receives data bytes and stores them in the RAM buffer. If it detects an error, the routine writes $FF to that buffer location.

```
                ORG $C000

        ; Equates here
        ; Init routine

   LDX   #BASE
   LDAA  #$30
   STAA  BAUD,X
   LDAA  #00
   STAA  SCCR1,X
```

```
                LDAA #$0C
                STAA SCCR2,X
                LDY  #BUFFER
                CLI
WAIT            BRA  WAIT

RECVISR         BRCLR SCSR,X,$0E,NO_ERR
                LDAA #$FF
                STAA 0,Y
                LDAA SCDR,X      ;clear flags
                BRA  NEXT
NO_ERR          LDAA SCDR,X
                STAA 0,Y
NEXT            INY              ;next buffer
                RTI


                ORG $FFD6


        FDB         RECVISR
```

Exercises

1. Write a program to load a printer from a queue. The queue is 256 bytes long.

2. Write a routine for the receiver to detect the three types of receiving errors.

5.7   Serial Peripheral Interface

The SPI is a synchronous system and is used for communication with peripherals such as
LCD drivers and DAC converters as well as with other microcontrollers. The SPI block
diagram is shown in Figure 5.18. It has four pins: MIS0, MOSI, SCK, and $\overline{SS}$.

SPI Features

- Full duplex
- Master or slave operations
- Programmable clock polarity and phase

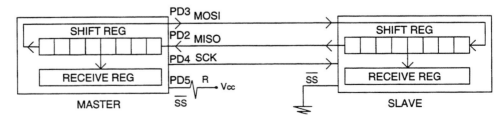

Figure 5.18

- End-of-transmission interrupt flag
- Write-collision flag protection

SPI Operation

Master and slave devices communicate by shifting serial data to each other's registers. On completion of an 8-bit shift, the following occur:

1. The SPF status flag is set.
2. An interrupt request may be generated if enabled.

When the SPI is configured as a master, Master In Slave Out (MISO) is the master "data input" line and Master Out Slave In (MOSI) is the master "data output" line. When SPI is configured as a slave, MISO becomes the "data output" line and MOSI becomes the "data input" line, as shown is Figure 5.18.

$\overline{SS}$ may be pulled high on the master or can be used to enable SPI slave devices. The serial clock (SCK) signal is used to synchronize the data in and out of the device through MOSI and MISO pins. SCK is initiated from the master devices and data are shifted on one edge of the clock and sampled on the opposite edge. The clock SCK could be in phase or 180° out of phase depending on the control bit clock polarity (CPOL) in the control register. We should mention that the data transfer occurs through the shift registers either in the master or the slave device. The MSB is shifted first and the LSB last. This is the opposite of the SCI operation, where the LSB is shifted first, then the MSB last. Table 5.7 shows the direction of each pin in a master-slave combination.

SPI Registers

There are three registers to control the SPI operation: the control register (SPCR), the status register (SPSR), and the data register (SPDR).

Control Register
SPCR selects the clock rate and phase of SPI and other control bits, as shown in Figure 5.19.

Status Register
When the data transfer is done, the SPIF flag bit is set. It is automatically cleared by reading the SPSR register followed by an access of the SPDT register. The WCOL flag is set if an attempt was made to write to the serial peripheral data register while data transfer was in progress. This is similar to the overrun flag operation in the SCI system. To clear

Table 5.7

Device Mode	Signals MISO	MOSI	SCK	$\overline{SS}$
Master	Input	Output	Output	Programmable
Slave	Output	Input	Input	Input

MISO output is three-stated until enabled by $\overline{SS}$.
Copyright of Motorola. Used by permission.

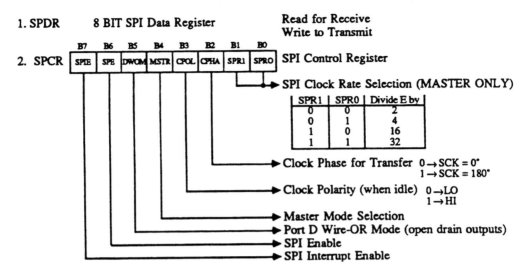

Figure 5.19
Copyright of Motorola. Used by permission.

the WCOL flag, SPSR is read followed by read or write to SPDR, after SPIF is set. The MODF flag is set if the SS signal goes low while SPI is configured as a master. To clear this flag, SPSR is read followed by a write to SPCR. (See Figure 5.20.)

Data Register

Writing to this register in a master mode will initiate transmission/reception of data. The data are then loaded directly to the 8-bit shift register. On completion of data transfer, the SPIF flag is set.

Program 4:   Using the SPI Subsystem of the M68HC11

This program shows how to use the M68HC11 as a master to output data to a slave device. The slave device is a Serial-In Parallel-Out (SIPO) register such as MC74HC595, as shown in Figure 5.21. PC0 is used to enable SIPO when this line goes low and disables it when it goes high. When SW is closed, the master SPI sends a byte of data. When the switch (SW) is pressed initially, 00 is sent to the slave, then it increments at the second pressing of the switch and so

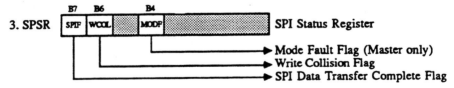

PINS: 1. $\overline{SS}$, SCK, MOSI, MISO, (Port D Pins 5-2)

Figure 5.20
Copyright of Motorola. Used by permission.

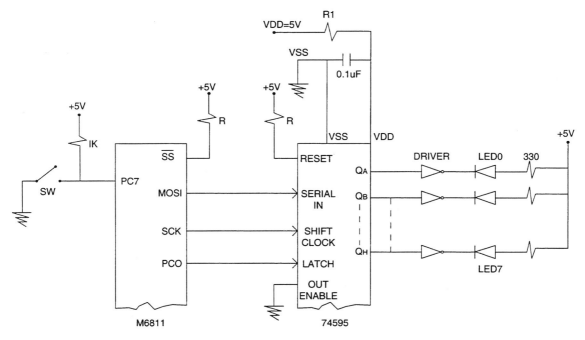

Figure 5.21

on. This means that the first time SW is pressed, 00 is shifted out serially to the SIPO, and LED0 to LED7 are off. The second time SW is pressed, SPI shifts out 01 hex and LED0 alone is turned on. The third time SW is pressed, only LED1 is turned on and the process continues.

```
BASE   EQU  $1000
PORTB  EQU  $04
PORTC  EQU  $03
DDRC   EQU  $07
DDRD   EQU  $09
SPCR   EQU  $28
SPSR   EQU  $29
SPDR   EQU  $2A


           ORG  $C000

; Init  routine

       LDX    #BASE
       LDAA   #$0F
       STAA   DDRC,X        ; 0—>3 output,4—>7 input
       LDAA   #01
       STAA   PORTC,X       ; enable line is high (SIPO disabled)
       CLRB                 ; clear counter
       LDAA   #$38
       STAA   DDRD,X        ; pd3,pd4,pd5 outputs
       LDAA   #$50
```

```
            STAA   SPCR,X          ; master, enabled

        ; main routine

MAIN        BRSET PORTC,X,$80,MAIN  ;loop till SW is closed
            JSR   DLY_20      ;Debounce the SW
            BCLR  PORTC,X,$0  ;enable line is active
            STAB  SPDR,X      ;start transmit
            JSR   DLY_500
            INCB              ; increment counter
LOOP        BRCLR SPSR,X,$80,LOOP ; is it complete yet ?
            BSET PORTC,X,$0      ; Yes, disable SIPO
            BRA   MAIN
            END
DLY_20      |                   ; time delay for 20 mS
            RTS
DLY_500     |                   ; time delay for 500 mS
            RTS
```

In the initialization routine, PC0 to PC3 are outputs while PC4 to PC7 are inputs. PD3, PD4, and PD5 are configured outputs. SPI is configured as a master; CPHA = 0 and CPOL = 0. The SIPO chip is disabled and the counter (ACCB) is cleared.

The main routine waits for the switch (SW) to close. If closed, it is debounced first (20 mS delay) and the SIPO is enabled. The data register keeps transmitting to the SERIAL IN line the incremented value of ACCB each time the SW is closed.

5.8   Programmable Timer

The timer is a 16-bit free-running counter with a prescalar, and its capabilities are provided by using timer registers. The timer functions are:

1. Input capture
2. Output compare
3. Periodic interrupts real-time interrupt (RTI)
4. Pulse accumulator

The timer is the basic counter for all the timing functions. The E-clock drives a prescalar, divisible by 1, 4, 8, or 16, which in turn drives a 16-bit counter (Figure 5.22).

The double-byte free-running counter (TCNT) has two read-only registers and can be read as a 16-bit value.

The prescalar bits PR0 and PR1 of Timer Mask Register 2 (TMSK2) select the prescalar rate, as shown Figure 5.23.

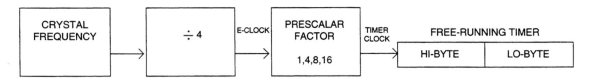

Figure 5.22

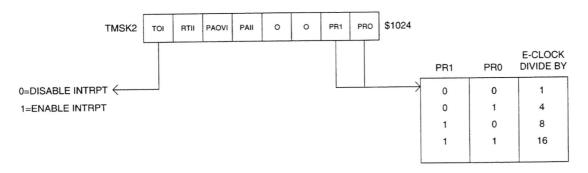

Figure 5.23

Timer Overflow

After the reset condition, TCNT is initialized to $0000. The counter then is incremented on each clock signal, counts up to $FFFF, and then rolls over to $0000 again. This rollover sets B7 of the timer flag register (TFLG2). This bit is used for polled operation to know if the timer has an overflow. If B7 of TMSK2 is enabled, the interrupt has occurred and an ISR is served. The TOF flag is cleared by writing a logic 1 to it, which must be done before the interrupt occurs.

EXAMPLE 3

Assume that a crystal frequency is 8 MHz. Calculate the rollover time for the timer if the PR1, PR0 are:

1. 00
2. 01

Solution

As noted, E-clock = 8/4 = 2 MHz and its cycle time = 0.5 µS.

1. If PR1, PR0 = 00, then E is divided by 1 and timer clock cycle = 0.5 µS. The rollover time = 65,536 × 0.5 µS = 32.77 mS. Thus, the timer takes 32.77 mS from $0000 to $FFFF.
2. If PR1, PR0 = 01, then timer clock = E/4 = 2 MHz/4 = 0.5 MHz. Cycle time = 2 mS. Then, the rollover time = 65,536 × 2 µS = 131.072 mS.

Timer Reset Conditions

The conditions after the timer is reset are as follows:

1. TCNT is initialized to 0000.
2. Timer overflow interrupts are disabled.
3. Timer overflow flag is cleared.
4. Prescalar is 1.

Real-Time Interrupt

For short time delays, the loop time delay method is used. But for longer delays, the RTI method is preferred. It is used to generate hardware interrupts at fixed periodic rates. The three registers associated with RTI are TMSK2, TFLG2, and pulse accumulator register PACTL, as Figure 5.24 shows. To generate a real-time hardware interrupt, these steps are followed:

1. Select the prescalar factor using bits RTR1, RTR0 as shown.
2. Enable RTII bit in TMSK2 register.
3. Clear RTIF by writing a logic 1 to it.
4. After the interrupt request has occurred and ISR is executed, RTIF is cleared again.

Program 5:   Demonstration of RTI

Assume that we would like to increment a counter from 00 to 0A hex. Each count takes 1 sec (10-sec counter). Using the RTI, write a program to accomplish that. Assume the prescalar is: (1) Divided by 4. (2) Divided by 8.

1. Prescaler is divided by 4.

> RTR1, RTR1 = 10
> Then the RTI clock = E ÷ (2**13)/4 = 61.03 Hz
> RTI clock cycle = 1/61.03Hz = 16.38 mS
> Number of counts = 1 sec/16.38 mS = 61

This means that there are 61 RTI interrupts to form 1 sec, as Figure 5.25 indicates. The vector addresses for the M68HC11 are $FFF0 and $FFF1 and for the EVB, $00EB and $00ED.

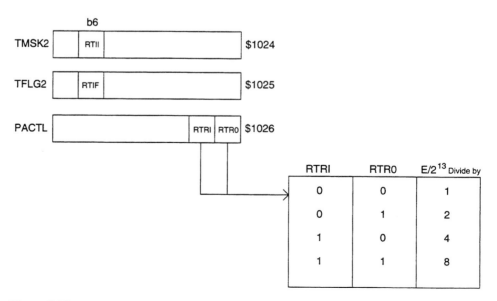

Figure 5.24

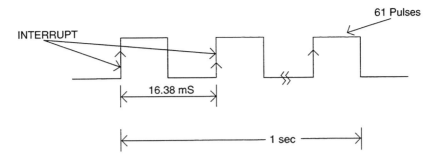

61 Pulses

INTERRUPT

16.38 mS

1 sec

Figure 5.25

Here is the program:

```
BASE    EQU    $1000
TMSK2   EQU    $24
TFLG2   EQU    $25
PACTL   EQU    $26


                ORG $0000


TICKCNT     RMB 1                   ;holds value for the time delay
COUNTER     RMB 1                   ;holds number of loops


                ORG $C000


            LDX     #BASE
            CLR     COUNTER    ;no. of  loops
            LDAA    #02
            STAA    PACTL,X    ;RTI clock divide by 4
            LDAA    #$40
            STAA    TMSK2,X  ; enable RTII
            STAA    TFLG2,X  ; clear flag


MAIN        LDAA    #61
            STAA    TICKCNT
            CLI              ; clear interrupt mask
LOOP        BRA     LOOP


        ; interrupt service routine


ISRVEC          DEC     TICKCNT
                BNE     CLRFLG
                LDAA    COUNTER
                INCA
                STAA    COUNTER
                CMPA    #10
                BNE     MAIN
HERE            BRA     HERE
                END
```

```
CLRFLG                 LDAA   #$40
                       STAA   TFLG2,X
                       RTI

                       ORG $FFF0

                       FDB    ISRVEC
```

2. Prescale RTR1, RTR2 = 11

RTI clock = E − (2**13)/8 = 30.5 Hz
RTI cycle time = 32.7 mS
No. of counts = 1 sec/32.7 =30, then proceed in programming as above.

When using the EVB evaluation board, the only change will be the interrupt vector address using the jump table, and this is valid for all programs in this book.

```
RTIVEC EQU    $EB                         ;LSB of pseudovector
BASE   EQU    $1000
TMSK2  EQU    $24
TFLG2  EQU    $25
PACTL  EQU    $26


                       ORG $0000

TICKCNT  RMB  1                           ;holds value for the time delay
COUNTER  RMB  1                           ;holds number of loops


                       ORG $C000

                 LDX      #BASE
                 CLR      COUNTER   ;no. of  loops
                 LDAA     #$7E      ;opcode for jump instruction
                 STAA     RTIVEC
                 LDD      ISRVEC
                 STD      RTIVEC+1  ;pseudovector from $EB to $ED
                 LDAA     #02
                 STAA     PACTL,X   ;RTI clock divide by 4
                 LDAA     #$40
                 STAA     TMSK2,X   ; enable RTII
                 STAA     TFLG2,X   ; clear flag

MAIN             LDAA     #61
                 STAA     TICKCNT
                 CLI                ; clear interrupt mask
LOOP             BRA      LOOP

           ; interrupt service routine

ISRVEC                 DEC    TICKCNT
                       BNE    CLRFLG
                       LDAA   COUNTER
```

```
                       INCA
                       STAA      COUNTER
                       CMPA      #10
                       BNE       MAIN
         HERE          BRA       HERE
                       END

           CLRFLG      LDAA      #$40
                       STAA      TFLG2,X
                       RTI
```

Input Capture

Port A can be used either for general purpose I/O or for timer functions such as input capture, output compare and pulse accumulator. The timer has three input capture pins (IC1 to IC3) and could have a fourth (IC4) with six associated internal 16-bit registers, TIC1 to TIC3. Each input capture function includes a 16-bit latch, an input edge-detection logic, and an interrupt-generation logic. The 16-bit latch captures the current value of the free-running counter when a selected edge is detected at the input, as shown in Figure 5.26. If an edge occurs, the following may happen:

1. The counter value goes into the input capture register.
2. The status flag ICAF is set.
3. An interrupt may occur, if the ICXI bit is set to 1.

The input captures are used mainly in measuring periods or pulse widths. These can include the activation of a pushbutton, pulse code detection, detection of the period of rotation of an engine, and accumulation of the "on" time of a process control valve.

To clear a status flag, write 1 to the ICX F bit in the TFLG1 register (see Figure 5.27).

Reset Conditions

The input capture status after reset is as follows:

1. Input capture functions are disconnected from input capture pins.
2. Interrupts are disabled.
3. Flag bits are cleared.

Figure 5.26

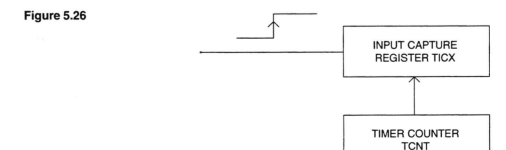

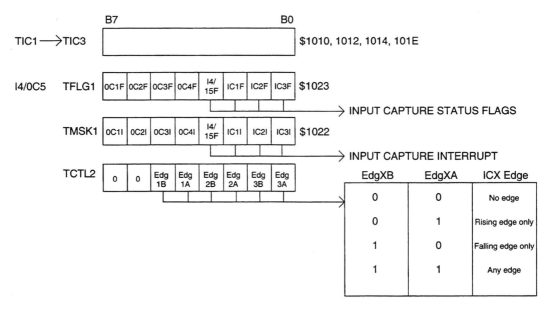

Figure 5.27

The input capture can be programmed to detect a positive-going or negative-going pulse or both using the bits of the TCTL2 register.

Program 6: Measurement of a Period Using Polling

Write a routine to measure a period using *polling*. We would like to measure a pulse period between two executive rising edges using IC1. The time of the E cycles between the two edges is the pulse period. Figure 5.28 displays the program.

```
BASE    EQU $1000
TIC1    EQU $10
TCTL1   EQU $20
TMS1C1  EQU $22
TFLG1   EQU $23
```

Figure 5.28

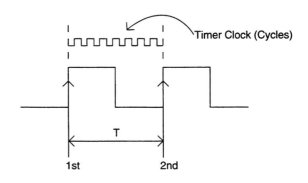

```
                        ORG  $0000

FRST    RMB   2               ; Holds the 1st edge reading
PERIOD  RMB   2               ; Holds the period reading

                        ORG  $C000
            ;Init routine

            LDX   #BASE
            LDAA  #$10
            STAA  TCTL2,X      ;rising edge on IC1
            LDAA  #04
            STAA  TFLG1,X      ; clear detection flg
LOOP1       BRCLR TFLG1,X,$04,LOOP1 ; edge detected yet?

        ;First edge detected

            LDD   TIC1,X     ; yes
            STD   FRST
            LDAA  #04        ; clear detect flg
            STAA  TFLG1,X
LOOP2       BRCLR TFLG1,X,$4,LOOP2 ; edge detected yet?

        ; Second edge detection

            LDD   TIC1,X     ;yes
            SUBD  FRST
            STD   PERIOD
            LDAA  #04
            STAA  TFLG1,X
            BRA   LOOP1
            END
```

To measure the frequency of a pulse, we usually take the average of several periods, say 50 periods.

Program 7:   **Measurement of a Period Using Interrupt**

Write a routine to measure the above period using *interrupts* on IC1.

```
        ; Same equates here

EDGMOD  RMB  1                         ;status byte

                        ORG  $C000

            ;init routine

            LDX  #BASE
            LDAA #$10
            STAA TCTL2,X     ; rising edge detection
            LDAA #04
            STAA TFLG1,X     ; clear detection flag
```

```
                    BCLR EDGMOD,$01 ; 0 in 1st detect, 1 in 2nd
                    CLI
LOOP                BRA    LOOP


        ;ISR routine
ISRADR              BRSET EDGMOD,$01,SECOND

        ; 1st edge detection
                    LDD   TIC1,X
                    STD   FRST
                    BSET  EDGMOD,$01 ; set the flag
                    BRA   EXIT


        ; 2nd edge detection

SECOND              LDD   TIC1,X
                    SUBD  FRST
                    STD   PERIOD
EXIT                LDAA  #04
                    STAA  TFLG1,X
                    RTI


                        ORG $FFEE

                    FDB   ISRADR
```

Exercise

Using the EVB board, make changes to Program 7.

Program 8:   Measurement of a Pulse Width Using Polling

Write a routine to measure a pulse width using *polling* on IC1. The pulse width is the number of E cycles between two consecutive rising and falling edges of a pulse.

```
                    ; Same equates

                        ORG $C000
        ;Init routine
                    LDX    #BASE
                    LDAA   # $10
                    STAA   TCTL2, X    ; rising edge
                    LDAA   #04
                    STAA   TFLG1,X     ; clear detection flag
LOOP1               BRCLR TFLG1,X,$04,LOOP1 ;edge detected yet?


        ;First edge detected

                    LDD    TIC1,X      ; yes
                    STD    FRST
                    LDAA   #4
```

```
             STAA   TFLG1,X
             LDAA   #$20
             STAA   TCTL2,X      ; falling edge
LOOP2        BRCLR  TFLG1,X,$04,LOOP2

     ; Second edge detection

             LDD    TIC1,X
             SUBD   FRST
             STD    PERIOD
             LDAA   #04
             STAA   TFLG1,X
HERE         BRA    HERE
             END
```

Exercises

1. Write a routine to measure a pulse period by taking the average of 50 pulse periods.

2. Write a routine to convert the above period value to a frequency.

3. Calculate the speed of an object as shown in Figure 5.29 if $d = .5$ ft.

$$V = \text{speed} = \frac{\text{distance}}{\text{time}}$$

4. In engine control systems, the crankshaft position sensor provides the necessary signals for:
   a. Fuel injection timing
   b. Engine speed (rpm) calculations
   c. Turbine and transmission speed (rpm) calculations

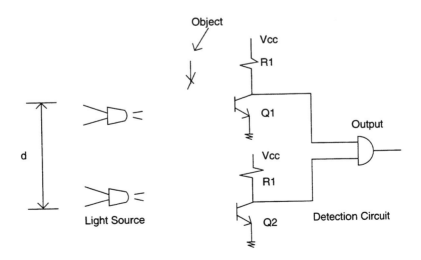

Figure 5.29

Figure 5.30

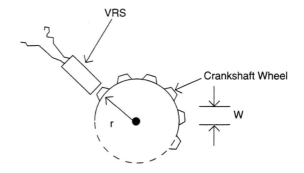

The most popular sensors used to detect the crankshaft position are the Hall effect and the variable reluctance sensors (VRS). As Figure 5.30 shows, rpm is calculated as follows:

$$\text{rpm} = \frac{60\ w}{2(rt)}$$

where

w = tooth width
r = tooth radius
t = pulse width

Write a routine to calculate this rpm.

Output Compare

The timer has four output compare pins (OC1 to OC4), and could have a fifth (OC5) with eight internal 16-bit associated registers, TOC1 to TOC4. When the value in the TOCX register matches the value of the free-running counter TCNT, the output sets the OCXF bit as shown in Figure 5.31. When a match occurs between the TCNT and TOCX, the following may occur:

1. OCXF is set.

2. An interrupt may occur.

The output compare function is mainly used for outputting waveforms (software controlled) to control actuators such as motor speed or solenoid valve close-hold-open states or is used to generate time delays for I/O functions.

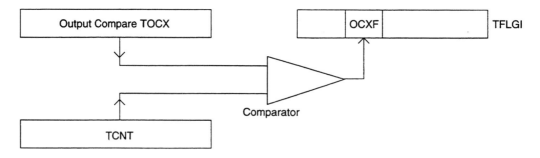

Figure 5.31

To clear OCXF, write 1 to the bit. If the OCXI is set, the interrupt request is enabled when a match occurs (see Figure 5.27). The user can independently program the automatic pin actions to occur for each output compare function. For OC5–OC2, a pair of control bits (OMX, OLX) in the timer control 1 register TCTL1 are used to control the automatic pin actions. (See Figure 5.32.)

Program 9: **Generating a Square Wave Using Polling**

Write a program to generate a square wave with a period of 4 mS and DC = 50% using *polling* on OC3.

```
BASE   EQU $1000
TOC3   EQU $1A
TCTL1  EQU $20
TMSK1  EQU $22
TFLG1  EQU $23
TCNT   EQU $0E

                     ORG $C000

         LDX    #BASE
         LDAA   #$20
         STAA   TFLG1,X     ; clear status flag
         LDAA   #$10
         STAA   TCTL1,X     ; output toggle on OC3
         LDD    TCNT,X
AGAIN    ADDD   #4000       ; 2 mS
         STD    TOC3, X
LOOP     BRCLR  TFLG1,X,$20,LOOP  ; a match occurred yet?
         LDAA   #$20            ;yes
         STAA   TFLG1,X     ; clear status flag
         BRA    AGAIN
         END
```

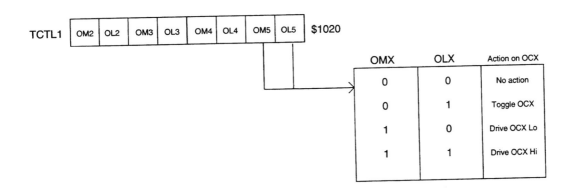

Figure 5.32

We assumed that the prescalar for the time is divided by 1. Thus the clock frequency is 2 MHz or 0.5 µS for cycle time. The value stored in TOC3 is the number of timer cycles. This number of cycles determines the output waveform period and its duty cycle, as Figure 5.33 illustrates.

$$\text{The number of timer cycles for } 1/2 \text{ period} = 2 \text{ mS}/0.5 \text{ µS} = 4000 \text{ cycles}$$
$$= 0FA0 \text{ hex}$$

This value is stored in TOC3. Every 4000 cycles, a match would occur between the free-running counter and the TOC3 register, and then status flag OC3F is set. Also, the TCTL1 register is programmed so that TOC3 changes its output level every 4000 cycles.

Multiple Output Compare

Output compare 1 (OC1) has a special feature that can control multiple output compares OC1 to OC5 when a match occurs between TOC1 and TCNT. The two associated registers controlling this feature are the output compare1 mask (OC1M) and output compare1 data (OC1D), as shown in Figure 5.34. If a logic 1 is written to a bit in OC1M, the corresponding output compare is affected by OC1. If a 0 or 1 is written to a bit in OC1D, the corresponding OCX will be low or high respectively when a match occurs.

Pulse Accumulator

The pulse accumulator (PACNT) is an 8-bit counter and can be configured to operate as a simple event counter or for a gated time accumulation. Bit 7 of port A (PAI) associated with the pulse accumulator can be configured to act as a clock (event-counting mode) or as a gate signal to enable a free-running E divided by 64 clock to the 8-bit counter (gated time

Figure 5.33

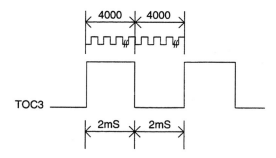

Figure 5.34

	OC1	OC2	OC3	OC4	OC5	OC6	OC7	OC8	
OC1M	OCM7	OCM6	OCM5	OCM4	OCM3	0	0	0	$100C
OC1D	OCD7	OCD6	OCD5	OCD4	OCD3	0	0	0	$100D

accumulation mode), as shown in Figure 5.35. Bit 5 of PACTL determines the counter mode. For the event-counting mode, PEDGE of the PACTL register selects which PAI edge is used to increment the PACNT register. For the gated accumulation mode, PEDGE selects which PAI state is used to inhibit counting. If the specified edge occurs, the following may occur:

1. PAIF is set.

2. The counter may be incremented.

3. An interrupt may occur.

4. The PAOVF may be set.

The PAI line should be configured as input when dealing with the PACNT (See Figure 5.36.)

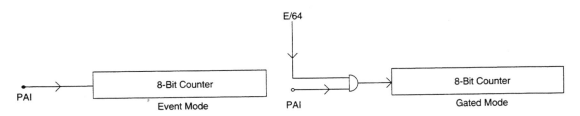

Figure 5.35

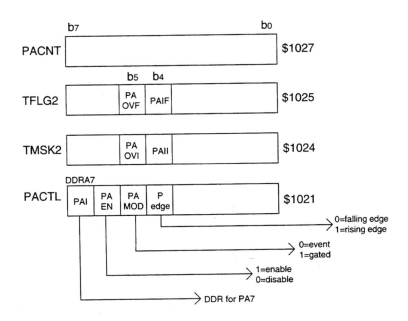

Figure 5.36

Program 10: Using the Pulse Accumulator in Event-Counting Mode

Write a routine to count the *negative* edges using the pulse accumulator.

```
; Equates here

                            ORG $0000

TEMP  RMB1
                            ORG $C000
                CLR   TEMP
                LDX   #BASE
                LDAA  #$30
                STAA  TFLG2,X     ; clear status flags
                STAA  TMSK2,X     ; enable interrupts
                LDAA  #$4F
                STAA  PACTL,X     ; event mode, falling edge, PA
                                  ; enable, PA is input
                CLI
LOOP            BRA   LOOP

ISREVT          LDAA  #$10
                STAA  TFLG2, X    ; clear status flag
                RTI

ISROVR          LDAA  #$20
                STAA  TFLG2,X     ; clear overflow flag
                INC   TEMP
                RTI

                        ORG   $FFDA

            FDB   ISREVT

                        ORG   $FFDC

            FDB   ISROVR
```

Note that the pulse accumulator is incremented automatically every time it detects a negative edge.

$$\text{Total pulses} = \text{contents of PACNT} + 255 \times (\text{contents of temp})$$

For example, suppose that contents of temp = 4 and the final contents of PACNT = F0 (240 dec). Then total pulses = $4 \times 255 + 240 = 1260$ pulses.

Exercises

1. Write a program to generate a square wave with 333 Hz and 30% DC. Use polling and interrupt modes.

2. In the biocontrol system, the MC68HC11 can be used to modulate the output signal on OC2 to provide an electrical shock to the muscle. The duration of a shock is proportional

to the dial output, as Figure 5.37 indicates. The dial output is an 8-bit word. Develop a routine to generate this PWM signal proportional to the dial output.

3. A photocell—shown in Figure 5.38—is connected to the pulse accumulator and used to count the number of filled bottles on a conveyer. Develop a routine to accomplish this task if it counts from 0000 to 9999.

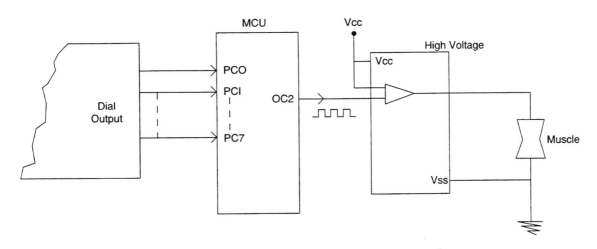

Figure 5.37

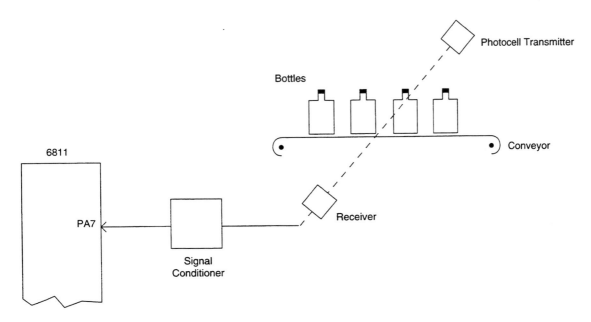

Figure 5.38

5.9   Analog-to-Digital Conversion

Figure 5.39 depicts a simple 3-bit analog-digital (ADC) converter. If an ADC has n bits, it generates 2^n output digital levels. In this example, the ADC has 3 bits and the outputs are $2^3 = 8$ digital levels, as shown in Fig 5.39b. These levels are $000, 001, \ldots, 111$. The way the ADC works is that it generates a smallest quantity of conversion called *step size* and adds these step sizes as a *stairway* to form the digital equivalent of the input analog. The step size can be calculated as follows:

$$\text{Step size} = \text{least significant bit (LSB)} = V_{Fs}/2^n$$

where V_{Fs} is called *full scale*. For example, if a V_{Fs} of this ADC is 5V, the step size is:

$$\text{Step size} = \text{LSB} = 5/8 = 0.625V$$

The maximum output of this ADC shown in Figure 5.39 is $7/8 \times 5 = 4.375V$. This means there is a difference between the actual reading (4.375V) as an output and the FS reading as an input (5V). For any ADC, the actual output is calculated as: Output = FS − LSB. In our example, the output $= 5 - 0.625 = 4.375V$.

To minimize this difference between the full scale and the actual output, the number of ADC bits has to increase. For example,

$$\text{An 8-bit ADC with } V_{FS} = 5V$$
$$\text{Step size} = 5/2^8 = 5/256 = 0.0195V$$
$$\text{The actual reading} = 5 - 0.0195 = 4.9804V$$

which is very close to the V_{FS} or the input voltage signal. If we increase the ADC bits further to be 10 bits, then:

$$\text{Step size} = 5/2^{10} = 0.00488V$$
$$\text{Actual reading} = 5 - 0.00488 \doteq 4.9951V, \text{ and so on}$$

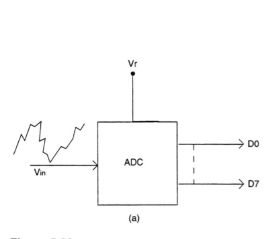

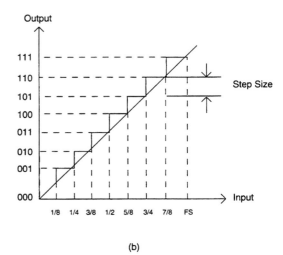

(a)   (b)

Figure 5.39

Resolution

Resolution is the smallest analog output change that can occur as a result of a change in the digital input. It can be given in percent of full scale, in millivolts for a given input range, or as the number of conversion bits.

For example, percent resolution of an 8-bit ADC is:

$$\text{percent resolution} = \frac{1}{2^n} \times 100 = \frac{1}{2^8} \times 100 = 0.391\%$$

where n = number of ADC binary input bits. Or we can say that

$$\text{resolution} = \text{LSB} = 195 \text{ mV (for 5V)}$$

or we can say that for 8-bit ADC, the resolution is simply 8 bits.

Accuracy

The accuracy is the actual analog output of an ADC compared to the expected output (full scale). The accuracy is expressed as +1/2 LSB. For the above 8-bit converter, we calculated its LSB as 392 mV. Then the accuracy is to ±19 m.v.

ADC Conversion Techniques

Counter or Ramp ADC

The counter will start counting from zero and is incremented by a clock pulse. The digital output of the counter is converted back to analog by the digital-to-analog converter (DAC) and is compared with the analog input voltage signal (V_{in}). If V_{in} is higher than the DAC output, the counter keeps incrementing, but if V_o increases the value of V_{in}, the counter stops immediately and the counter output is the ADC output.

The disadvantage of this ADC is that it is very slow since it depends on the up-counter. For example, in 10-bit ADC, the conversion time is about 1024 cycles. Also, it is not accurate, especially in converting sudden peaks.

Successive Approximation

The structure of this technique contains a register called the successive approximation register (SAR). It does not use the counting method but modifies the contents of the register bit by bit starting with the MSB bit until the register data are the digital equivalent of the analog voltage.

The SAR uses a trial-and-error method. Assume a 4-bit converter with four outputs Q_o to Q_3 at the SAR register with a step size of 1V for simplicity, and $V_{in} = 10.4$V. Let's see the sequence of operations of the SAR register (Figure 5.40):

1. Clear all outputs Q_o, Q_1, Q_2, Q_3. This means Q = 0000.
2. Set MSB = 1—that is, Q = 1000 where Q_3 = MSB. This means that V_{in} = 8V and is compared with V_{in}. If it is less, this bit is kept set; if it is higher, the bit is reset. Since $V_{in1} < V_{in}$, then MSB is kept set. Now the output is 1000.
3. Set the next lower bit Q_2 = 1; now output Q = 1100 = 12V, which is higher than V_{in}. In this case, Q_2 is reset and the output is 1000, or 8V.

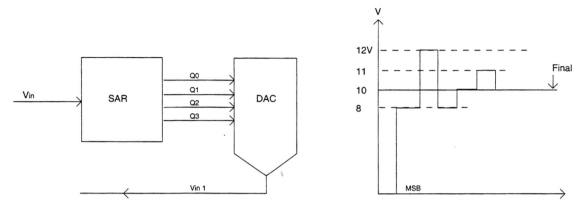

Figure 5.40

4. Set $Q_1 = 1$ or $Q = 1010 = 10V$, which is lower than V_{in}, then Q_1 is kept set. The output now is 1010.

5. Set $Q_0 = 1$ or $Q = 1011 = 11V$, which is higher than V_{in}; then Q_0 is reset. The output now is 1010 and is the final reading for 10.4V.

The advantage of this ADC is that it is very fast. For 8-bit ADC, the conversion takes 8 cycles only. For 10-bit ADC, the conversion takes 10 cycles only and so on. Also it is very accurate. This type of converter is used in most *microcontrollers* and is the most popular conversion technique on the market.

Exercises

1. Compare the maximum conversion times of the 10-bit ADC ramp type and the 10-bit ADC successive approximation type. Assume that the clock frequency is 1 MHz.
2. For a 12-bit ADC and $V_{in} = 10V$, calculate:
   a. Step size
   b. Percent resolution and resolution in mV.
   c. Accuracy
   d. Output voltage.

Dual-Slope ADC

If analog voltage V_{in} is applied to the integrator, it causes the integrator capacitance (C) to charge in a direction that depends on the polarity of the input voltage for a fixed time (t1). After the capacitor is charged, the analog switch is switched to the V_{ref} and allows the capacitor to discharge. While C is discharging, the counter starts to count up. When the capacitor charge reaches 0 (complete discharge), the counter stops through the control logic circuit and counter outputs are now the digital equivalent. This kind of ADC is expensive and slower than successive approximation, but it is very accurate.

Flash ADC

Flash ADC is considered the highest speed ADC available. It has n bits with m comparators. The number of comparators (m) always equals $2^n - 1$. The major drawback is that this ADC is expensive—for example, an 8-bit ADC requires $2^n - 1$ comparators, i.e., $2^8 - 1 = 255$ comparators!

ADC Converter in the M68HC11

The ADC built in the M68HC11 uses a conversion technique similar to the successive approximation just explained. It uses a network of capacitors connected to the comparator. For each conversion, the system switches into three modes: a sample mode, a hold mode, and an approximation mode.

During the sample mode, the capacitors are charged and the charge is proportional to the input analog signal Vx. During the hold mode, the charge remains conserved. In the approximation mode, different capacitors are switched to modify the input voltage using the successive approximation technique.

As mentioned earlier, port E (pins 43–50) can be used as a general input I/O port (PE0–PE7) or as ADC converter channels (AN0 to AN7) in the PLCC version or channels (AN0 to AN3) in the DIP version. The converter consists of these channels, a sample and hold circuit, and an 8-bit successive approximation converter. It makes use of high-reference voltage (Vrh) and low-reference voltage (Vrl). An input voltage equal to Vrl is converted to 00 and an input voltage equal to Vrh is converted to $FF.

The registers associated with the ADC converter are shown in Figure 5.41. There are two conversion modes: *single-channel* and *multichannel* modes. In the single-channel mode the MULT bit equals 0 and one channel is selected at a time through the CA – CD bits of the ADCTL registers, as shown in Table 5.8. Four conversion readings of this channel are carried out, one every 16 µS (32 cycles) if the crystal is 8 MHz. The first conversion goes to the ADR1 register, and 16 µS later the second goes to ADR2 and so on to ADR4, where ADR1 to ADR4 are A/D result registers.

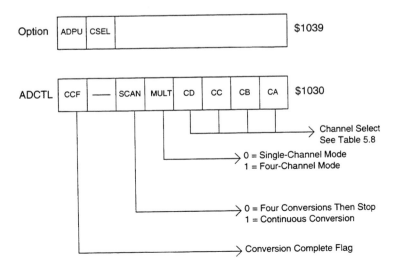

Figure 5.41

Table 5.8

ADCL Bits 3-0:	Single-Channel Selection				
	CD	CC	CB	CA	Channel
	0	0	0	0	AN0
	0	0	0	1	AN1
	0	0	1	0	AN2
	0	0	1	1	AN3
	0	1	0	0	AN4
	0	1	0	1	AN5
	0	1	1	0	AN6
	0	1	1	1	AN7
	1	0	0	0	Reserved
	1	0	0	1	Reserved
	1	0	1	0	Reserved
	1	0	1	1	Reserved
	1	1	0	0	V_{RH} PIN
	1	1	0	1	V_{RL} PIN
	1	1	1	0	$(V_{RH}-V_{RL})/2$
	1	1	1	1	Reserved

On completion of the fourth conversion, the conversion complete flag (CCF) of ADCTL is set. If the SCAN bit of ADCTL is 0, this conversion process stops and a reading of one result register is taken. But if the SCAN bit is set, the conversion process continues. In the *multichannel* mode, the conversion process is the same as in the *single* mode, except the channel selection is different. The selection bits CC and CD of ADTL select a group of four channels at a time with four analog inputs. For example, if both the CC and CD bits are 0, the conversion result of channel AN0 goes to ADR1 and of channel AN1 goes to ADR2 and so on. (See Table 5.9.)

Program 11: A/D Conversion in Single- and Multi-Channel Modes

Write a program to convert Vrh to a digital value using the single channel and multi-channel modes. Also, use the one-conversion and continuous conversion scan modes. Store the results in temp locations 0000–0003.

```
BASE  EQU $1000
ADCTL EQU $30
ADR1  EQU $31
ADR2  EQU $32
ADR3  EQU $33
ADR4  EQU $34
OPTION EQU $39
TEMP  EQU $0000
```

1. Single-Mode, 1 Conversion

```
        ORG $C000

    LDX   #BASE
    LDAA  #$80          ;time on ADC
```

Table 5.9

ADCTL Bits 3-2:	CD	CC	Channel	Register
			MultiChannel Selection	
	0	0	AN0	ADR1
			AN1	ADR2
			AN2	ADR3
			AN3	ADR4
	0	1	AN4	ADR1
			AN5	ADR2
			AN6	ADR3
			AN7	ADR4
	1	0	Reserved	ADR1
			Reserved	ADR2
			Reserved	ADR3
			Reserved	ADR4
	1	1	V_{RH} PIN	ADR1
			V_{RL} PIN	ADR2
			$(V_{RH}-V_{RL})/2$	ADR3
			Reserved	ADR4

```
              STAA   OPTION, X
              LDAA   #$C
              STAA   ADCTL,X,Vrh is selected, single mode
              LDY    #TEMP
LOOP          BRCLR  ADCTL,X,$80,LOOP ;conversion complete yet?
NEXT          LDAA   ADR1, X              ; yes
              STAA   0,Y
              INX
              INY
              CPX    #$1035
              BNE    NEXT
DONE          BRA    DONE
```

2. Single-Mode, Continuous Conversion

```
; Same equates
                        ORG $C050

        LDX    #BASE
        LDAA   #$80
        STAA   OPTION,X          ; turn on ADC
        LDAA   #$2C
        STAA   ADCTL,X ; Vrh is selected, single mode,continous
        LDY    #TEMP
AGN     LDAB   #26
LOOP    DECB
        BNE    LOOP      ; time delay for 64 µS for 4 conversions
```

```
NEXT       LDAA   ADR1,X
           STAA   0,Y
           INX
           INY
           CPX    #$1035
           BNE    NEXT
           BRA    AGN
```

3. Multimode, Single Conversion

```
; Same equates
                        ORG $C100

           LDX    #BASE
           LDAA   #$80
           STAA   OPTION,X          ; turn on ADC, single mode
           LDAA   #$1C
           STAA   ADCTL,X
LOOP       BRCLR  ADCTL,X,$80,LOOP ; conversion complete yet?
           LDAA   ADR1,X               ; yes
           STAA   TEMP
DONE       BRA    DONE
```

4. Multimode, Continuous Selection

```
; Same equates
                        ORG $C150

           LDX    #BASE
           LDAA   #$80
           STAA   OPTION,X
           LDAA   #$3C
           STAA   ADCTL,X
LOOP       BRCLR  ADCTL,X,$80,LOOP
           LDAA   ADR1,X
           STAA   TEMP
           BRA    LOOP
```

Exercises

1. For an 8-bit ramp ADC, if the clock frequency = 100 KHz, Vfs = 10 V. How long does it take to make the conversion if V_{in} = 7 V?

2. In M68HC11 system, the crystal frequency is 12 MHz. What is the conversion time if it is in the single-conversion mode?

5.10   Memory Addressing Decoding

An address decoder is a device or digital circuit that can address a particular area of memory or point to it by MCUs and MPUs. There are two types of address decoding, full decoding and partial decoding. Four main methods are used in address decoding:

1. Using random logic
2. Using data decoders
3. Using PROMS
4. Using PGA, PLA, or PAL

We will explain the first method in detail; then the rest will be self-explanatory.

Random Logic

Random logic describes a system using small-scale logic such as AND, OR, NAND, NOR, and inverters. The advantage of this method is its custom design to fit a particular system. The disadvantage could be its high cost because it uses many logic components.

Full Decoding

Full-address decoding means that each addressable location in a memory chip responds to a single address on the system address bus.

To address any byte in memory like the one shown in Figure 5.42, there are two steps in decoding. The first step is to construct an address decoding table that contains *chip select* to select a particular memory chip and *word select* to select an address inside the

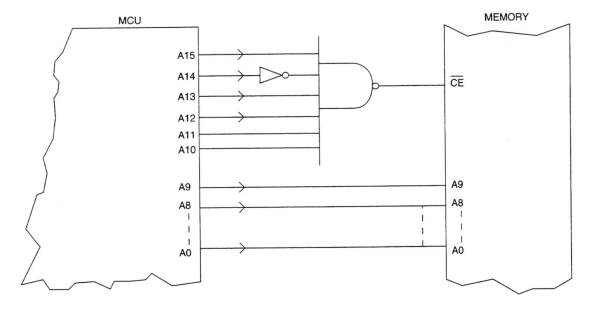

Figure 5.42

selected chip. To select a memory chip, the few higher address lines of the MCU are first used to decode this address. In this example A10 to A15 are used with the NAND gate to select the chip. Lower address lines (A_o to A9) are used to select the location inside this chip. Select enable ($\overline{CE}$) or, as it is sometimes called, chip select ($\overline{CS}$) is an active low in this example. This means that the NAND output should be low to activate this line. Thus:

- A15 should be high.
- A14 should be low.
- A13 should be high.
- A12 should be high.
- A11 should be high.
- A10 should be high.

Second, we display all address lines as follows:

CHIP SELECT						WORD ADDRESS									
A15	14	13	12	11	10	9	8	7	6	5	4	3	2	1	A0
1	0	1	1	1	1	X	X	X	X	X	X	X	X	X	X

A10 to A15 are always 101111 for a particular chip. But A0 to A9 are variable, which is why we denoted them as X (do not care). To select the first location, A0 to A9 must be 0s, as shown:

A15	A14	A13	A12	A11	A10	A9	A8	A7	A6	A5	A4	A3	A2	A1	A0
1	0	1	1	1	1	0	0	0	0	0	0	0	0	0	0

which is BC00 in hex.

To select the second location, the address lines should be as follows:

A15															A0
1	0	1	1	1	1	0	0	0	0	0	0	0	0	0	1

which is BC01 in hex.

To select last location, the address lines should be like this:

A15	A14	A13	A12	A11	A10	A9	A8	A7	A6	A5	A4	A3	A2	A1	A0
1	0	1	1	1	1	1	1	1	1	1	1	1	1	1	1

which is BFFF.

So the address range of this memory chip is:

BC00—>BFFF.

As a rule, to design a decoder for a memory chip, follow these steps. First, find out the high-address lines that are going to select the chip itself. Second, the rest of the address lines are considered all 0s one time to select the first location, and then considered all 1s to select the last location. The address range (space) of the chip will be the total address lines that select the first location and the total address lines that select the last location. The following question now arises: What is the capacity (size) of this memory? As mentioned in the first chapter, we should look at the number of address lines. The address lines that select the locations are A0 to A9 or 10 lines. Then the memory address = 2^{10} = 1K.

Partial Decoding

In full decoding, we used all address lines A0 to A15. In partial decoding we do not have to use all the address lines, especially the select chip address lines. Address lines A10, A11, A12, and A13 are not used here and considered as "don't care." Now the first location address looks like this:

A15	A14	A13	A12	A11	A10	A9	A8	A7	A6	A5	A4	A3	A2	A1	A0
0	1	X	X	X	X	0	0	0	0	0	0	0	0	0	0

So, there are many possibilities for the first location. It could have the address 8000 or A000 or B000 and so on. The advantage of partial decoding is to save some logic gates.

Using Data Decoders

Each decoder chip has *m* input lines, which are decoded to *n* lines where

$$n = 2^m$$

Some popular decoders are the 74LS154 (4-to-16), 74LS138 (3-to-8), and 74LS139 (2-to-4). These decoders have active low outputs suitable for many address decoding applications because most memory chips have active low chip select inputs. E1 and E2 are active low enable inputs, while E3 is an active high input. To get the outputs of this decoder, E1 and E2 should be low while E3 should be high. If E1 or E2 is high or E3 is low, there will be no outputs (outputs are floating).

PROM Decoders

PROMS can be programmed with the truth table directly relating addresses to device select outputs. Figure 5.43 shows a PROM-based address decoder. The PROM (32x8) is

Figure 5.43

From A. Clements, *Microprocessor Systems Design.* Copyright 1987 by PWS-KENT Publishers, Boston, MA. Reprinted with permission.

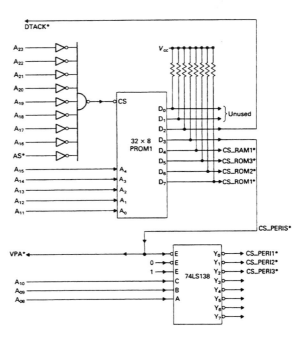

used to decode several RAMs and ROM memory chips. Address lines A10 to A15 are used to enable the PROM.

Using FPGA, PLA, and PAL

These devices are general-purpose logic elements and are configured by the programmer. The 82S102 and 82S103 are examples of the field programmable gate array (FPGA). The FPGA consists of inputs to NAND gates and outputs from EOR gates. The inputs pass through fusible links to NAND gates. The fusible links could be open-circuit (blown) or short-circuit (left intact) during programming. The outputs can be programmed to be active low or active high.

The field programmable logic array (FPLA) and the programmable array logic (PAL) are similar to the FPGA in programming techniques, although their architecture differs.

6 General Interfacing

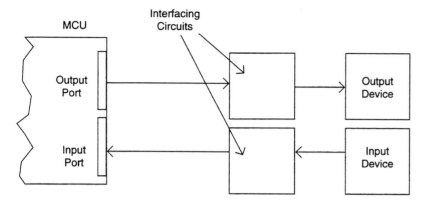

Figure 6.1

After you become familiar with the on-chip ports, we are going to perform several practical experiments to learn how to interface the M68HC11 to the real world. We have done these experiments using both Motorola's evaluation boards, the EVM 6811 and EVB 6811. You can use any development system you like as long as it does the job.

In Chapter 4, you became familiar with programming techniques using the assembler directives. These programs did not input or output any data to or from the MCU but performed using its internal registers and memory. In real life, the MCU deals with outside devices using its internal registers, memory, timer, A/D, serial ports, and I/O ports. Usually there is an electronic interfacing circuit between the MCU and the outside-world devices, as shown in Figure 6.1.

The outside devices could be input or output devices. The input devices could be digital or analog. Digital input devices include switches, keyboards, digital integrated sensors that send 0 V (logic 0) or 5 V (logic 1), and so on. There are many analog input devices—for example, all types of transducers. A transducer is an analog device that converts some form of energy to an electrical signal. For instance, transducers can measure position, motion, force, pressure, flow, level, temperature, light density, and so forth. Output devices may include relays, DC and AC motors, stepper motors, all types of displays, solenoids, and other actuators.

Let's explain the interfacing circuits for inputs and outputs.

6.1   Input Analog Devices

The analog signal from any transducer must first be conditioned by a signal conditioning circuit to meet the input requirements of an ADC input. Different circuits such as amplifiers, filters, bridges, and comparators are used for analog signal conditioning.

Signal Conditioning Circuits

Operational Amplifiers

Operational amplifiers (op-amps) are used to amplify a signal or a difference between two signals. We will not explain op-amp functions and features in detail in this book; you can check any textbook on linear devices for more discussion.

The most commonly used op-amp circuits are described briefly here.

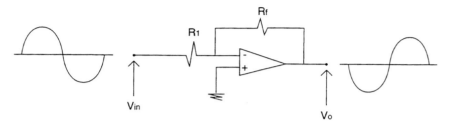

Figure 6.2

In an *inverting op-amp*, the input voltage is inverted as shown in Figure 6.2.

$$Vo = \text{output voltage} = \frac{-Rf}{R1}\, V_{in}$$

where (–Rf/R1) is the closed-loop gain.

In a *noninverting op-amp*, output voltage is calculated as

$$Vo = Vi\,(1 + \frac{Rf}{R1})$$

where (1 + Rf/R1) is the closed-loop gain.

In a *differential op-amp*, output voltage is calculated as

$$Vo = (V2 - V1)\,\frac{R2}{R1}$$

where V1 and V2 are the op-amp inputs.

In a *voltage follower*, output voltage is calculated as

$$Vo = vi$$

It is used as a buffer because it has a very high input impedance.

In a *current-to-voltage converter*, output voltage is calculated as

$$Vo = -I_{in}\, Rf$$

The current source I_{in} is converted to voltage Vo.

An *instrumentation amp* has a high gain with higher immunity to noise. Output voltage is calculated as

$$Vo = (Vb - Va)\,\frac{R2}{R1}\,(1 + \frac{2R4}{R3})$$

Filters

Filters are required in signal conditioning circuits to eliminate unwanted frequencies. There are two types of filters: passive and active. Passive filters are designed using standard discrete components such as resistors, capacitors, and inductors.

Active filters use op-amps in their design and have advantages over passive filters, although they are more expensive. Filters generally can be categorized as low pass, high pass, bandpass, and band reject. See the frequency response curves in Figure 6.3 for each type.

$$fc = cutoff\ frequency$$
$$fcl = lower\ cutoff\ frequency$$
$$fch = higher\ cutoff\ frequency$$

As you can see in Figure 6.3, fc, fcl, and fch occur when the gain reaches 0.707 of the maximum.

The cutoff frequency for a *low-pass active* filter is:

$$fc = \frac{1}{2\pi RC}$$

We can calculate the desired fc by choosing the right values of R and C.

The cutoff frequency for a *high-pass active* filter is

$$fc = \frac{1}{2\pi RC}$$

The central frequency for a *bandpass active filter* is

$$fo = \frac{1}{2\pi C\ \sqrt{R_1 R_2}}$$

where

$$Bandwidth = bw = fch - fcl$$

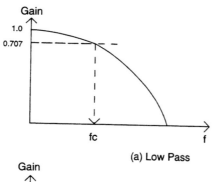

(a) Low Pass

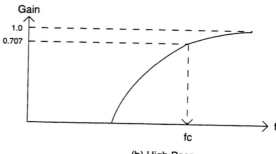

(b) High Pass

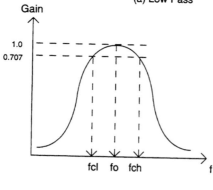

(c) Bandpass

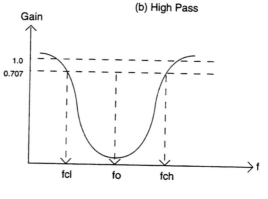

(d) Band Reject

Figure 6.3

The *reject pass active filter* is a combination of low-pass and high-pass filters. The central frequency for a reject pass active filter is

$$fo = \frac{1}{2\,(R1C1)}$$

where R1, C1 are of a low-pass filter.

Bridge Circuits

Bridge circuits (both AC and DC) are commonly used with variable resistance transducers. We are going to explain DC bridges only since they are more popular. These bridges are called *wheatstone bridges*. The wheatstone bridge shown in Figure 6.4 is balanced when

Vo = 0 V and R1R4 = R2R3

It is used to measure the small resistance changes caused by a resistive transducer. The transducer (R4) is connected as shown in Figure 6.4. The following steps are taken to measure the change of a transducer resistance:

1. R3 is used to balance the bridge initially where Vo = 0.
2. When the transducer starts to work, its resistance (R4) changes and Vo ≠ 0. Vo could be in millivolts.
3. This output is amplified by a difference or an instrumentation amplifier and its output is directly proportional to the resistance change.

   Transducers using wheatstone bridges include strain gauges, thermistors, resistance temperature detectors (RTDs), and almost all resistive transducers.

Exercises

1. Calculate the cutoff frequency of a low-pass active filter if R = 10 K, C = 0.001 μF.
2. In the wheatstone bridge of Figure 6.4, R1 = R2 = 100 ohms and R3 = R4 = 200 ohms.
   a. Determine whether the bridge is balanced initially.
   b. If the transducer resistance (R4) is changed to 200.001 ohms, calculate the value of Vo if V = 10 V.
   c. If the gain of the amplifier (to amplify Vo) is 100, what is the value of its output?
   d. What new value of R3 is required to rebalance the bridge?

Figure 6.4

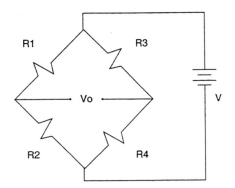

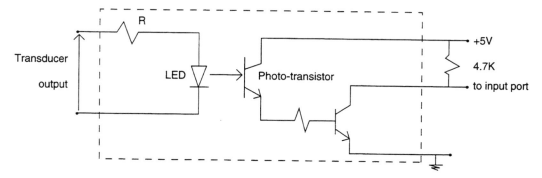

Figure 6.5

Opto-Couplers

An opto-coupler—as shown in Figure 6.5—consists of a light-emitting diode LED powered by the transducer, and a phototransistor or a thyristor acting like a detector (switch). Both the input and the detector are included in one package. Since there is no direct connection between the grounds of the two sides of the isolator, the common mode signal is eliminated and there is a high electrical isolation (several megaohms) between the two sides. Therefore, the LED is driven by a low voltage from an MCU or a logic circuit and the detector side could be a part of a high-voltage AC or DC circuit.

6.2   Digital Input Devices

Mechanical Switches and Keypads

Mechanical switches could be *single-pole single-throw* (spst) or *single-pole double-throw* (spdt) or any combination. As shown in Figure 6.6, when the switch is open, Vo = 5 V; when it is closed, Vo = 0 V, that is, Vo is either a logic 0 or a logic 1. One problem with using a mechanical switch is that when it is closed, it "bounces" for a few milliseconds, which lets the microcontroller see that it has been pressed several times, and gives the wrong information. There are two ways to debounce a switch:

1. Using an R-S flip-flop.
2. Using a software time delay (usually 20 mS).

    The second method is the one used in software programming. Some keypads use a mechanical switch matrix and have the same problem of bouncing. To debounce the keypad, we use a time delay loop (or software timer).

Figure 6.6

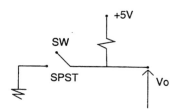

Digital Integrated Sensors

An electronic switch can be activated by the detection of a physical event such as the *Hall effect* and other digital sensors. These sensors have an output of 0 V or 5 V.

6.3   Output with Two State Actuators

The output lines from an MCU output port can supply only small amounts of power. The high-level output signal has voltage between 2 V and 5 V, and a low-level output signal has voltage less than 1 V. The output port is said to be *sourcing* current if it flows out of the port into the load, as shown in Figure 6.7a. It is said to be *sinking* if the current flows through the load and into the output port, as indicated in Figure 6.7b.

Most actuators require more than 100 mW. Since most types of digital logic circuits source or sink 20 mA (100 mW), the ports by themselves are not enough to drive the actuators. We are only going to discuss some types of driving circuits.

Integrated Circuits

Integrated circuits containing transistor switches can be used when the current is under 4 A. An example is "MC34151" made by Motorola.

Discrete Solid-State Switches

Power transistors and power MOSFETS are used for higher currents. The circuit in Figure 6.8 uses a power *Darlington transistor,* which has a rated current of 5 amps.

Electromechanical Relays

Electromechanical relays cannot be driven directly by the MPU output port. Therefore, they require a relay driver circuit. Relays have some advantages; they can switch DC or

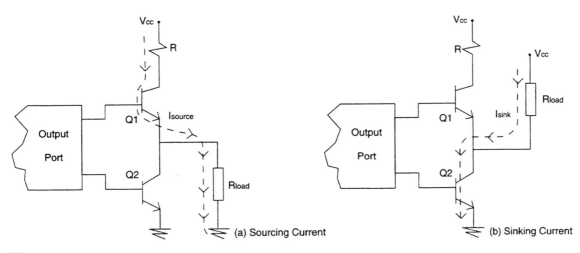

Figure 6.7

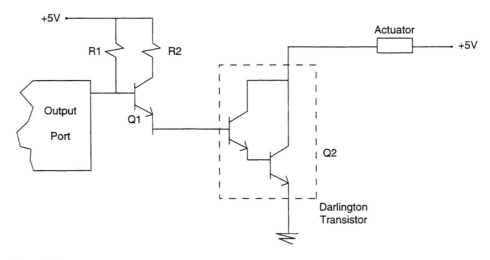

Figure 6.8

AC current and carry heavy currents. In terms of disadvantages, they switch in milliseconds instead of microseconds and suffer from contact bouncing. Because of these disadvantages, solid-state relays are replacing the electromechanical ones.

The relay coil must be shunted by *a freewheeling diode* to protect it from damage when the driver is turned off.

Solid-State Relays

Solid-state relays use members of semiconductor power switches called *thyristors*. These members are silicon controlled rectifiers (SCRs) and triacs. They are used if the relay is to switch AC current. But if it is to switch DC, power transistors and power MOSFETS are used.

These relays also use opto-couplers to send the signal to the solid-state switch. Since the solid-state relays use solid-state switches, there will be no contact bounces, switching delays, or contact sparks as happens with electromechanical relays.

6.4   Output System with Continuous Actuators

Not all systems can be controlled with two state actuators. Sometimes the output of an actuator must be continuously variable.

Power Amplifiers

Power op-amps with a medium power rating are used to drive continuous actuators such as DC motors or valves. Other power amplifiers called *servo amplifiers* have an even higher rate of power than power op-amps.

Pulse-Width Modulated Amplifiers

Pulse-width modulation (PWM) is used to drive motors using PWM amplifiers. These amplifiers' power control section consists of solid-state switches. The switches are turned on or off at a constant frequency. The duty cycle of these square pulses can be varied,

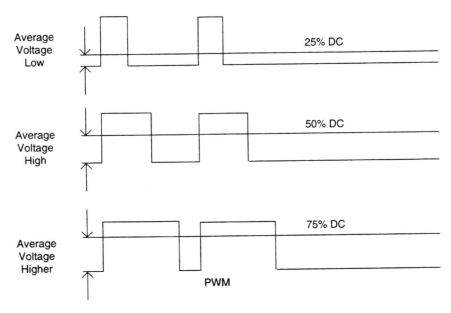

Figure 6.9

which will vary the average voltage applied to the load (Figure 6.9). The output average voltage depends on the pulse duty cycle. The higher the duty cycle, the higher this voltage. The advantages of PWM are numerous; power dissipation is low, and it is easy to implement in the MCU.

6.5   Introduction to Light-Emitting Diodes

As with diodes, current passes through an LED in one direction only. When a positive voltage is applied to the anode, current flows from the anode to the cathode and electrons flow in the opposite direction and migrate across an energy gap in the LED, causing it to light up.

The size of the energy gap determines the voltage drop across the LED. LEDs can be designed to emit light from ultraviolet to infrared. The most common LEDs emit a red light using a gallium arsenide phosphide (Ga Asp) chip, yellow light using (Ga Asp on Gap), and green light using (Gap). A tinted case can also vary the color.

Table 6.1 shows typical forward voltage drops for different colors of LEDs. Typical LED operating currents are between 10 and 30 mA. For a bright display with low power consumption, look for types labeled "High efficiency."

One disadvantage to LEDs is that their light is hard to detect in bright light—outdoors, for example. Therefore, a tinted transparent sheet of plastic is mounted over the display to make it more visible.

Generally, there are two methods of connecting a group of LEDs as indicator patterns or as seven-segment displays. They can be connected as common-cathode or as common-anode LEDs.

In the *common-cathode* type, the LEDs' cathodes are connected to ground. To turn a particular LED on, a +5 V (logic 1) must be applied to its anode as shown in Figure 6.10a.

Table 6.1

LED Color	Typical Forward Voltage (Volts)
Red	1.6
Green	2.0
Yellow	2.0
Blue	3.2

In the *common-anode* type, the LEDs' anodes are connected to +5 V. To turn a particular LED on, a 0 V (logic 0) must be applied to its cathode as shown in Figure 6.10b. The resistors shown are *current-limiting* resistors to provide the right current.

Bicolor LEDs

Bicolor LEDs have both a red and a green LED in a single package. Some have two leads, while others have three leads. By turning on one, both, or neither, you can use a single indicator to show as many as four states.

These bicolor LEDs can be common-anode, common-cathode, parallel connect, one-line control, or two-line control as shown in Figure 6.11. The truth table attached to each type should be examined.

Driving Design Examples

Driving LEDs

To turn any LED on, its anode voltage (Va) has to exceed its cathode voltage (Vc) by the voltage drop between the anode and the cathode. For example, if a voltage drop (in the forward direction) between Va and Vc = 1 V, then to turn an LED on, (Va − Vc) must exceed 1 V. Va has approximately 5 V, and if we provide the cathode with low voltage (Vc = 0), then the LED should light. This voltage is called *turn-on* voltage. Above this voltage a small resistance starts to appear. The equivalent circuit of an LED

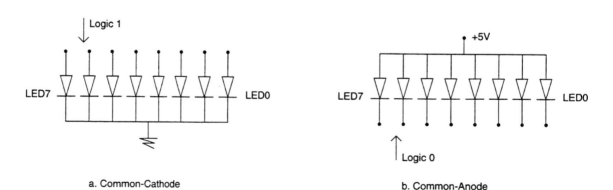

a. Common-Cathode   b. Common-Anode

Figure 6.10

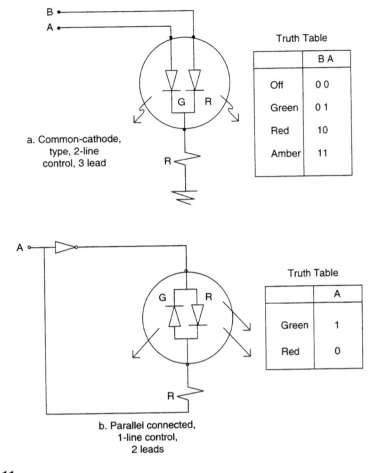

Truth Table

	B A
Off	0 0
Green	0 1
Red	1 0
Amber	1 1

a. Common-cathode, type, 2-line control, 3 lead

Truth Table

	A
Green	1
Red	0

b. Parallel connected, 1-line control, 2 leads

Figure 6.11

is shown in Figure 6.12 and consists of a battery source ($V_{turn-on}$) and a resistance (rd). The voltage drop equals:

$$V_d = V_{turn-on} + I_d rd$$

$V_{turn-on}$ and rd differ from one device to another. LEDs can be driven either by discrete transistors or by IC inverting and noninverting TTL drivers. The driver could be sinking or sourcing current.

Figure 6.12

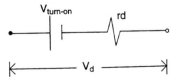

Example 1

Figure 6.13 shows discrete transistor drivers. In the sinking current driver, the transistor should be in the saturation mode (Vce is low) to let the driving current (I_d) pass through the LED from Vcc to ground. In a sourcing current driver, the transistor must be off (Vce is high) to let the driving current (I_d) pass through the LED from Vcc to ground. The LED current is limited to 20 mA. Assume rd = 20 ohms, Vce(sat) = 0.7 V, and $V_{turn-on}$ = 1.5 V. Calculate Rl in both cases.

Solution

1. Sinking Current Driver

$$Vcc = V_{ce(sat)} + V_d + I_d Rl$$

where,

$$V_d = V_{turn-on} + I_d rd$$

Then,

$$Vcc = V_{ce(sat)} + V_{turn-on} + I_d(Rl + rd)$$

$$Rl = \frac{Vcc - Vce - V_{turn-on} - I_d rd}{I_d}$$

$$= \frac{5 - 0.7 - 1.5 - 20 \times 0.02}{0.02 \text{ A}}$$

$$= 120 \text{ ohm}$$

2. Sourcing Current Driver

$$Vcc = I_d Rl + V_d$$

$$Rl = \frac{Vcc - V_d}{I}$$

where

$$I = I_d + I_{ce}$$

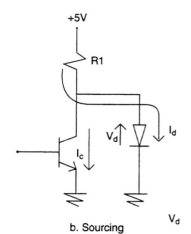

a. Sinking   b. Sourcing

Figure 6.13

Since $$I_d \gg I_{ce}$$

then $$I = I_d$$

$$Rl = \frac{5 - 1.5 - 20 \times 0.02}{0.02 \text{ A}}$$
$$= 150 \text{ ohms}$$

Notice the difference between the two circuits. In the source current circuit, there is no Vce drop voltage that increases Id. To limit this current to 20 mA, Rl has to increase.

Driving Induction Loads

Induction loads such as incandescent lamps, relays, solenoids, motors, and so on have enough inductance to cause destructive reverse voltage transients during switching and can damage the integrated circuit drivers. Some of these inductive loads have a turn-on peak surge current of up to 10 times their steady-state value. There are several ways to limit these surge currents.

The most popular method uses a clamp diode (or a transistor), as shown in Figure 6.14. This clamp diode is included in many peripheral integrated driving circuits. If it is not included, then it has to be added externally to the driver.

6.6   Practical Interfacing Programs

In this section, we want to interface some of the I/O (input/output) devices to the M68HC11 using some of the signal conditioners and drivers mentioned. We have selected LEDs and switches to be used in some of these programs as simulators. LEDs simulate real-world output devices such as solenoids, relays, motors, and so on. Switches can also simulate real-world input devices like digital transducers, keypads, and mechanical switches. The code is the same for both real-world I/O devices and the simulators. The only difference is the electronic interfacing circuits such as signal conditioning and the driver circuits. You are encouraged to connect each one to your emulator and run these programs. You are also encouraged to do each of the exercises. Practice makes perfect! You might write your own programs for these exercises. That is a great help too! ·

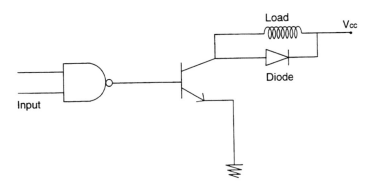

Figure 6.14

Program 1: Turning ON and OFF Each LED Light (Walking One)

In this experiment, we would like to interface six LEDs to the MCU. As noted earlier, we have to use a driver circuit between the MCU and the LEDs. This driver is the 7407, which has noninverting drivers. These drivers are going to drive the signal coming out of the I/O port to turn on and off the LEDs. We choose Port B as an output port.

To turn on any LED, we have to generate a low-level voltage (logic 0) at the output lines PB0 to PB5. If the drivers are inverting, we have to generate high-level voltage (logic 1) on these output lines. The pull-up resistors (330 ohm) are current-limiting resistors.

Using the circuit in Figure 6.15, we want to turn each LED off starting from LED0 while the other LEDs are on. Usually to turn on or off any display we prefer to keep it in its current state for enough time (time delay) to give the human eye a chance to capture it. If we do not have this time delay in the program, the display could blink so fast that the eye could not capture it.

To turn each LED off, we take advantage of the "ROTATE" instruction. The idea is to output a logic 1 at each LED cathode starting from LED0. When a logic 1 is at the cathode of an LED, it turns it off while the others are on. The logic 1 will stay at the cathode long enough to be noticed, then moves again to the next LED and so on.

Here is the program:

```
LED1.asm


BASE    EQU $1000          ;base address for all registers
PORTB   EQU $04


                    ORG $C000


            LDX   #BASE
            LDAA  #$01
REPEAT      STAA  PORTB,X         ;LED is off
            JSR   DLY             ;provide time delay
```

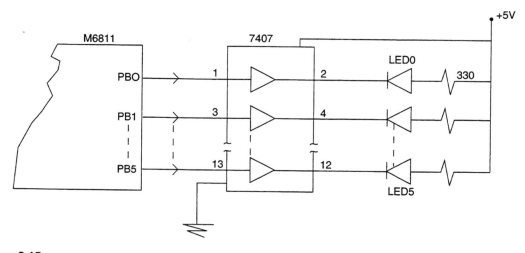

Figure 6.15

```
                        ROLA                    ;move "1" to next bit
                        BRA   REPEAT            ;keep shifting "1"
                        END

DLY                     LDY   #$FFFF            ;time delay base
AGN                     DEY
                        BNE   AGN
                        RTS
```

To calculate the time delay, DEY = 4 cycles, BNE = 3 cycles. Total = 7 cycles. In EVM, the crystal frequency is 8 MH—that is, the clock frequency is 2 MHZ. Thus the cycle period is 0.5 µS.
Now the time delay is T = 65,535 × 7 × 0.5 µS = 229.37 mS.

Exercises

1. Modify the above program so that you turn on each LED while the others are off (walking "zero").

2. Modify Exercise 1 so that the walking zero direction will be from left to right with a time delay of 100 mS.

Program 2:   Flashing LEDs (On and Off Constantly)

In this experiment, we are going to let all LEDs flash (on and off) every one second. This routine could be used in a warning application.
Here is the program:

Method 1: Using the CPU Time Delay

```
BASE   EQU $1000
PORTB  EQU $04
                              ORG $C000

                        LDX   #base
                        LDAA  #00
repeat                  STAA  PORTB,X           ;LEDs are on
                        JSR   DELAY             ;time delay
                        COMA                    ;LEDs are off
                        BRA   REPEAT            ;keep repeating
                        END

DELAY                   LDAB  #3                ;initial outer loop
NEXT                    LDY   #$FFFF            ;initial inner loop
AGAIN                   DEY              } Inner
                        BNE   AGAIN      } Loop
                        DECB
                        BNE   NEXT
                        RTS
```

Inner loop time = 65535 × 3.5 = 229 mS.
Outer loop time = 229 × 3 = 1 S.

Method 2: Using Real-Time Interrupt

As mentioned earlier, the RTI is very useful in long delays. Using the EVB board, the program looks like this:

```
RTIVEC   EQU $EB
TMSK2    EQU $24
TFLG2    EQU $25
PACTL    EQU $26
BASE     EQU $1000
PORTB    EQU $4


                 ORG $0000

TICKCNT RMB 1                              ;holds count value for RTI

                     ORG $C000

                 LDX  #BASE
                 LDAB #$FF
                 LDAA #$7E
                 STAA RTIVEC
                 LDD  #ISRVEC
                 STD  RTIVEC+1
                 STAB PORTB,X     ;turn off LEDs
                 LDAA #2          ;scalar div by 4
                 STAA PACTL,X
                 LDAA #$40
                 STAA TMSK2,X     ;enable interrupt
                 STAA TFLG2,X     ;clear flag

START            LDAA #61
                 STAA TICKCNT     ;no. of count value
                 CLI
LOOP             BRA  LOOP

         ;INTERRUPT SERVICE ROUTINE

ISRVEC           DEC  TICKCNT     ;decrement counter every 16.38 mS
                 BNE  CLRFLG      ;clear the flag every cycle
                 COMB             ;toggle ACCB after 1 S
                 STAB PORTB,X     ;on, off, on, ....
                 BNE  START       ;start all over again
                 END

CLRFLG           LDAA #$40        ;clear the flag
                 STAA TFLG2,X
                 RTI
```

RTI clock = $E \div (2^{13})/4 = 16.38$ mS. This means that we get an interrupt every 16.38 mS. To accomplish 1 S, we have to get these interrupts for 61 times. For this reason, the counter TICKCOUNT counts 61 times.

Program 3: Interfacing Binary Switches

If a switch is closed, the corresponding LED turns on (Figure 6.16).

```
BASE    EQU  $1000
PORTB   EQU  $04
PORTC   EQU  $03
DDRC    EQU  $07


               ORG  $C100

        LDX   #BASE
        LDAA  #$FF
        STAA  PORTB,X      ;turn off LEDS
        LDAA  #00
        STAA  DDRC,X       ;portc is input
REPEAT  LDAA  PORTC,X      ;get status of switches
        STAA  PORTB,X      ;turn on corresponding LED(s)
        BRA   REPEAT       ;keep going
        END
```

Exercises

1. Modify Program 3 so that when SW0 is closed, all LEDs start flashing.

2. If the SPST switches are replaced by "pushbuttons" (PBs), modify the above program so that when any PB is pressed, the corresponding LED turns on and remains on.

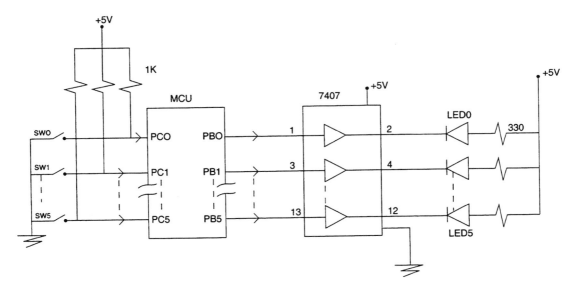

Figure 6.16

Program 4:   Inspector Tour

As Figure 6.17 indicates, the inspector has to press keys 0,1, . . . ,6 in that sequence to activate the light LED0. These keys in real life are located apart from each other; each key may exist in its own building. The LED0 light could also be a signal to store information in a monitor or a signal to printer. The clear (CL) switch is used to clear this light. Design a program to implement this using only three keys.

Method 1: Using Polling for the Clear Switch

```
BASE   EQU $1000
PORTC  EQU $3
PORTB  EQU $4
DDRC   EQU $7


                         ORG $C000

                 LDX   #BASE
                 LDAB #$FF
                 STAB PORTB,X        ;turn off LEDs
                 LDAA #00
                 STAA DDRC,X         ;port c is input
LOOP1            BRSET PORTC,X,$0,LOOP1 ;is sw0 closed?
LOOP2            BRSET PORTC,X,$1,LOOP2 ;AND is sw1 closed?
LOOP3            BRSET PORTC,X,$2,LOOP3 ;AND is sw2 closed?
                 LDAA  #$FE          ;yes, turn on LED0
                 STAA  PORTB,X

CLEAR            BRSET PORTC,X,$80,CLEAR ;is clear sw closed?
                 LDAA  #$FF              ;yes, turn off LED0
```

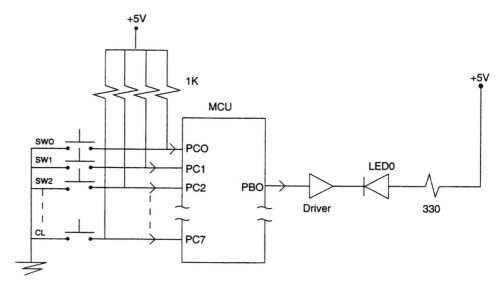

Figure 6.17

```
                       STAA    PORTB,X
HERE                   BRA     HERE
                       END
```

Method 2: Using an Interrupt for the Clear Switch

CL should be connected to the IRQ pin.

```
BASE   EQU  $1000
PORTC  EQU  $3
PORTB  EQU  $4
DDRC   EQU  $7

                       ORG  $C000

              LDX    #BASE
              LDAB   #$FF
              STAB   PORTB,X        ;turn off LEDs
              LDAA   #00
              STAA   DDRC,X         ;portc is input

LOOP1         BRSET  PORTC,X,$1,LOOP1
LOOP2         BRSET  PORTC,X,$2,LOOP2
LOOP3         BRSET  PORTC,X,$4,LOOP3
              LDAA   #$FE           ;turn LED0 on
              STAA   PORTB,X
              CLI                   ;enable interrupt
WAIT          BRA    WAIT           ;wait for interrupt

      ;  Interrupt Service Routine

ISR           LDAA   #$FF           ;turn off LED0
              STAA   PORTB,X
              RTI

                     ORG  $FFF2

              FDB    ISR
```

Exercises

1. See Figure 6.18. We are going to use keys sw0, sw2, sw1 in this sequence to unlock a padlock. If the wrong key is pressed, the alarm light (or buzzer) turns on for a short time and the user has to try the keys again in the same sequence. Design a program to implement this.

2. Modify the above program so that the total number of times the keys are pressed wrongly is three (instead of one), after which the alarm is turned on.

3. Also modify the padlock program so that the alarm light flashes for three seconds, then stops.

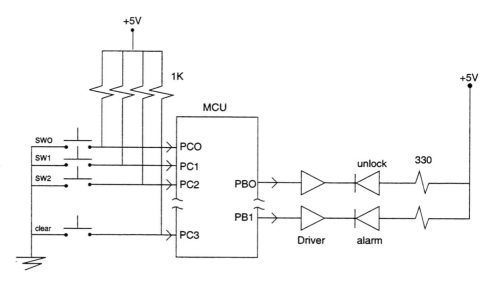

Figure 6.18

4. As Figure 6.19 illustrates, the normal status (if no fault) of this annunciator is that the green light is on while the others are off. In case there is a fault (fault switch is closed), the green light is off while both red and amber lights (could be a buzzer) are on. There are two possible situations that result from pressing the acknowledge switch:
   a. If there is no fault at this time (the fault is cleared)—that is, fault switch is open—the R and A lights are off and the G light is on.
   b. If the fault still exists at this time—that is, the fault switch is still closed—the R and A lights are still on and G light is off. Design a program to implement this.

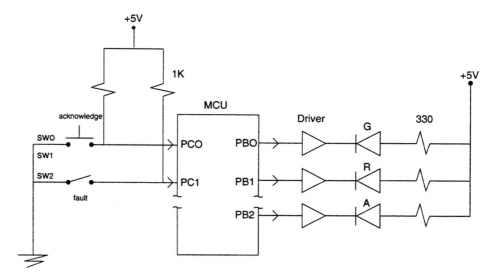

Figure 6.19

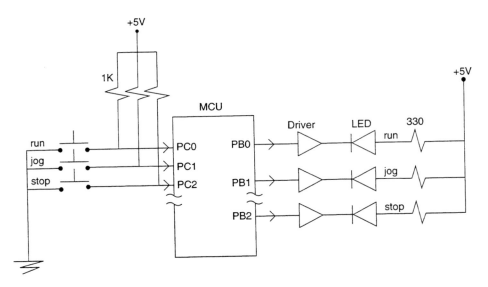

Figure 6.20

5. Modify the above program so that when pressing the acknowledge switch, the red light is turned off and the amber light (buzzer) is still on if the fault still exists.

6. As Figure 6.20 shows, we simulate the status of a motor by three lights: green when it is running, amber when it is jogging, and red when it stops. If the run switch is pressed, the motor will run and its green light is on. When the stop switch is pressed, the motor stops and its red light is on. If the jog switch is on from the stop status, the motor jogs (runs as long as the jog switch is pressed), and its amber light is on. Design a program to implement that.

7. Generating a tone from an 8-ohm speaker with 8-ohms is simple. All we have to do is to generate a square wave in the audible frequency range. The speaker has to be driven by a driver to supply enough current to it. In Figure 6.21, a 7400 NAND gate is used with a capacitor of 100 μF to protect the NAND gate from damage.

The frequency is 5 KHz since T = 200 μS = 0.2 mS. Develop a code to generate this tone.

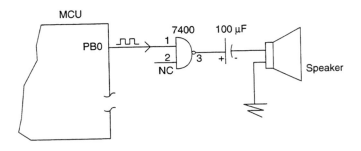

Figure 6.21

Figure 6.22

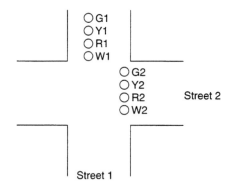

8. In this problem, we simulate a traffic light function using colored LEDs. The traffic light is located at the cross-section of two streets: street 1 and street 2. Each street has four lights: green, yellow, red for cars, and white for pedestrians, as shown in Figure 6.22.

The light schedule table is shown as Table 6.2. This is called a *look-up* table. The look-up table shows the status of every light at the two streets and the hex equivalent code for each row. The time delay subroutine between each status is set for 20 sec.

Program 6.5: Generating a Ramp Wave Using DAC

Many MCUs have built-in analog-to-digital converter (ADC) ports but do not have digital-to-analog converter (DAC) ports. Therefore, we have to use external DAC chips and interface them to the MCU. Figure 6.23 shows the DAC MC1408 interfacing to the MC6811 and its connections. This device is a current output device. Thus a *current-to-voltage converter* (op-amp 741c) must be used to convert the output current to an output voltage V_o.

Let's set R1 = Rf = 5 K. The MCU outputs 00 hex, then increments its output in small steps called *step sizes* until this output reaches a maximum value of FF hex, when it drops suddenly to 00. The waveform is then repeated. (See Figure 6.24.)

```
BASE   EQU $1000
PORTB  EQU $4
```

Table 6.2

street #1				street #2				
G1	Y1	R1	W1	G2	Y2	R2	W2	Code in Hex
1	0	0	1	0	0	1	0	92
1	0	0	1	0	0	1	0	92
0	1	0	0	0	0	1	0	42
0	0	1	0	1	0	0	1	29
0	0	1	0	1	0	0	1	29
0	0	1	0	0	1	0	0	24

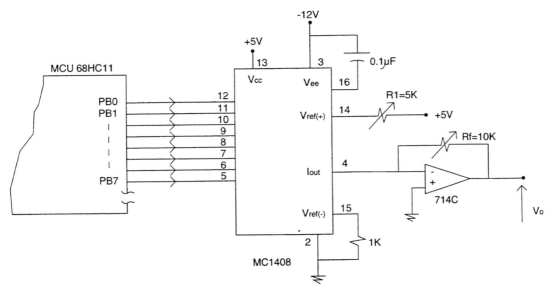

Figure 6.23

```
              ORG  $C000

              LDX  #BASE
              LDAA #00
REPEAT        STAA PORTB,X
              JSR  DLY      ;calculate step size
              INCA          ;next step
              BRA  REPEAT
```

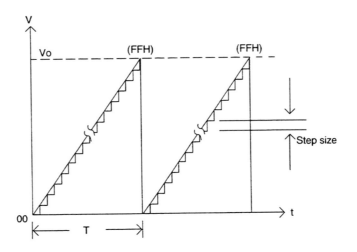

Figure 6.24

```
DLY                LDAB   #20          ;step size
AGAIN              DECB
                   BNE    AGAIN
                   RTS
```

Assume a crystal of 8 MHz. Each step has a duration $t = 5$ cycles $\times 0.5\ \mu \times 20 = 50\ \mu S$

$$T = \text{waveform duration} = 255 \times t$$
$$= 255 \times 50\ \mu S$$
$$= 12.750\ mS$$

$$F = \text{frequency} = 1/T = 1/12.750\ mS = 78\ Hz.$$

Exercises

1. Modify the above program to generate a ramp waveform with frequency of 90 Hz.
2. Write a program to generate a square wave with T1 = 30 mS and T2 = 45 mS. Calculate the frequency. Use both the DAC and the timer output compare methods. Compare them.
3. Develop an algorithm to generate a trapezoidal waveform. T1 = T3 = 2T2 = 10 mS.

6.7   Keypad Interfacing Techniques

A keypad consists of a number of keyswitches arranged in a matrix of rows and columns. In small matrices, rows and columns are connected directly to the MCU I/O lines, while in large matrices, decoders are used to save the number of I/O lines used.

Consider the hex keypad $(0,1, \ldots ,f)$ 4×4 matrix shown in Figure 6.25. It consists of four rows and four columns. The rows have high logic since they are connected to +5 V. To identify a pressed key, the technique is to send a logic 0 on each column, one at a time, then find out which row has a low logic. The combination of both low-logic column and low-logic row will identify which key is pressed.

For example, if we send a logic 0 signal on col 0 first, we then try to find out which row became low. If a low-level signal is detected at row0, this means that key0 is pressed. If this signal is detected at row1, it means that key4 is pressed. The process is repeated for all four rows. Next we send this low-level signal on col 1 and try to find out which row received this signal and thus which key is pressed. The same is repeated for col 2 and col 3.

Switch Bounce

Mechanical switches tend to bounce, as we mentioned at the beginning of this chapter. When the switch makes or breaks, it makes a number of very rapid make and break actions before the contacts become stable, as shown in Figure 6.26. This bouncing tells the MCU that the switch has been pressed a few times instead of once which means that false information has been sent to the MCU and misled it. To debounce a switch, a software time delay is used that provides a time delay (usually 20–30 mS) longer than the duration of the switch bouncing action.

Keypad Decoders

If the number of keyswitches is high, a decoder is used. For example, if we want to use a keypad 4×6 (24 switches), then utilizing the direct connection, we have to use twelve

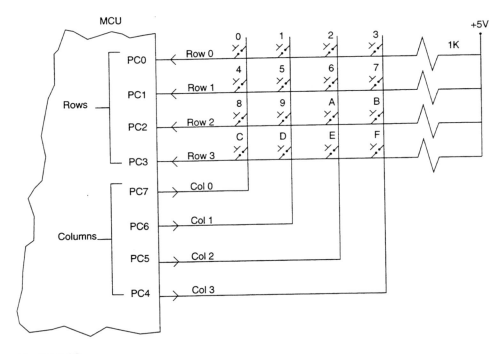

Figure 6.25

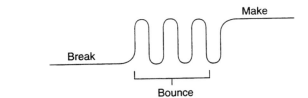

Figure 6.26

I/O lines. But by using a three-to-eight decoder, we can use only seven I/O lines, saving five lines (Figure 6.27). Using the truth table shown in Table 6.3, we can send high-logic signals on the decoder's input lines A, B, C to get high logic on its output lines Q1 to Q6. The technique of scanning is the same as explained previously.

Program 6:   Keypad Interfacing

We use the same circuit in Figure 6.25 to write a program to scan and detect pressed keys. PC0 to PC3 are configured as inputs, while PC4 to PC7 are configured as outputs. The technique, as mentioned before, is to output a low logic first on PC7 (col 0), then to find out if a key is pressed. If a key is pressed, it must be debounced first, then identified. Next we output a low logic on PC6 (col 1) and repeat the same steps as with col 0 and so on. Here is the program:

```
KEYPAD.ASM

BASE   EQU $1000
```

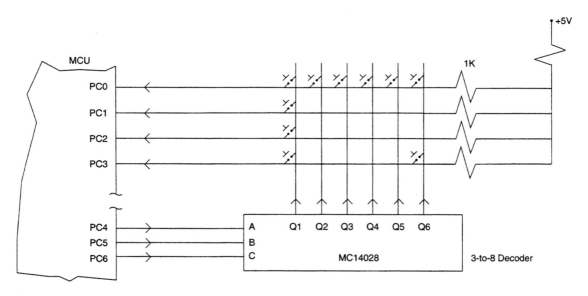

Figure 6.27

```
                    PORTC EQU $03
                    PORTB EQU $04
                    DDRC  EQU $07


                                        ORG $0000


                    FOUND RMB 1
                    TEMP  RMB 1


                                        ORG $C000


                              LDX  #BASE
                              LDAA #$F0
                              STAA DDRC,X        ;pc0-pc3 input,pc4-pc7 output

                    EXCITE            LDY  #KEYID        ;Y points to keypad table
```

Table 6.3

CBA	Q1	Q2	Q3	Q4	Q5	Q6
001	1	0	0	0	0	0
010	0	1	0	0	0	0
011	0	0	1	0	0	0
100	0	0	0	1	0	0
101	0	0	0	0	1	0
110	0	0	0	0	0	1

```
                  LDAA #$7F          ;put zero to col0
                  LDAB #4            ;set counter for rows
                  STAB TEMP
                  STAA PORTC,X
                  JSR  GETKEY

                  LDAA #$BF          ;put zero to col1
                  LDAB #4
                  STAB TEMP
                  STAA PORTC,X
                  JSR  GETKEY

                  LDAA #$DF          ;put zero to col2
                  LDAB #4
                  STAB TEMP
                  STAA PORTC,X
                  JSR  GETKEY

                  LDAA #$EF          ;put zero to col3
                  LDAB #4
                  STAB TEMP
                  STAA PORTC,X
                  JSR  GETKEY

                  BRA  EXCITE

GETKEY            BRSET PORTC,X,$0F,RETURN; any key pressed?

DEBOUNCE          PSHX               ;yes
                  LDX  #6666         ;20 sec debounce
LOOP              DEX
                  BNE  LOOP
                  PULX

                  LDAA PORTC,X       ;get keypad status
REPEAT            LSRA
                  BCC  KFOUND        ;key found yet?
                  INY
                  CPY  #16           ;reach end of keys yet?
                  BNE  NEXT_KEY      ;no
                  BRA  EXCITE        ;start all over again

NEXT_KEY          DEC  TEMP          ;row counter
                  BNE  REPEAT
RETURN            RTS                ;return to next scan

KFOUND            LDAA 0,Y           ;get the key ID
                  STAA FOUND         ;store it here
                  BRA  EXCITE        ;start all over again
                  END
```

```
                              ORG   $C100

KEYID            FCB $0,$4,$8,$C,$1,$5,$9,$D
                 FCB $2,$6,$A,$E,$3,$7,$B,$F
```

Exercises

1. Write a routine to display a key ID using 4 LEDs (0 to F).
2. Using a 20-mS sampling time, observe the input data of the keypad and accept it if three consecutive sample results are the same as shown in Figure 6.28.

Program 7:   Start/Stop a DC Motor Using the Keypad

This routine is used to start and stop a DC motor as shown in Figure 6.29. The 100 = ohm resistor is used to damp any rush current when the motor starts rotation. This routine can also be used in many applications using a keypad, including turning on/off lights, flashing lights, acquiring data, and so on.

```
;Key 0 is assigned to start the motor (full voltage)
;Key 1 is assigned to stop the motor

BASE  EQU $1000
PORTC EQU $03
PORTB EQU $04
DDRC  EQU $07

                              ORG $0000

FOUND RMB  1
TEMP  RMB  1

                         ORG $C000
```

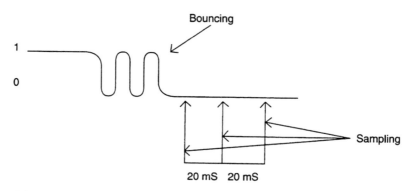

20 mS   20 mS

Figure 6.28

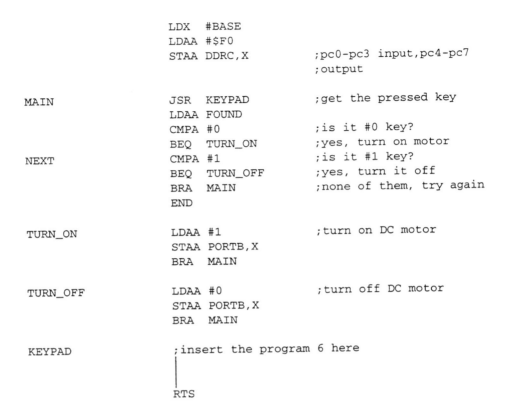

```
                    LDX   #BASE
                    LDAA  #$F0
                    STAA  DDRC,X        ;pc0-pc3 input,pc4-pc7
                                        ;output

MAIN                JSR   KEYPAD        ;get the pressed key
                    LDAA  FOUND
                    CMPA  #0            ;is it #0 key?
                    BEQ   TURN_ON       ;yes, turn on motor
NEXT                CMPA  #1            ;is it #1 key?
                    BEQ   TURN_OFF      ;yes, turn it off
                    BRA   MAIN          ;none of them, try again
                    END

TURN_ON             LDAA  #1            ;turn on DC motor
                    STAA  PORTB,X
                    BRA   MAIN

TURN_OFF            LDAA  #0            ;turn off DC motor
                    STAA  PORTB,X
                    BRA   MAIN

KEYPAD              ;insert the program 6 here


                    RTS
```

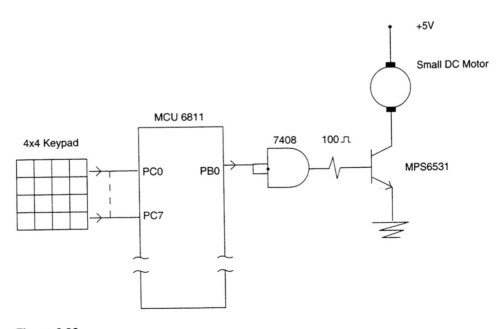

Figure 6.29

6.8   Thyristor-Based Power Interfaces

Thyristors are a family of high-power semiconductor switches. They include silicon-controlled rectifiers (SCRs), silicon unilateral switch (SUS), and triacs. Unlike transistors, they are pulsed on and will remain on after the switching pulse is removed. This means that we can switch large amounts of power on with little energy.

Thyristor Devices

The SCR is the most widely used thyristor. It is a PNPN structure and has three terminals: an anode, a cathode, and a gate. After the SCR, the most commonly used thyristor is the triac. It also acts as a switch and its gate controls the switching state. The triac has three terminals: main terminals MT1 and MT2 and a gate G (Figure 6.30).

Zero-Crossing Voltage

A better way of avoiding switching at high currents is to use *zero-voltage switching* (ZVS) ICs. With these ZVSs, the thyristor fires when the voltage across it is near zero. An example of ZVS is MO3031, manufactured by Motorola.

Since the ZVS turns only at or near the zero-voltage point of the load voltage, only complete *half* or *full* cycles are applied to the load. The average voltage across the load depends on how many half or full cycles are applied during a certain time (Figure 6.31).

This technique (also called zero-crossing voltage) is suitable in industrial applications such as solid-state relays (SSRs), heat, light, and valve control. It is not useful in controlling the speed of motors, since motors have the tendency to slow down during the time that power is not applied.

In this section, we are interested only in opto-triac drivers. We mentioned the types of opto-couplers (opto-isolators) and how they operate at the beginning of this chapter. One of these opto-couplers is the triac driver. Motorola produces a series of these opto-isolator triac drivers. Some opto-couplers are nonzero crossing drivers, while the others are. As examples, MOC3010 and MOC3011 are nonzero crossing devices, while MOC3031 and MOC3041 are zero-crossing devices.

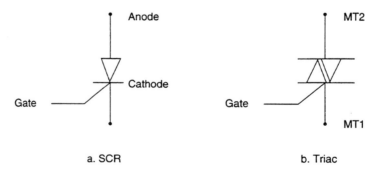

a. SCR   b. Triac

Figure 6.30

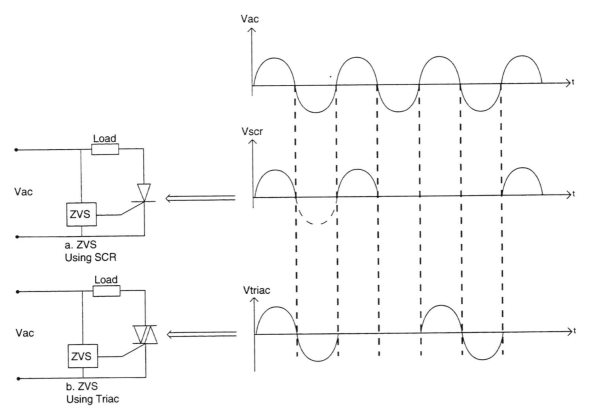

Figure 6.31

The schematic and pin diagram of MOC3011 are shown in Figure 6.32. It consists of an LED and a light-activated triac. The gallium arsenide (Ga As) LED has nominal 1.3 V forward drop voltage. The detector has a minimum blocking voltage of 250 Vac in either direction. It can pass 100 mA with a 3V voltage drop.

The MOC30XX is driven by sinking LED current with an active low. R1 is a limited-current resistor. R2 is used to limit the peak gating current to less than 1A to protect the MOC3011 and the triac itself.

Figure 6.32

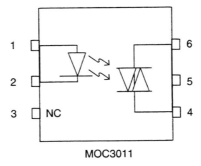

MOC3011

Program 8:   AC Load Control

In this program, the MCU will turn on/off eight high-voltage lamps using the zero-crossing voltage technique. The circuit is shown in Figure 6.33. The switches connected to port C (PC0 to PC3) indicate the status of the lamps. For example, SW0 indicates the status of lamp1 connected to PB0 and sw1 indicates the status of lamp2 and so on. When sw is pressed, port C switches are read, and if a switch is closed, it causes the corresponding port B to go low on the next zero-crossing and turns its lamp on.

The circuit consists of two parts:

1. Power side, which includes the 110 Vac power, lamps, and the triggering triacs.
2. Control side, including the detector transistor, which it detects when the power line goes to zero.

The zero-crossing detection should occur at or near the zero voltage. Let's assume that it occurs at ±5° of the input power line zero-voltage point, as shown in Figure 6.34a.

In the 60-Hz system, the power cycle period T is

$$T = \frac{1}{60} \text{ Hz} = 16.66 \text{ mS}$$

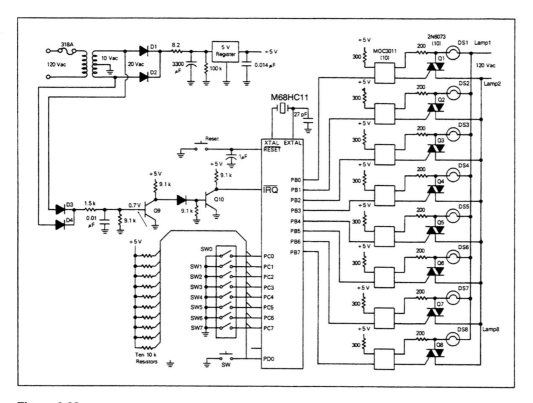

Figure 6.33
Copyright of Motorola. Used by permission.

Figure 6.34

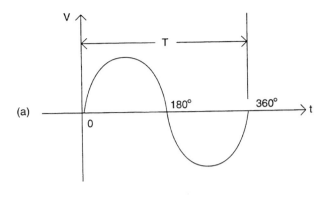

(a)

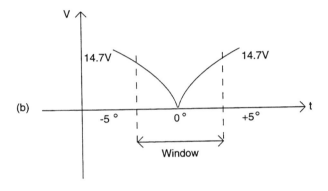

(b)

The power window 5° is calculated in terms of time as follows:

$$t = \frac{5°}{360°} \times 16.66 \text{ mS} = 231.5 \text{ μS}$$

As the circuit diagram shows, the line voltage is stepped down and full rectified to give an output of 10 Vac (rms).

The rectified voltage is fed to an AC detector, which consists of a filter and two transistors (Q1 and Q2). The output of the detector is fed to the $\overline{\text{IRQ}}$ pin of the MCU. Here's how the detector works.

When the input voltage of the 10 Vac goes high above 0.7 V (Vce of the transistor), Q1 turns on, which shorts out Q2. Q2 is off and IRQ is high. When the input goes below 0.7 V, Q1 is off, which turns Q2 on, and pin IRQ is low.

Let's calculate the control window:

$$Vt = Vp \sin wt$$
$$0.7 \text{ V} = 10\,2 \sin (2\pi f)t$$
$$t = 131.3 \text{ μS}$$

which is safe because it is within the power window (231.5 μS) calculated previously.

See the programming flowchart in Figure 6.35.

Exercise

Modify the above routine to accomplish a soft start for the lamps. Each half cycle of the IRQ full-wave input is divided into 32 time intervals. The duty cycle of each lamp is programmed initially and then incremented toward 100% power at one-second intervals.

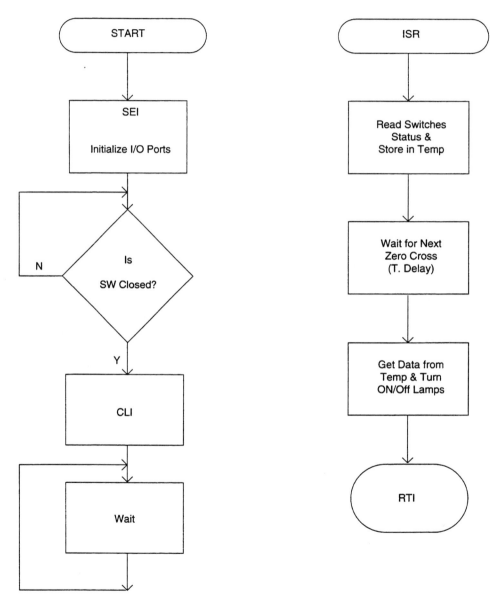

Figure 6.35

7 Microcontroller-Based Control Systems

Control systems are often divided into two major groups: servomechanism control and process control systems. In servo control, the position, speed, or acceleration of an object is made to follow closely the set-point command. The output responds fast, often within a very short time, to a change in a set point given a constant load. The process control system regulates one or more variables. Such variables include temperature, liquid, flow rate, pressure, force, and so on. The set-point changes occur only occasionally. The output response is slow, often taking minutes or hours.

A *servomechanism* is an electromechanical system consisting of devices designed specifically for accurately positioning or controlling motion. One servomechanism device is a *servomotor*. A servomotor is a precise electric motor whose function is to cause motion in proportion to a command signal.

There are two general types of servomotors: AC and DC. DC servos are used for high-power applications and are easy to interface to microcomputers, but the brushed-type generates radio frequency (rf) interference. AC servos require less maintenance and are more stable but lack large torque capability. We will concentrate on DC servos.

7.1   Modes of Control

Generally, modes of control include open- and closed-loop control. The latter is subdivided into on/off control, proportional control, integral control, proportion plus integral control, and PID control. We will control the speed of a DC motor using the above techniques.

Open-Loop Control

The simplest form of control is open-loop control. In this system, the output will follow the input as long as all the system variables are constant. If there is any disturbance—changes in load, amplifier gain, and so on—the output function is unpredictable because there is no way (such as feedback) to know these variations. Figure 7.1 depicts an open-loop motor control circuit.

A digital-to-analog converter (DAC) converts the digital command of the microcomputer to an analog signal. This signal is fed to the servoamplifier to amplify the command signal and feed it to the actuator. The *servoamplifier* is a power op-amp. There are several problems with the open-loop control system. First, some actuators do not perform as well at voltages other than those specified for them. Second, the open-loop system cannot guarantee the desired output from a process subject to disturbances. The DAC could be bipolar to change the motor direction. If you try to break down the motor, the speed will drop dramatically.

Closed-Loop Control

In the closed-loop system, the output variable is measured, fed back, and compared to the desired input function. Any difference between the two is a deviation (called *error voltage*) from the desired result. This deviation is amplified and used to correct the error.

On/Off Control

The on/off controller's output is either fully on or fully off. A block diagram of speed control using the on/off closed-loop is shown in Figure 7.2. To implement this control, the following algorithm is used:

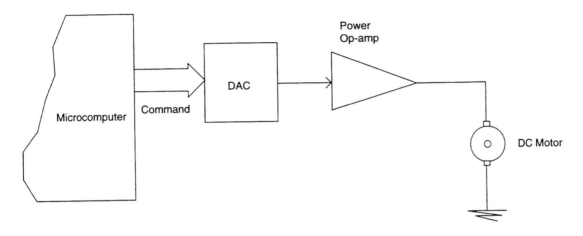

Figure 7.1

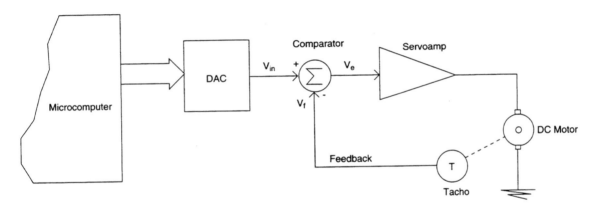

Figure 7.2

$$\text{If } V_{in} \geq V_f, \text{ then } V_e = \text{maximum}$$

or

$$\text{if } V_{in} < V_f, \text{ then } V_e = \text{minimum}$$

Figure 7.3 describes Algorithm 1.

Proportional (P) Control

The output voltage is given by

$$V_0 = K_p \times V_e + V_c$$

where V_c is the output with no error and is ignored in our discussion. K_p is the proportional gain and generates a speed command proportional to a speed error. K_p is a variable; its value is determined (called *tuning*) by trial-and-error techniques with the motor connected to the

Algorithm 1

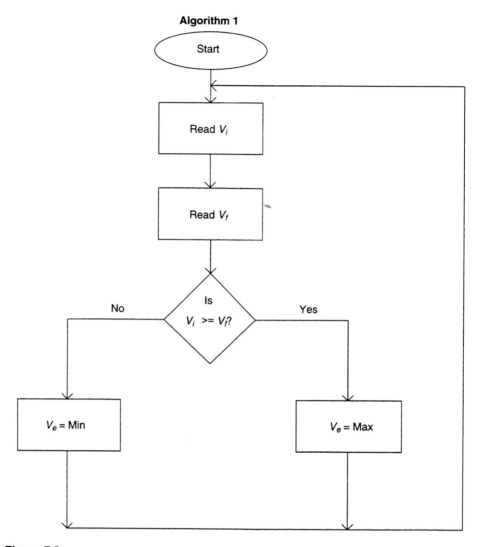

Figure 7.3

actual or simulated load. By increasing K_p higher than 1, the speed increases until K_p reaches a certain value at which the motor produces rough running. The algorithm for this type of control is illustrated in Figure 7.4 (Algorithm 2). If you try to load the motor shaft, the speed may drop by 20 to 30%, which is better than with the open-loop performance.

Integral (I) Control

The output voltage is given by

$$V_0 = K_i \int V_e dt + V_n$$

where V_n is the initial value of V_0 and is ignored in our discussion; K_i is integral gain.

Figure 7.4   **Algorithm 2**

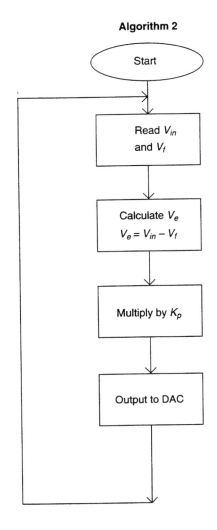

The term $\int V_e dt$ in mathematics means the sum of error voltages $V_{e1}, V_{e2}, \ldots, V_{en}$ to date. That is,

$$V_0 = K_i \times (\text{sum of all error voltages to date})$$

Using the integral control by itself is not the correct thing to do, because it is hard to control the speed in this mode. It is better to use both proportion and integral modes together. Again K_i is determined as explained in the case of K_p.

Exercise

Develop a flowchart for integral control.

Proportional and Integral (PI) Control

In this case, the output voltage is given by

$$V_0 = K_p \times \text{current } V_e + K_i (\text{sum of error voltages to date})$$

Figure 7.5   **Algorithm 3**

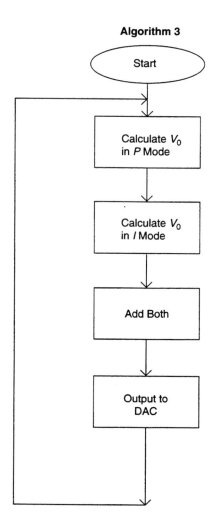

This combination of control provided excellent speed control over the entire range (low and high speeds). K_p and K_i are determined by trial and error, to obtain the best performance. Using the flowchart in Figure 7.4 and the flowchart in the preceding exercise, we can calculate V_0 with the aid of Algorithm 3 (see Figure 7.5).

Derivative (D) Control

The output voltage is given by the following equation:

$$V_0 = K_d \times dV_e/dt$$

where dV_e/dt is the derivative of error voltage V_e and equals the rate of change of V_e with respect to time (T). That is,

$$dV_e/dt = \{V_{e1} - V_{e2}\}/T = \{V_e(\text{current}) - V_e(\text{old})\}/T$$

$$V_0 = K_d\{V_e(\text{current}) - V_e(\text{old})\}$$

Figure 7.6   **Algorithm 4**

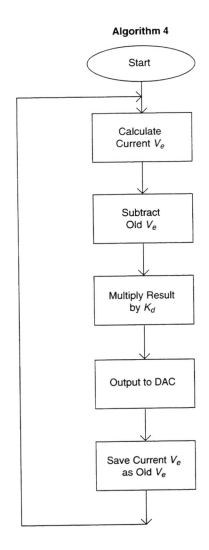

where T is constant. The derivative control cannot be used by itself but must be used with the other terms. It is used to provide damping. The appropriate algorithm is presented in Figure 7.6, Algorithm 4.

PD Controller

The output voltage is calculated as follows:

$$V_0 = K_p \times V_e + K_d \times dV_e/dt$$

The advantage of this mode is that the system has a fast response and is well damped at higher values of K_d.

Proportional Integral Derivation (PID) Control

One of the most powerful but complex controller mode operatives combines the previous three modes: the proportional, integral, and derivative modes. This system can be used for

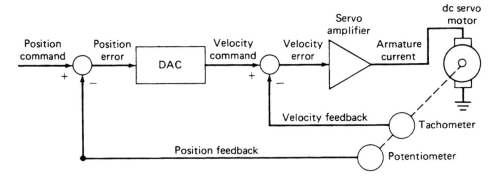

Figure 7.7

virtually any process condition. It provides fast response, no overshoot responses, and good speed regulation.

$$V_0 = K_p V_e + K_i \int V_e \, dt + K_d \, dV_e / dt$$

Exercise

Develop a flowchart to implement a PID control system.

Position Control

The closed-loop motion control system consists of a controller (computer) sending out position commands in the form of rules. These rules command a drive (DC motor in our case) to move an object a certain distance in the next increment of time. The pulse count from a position transducer (optical encoder) indicating the drive position is then subtracted from the position command (Figure 7.7).

The difference between the position command and the feedback position is called the *following error*. This error is converted to voltage via the DAC and is proportional to the position difference. This voltage then powers the motor, generating a new position. Each time the optical encoder moves a certain distance corresponding to one pulse (forward or reverse), a counter increments or decrements by 1. The servoamplifier drives the DC motor directly. The position loop also includes the speed loop. The P, PI, PD, or PID control can also be implemented in the position control system, as explained earlier for the case of speed control.

7.2   Servosystem Components

Servosystem components consist of the following:

1. Input components, which produce a reference input signal
2. Error-determining components, such as comparators
3. Control components—for example, amplifiers
4. Output components, including actuators
5. Feedback components, like potentiometers and encoders

Switching Amplifiers in Digital Systems

A circuit diagram of the H type is shown in Figure 7.8. There are two methods of switching the transistors to provide the desired polarity. These methods are called the *bipolar* and the *unipolar* drive methods. In the bipolar drive method, we turn on one pair of transistors at a time. For example, if we want the current to pass through the motor to the right, Q1 and Q4 are turned on only and the motor turns in one direction. If we wish the current to pass to the left, Q2 and Q3 are turned on simultaneously and the motor turns in the other direction. This method has a limitation. When a pair of transistors are turned off, it takes a certain time (about 5 mS) for the transistors to get to the off state. This time delay limits the switching frequency to about 20 KHz.

In the unipolar drive method, Q1 is turned on and a PWM voltage is applied to Q4 (Figure 7.9). The advantage of PWM amplifiers is that the power dissipation in the total system is decreased.

Tachometers

A tachometer is a device that measures the angular speed of a rotating shaft. Tachometers in industry use one of two basic measuring methods:

1. Representing angular speed by the magnitude of the voltage generated
2. Representing angular speed by the frequency of the voltage generated

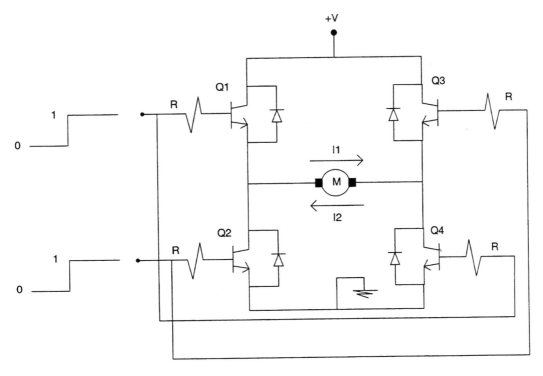

Figure 7.8

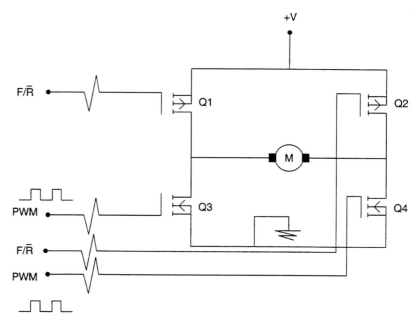

Figure 7.9

We are assuming that all the tachometers dealt with in this book are of the first type. They are simply DC generators.

Encoders

The types of encoders include potentiometer and optical encoders. Optical encoders may be absolute or incremental.

Potentiometer Encoders

Precision potentiometers are simply rotating devices for obtaining shaft position information. A potentiometer converts the mechanical position of a shaft to an electrical signal. It consists of a resistive element with a movable slider in contact with the element. As the shaft rotates, the slider rotates and the resistance varies between the end of the resistive element and the slider.

Some potentiometers have a continuous rotation with no internal stops. Others— called *multiturn* potentiometers—have 10, 15, or more revolutions before hitting the internal stops.

Optical Encoders

Absolute Encoders. An absolute encoder has a binary coded disk such as the one shown in Figure 7.10. An LED is mounted on one side of each track and a phototransistor (detector) is mounted on the other side. Since the disk has four tracks, there are a total of four LEDs on one side of it and four phototransistors on the other. The output of these phototransistors follows a special binary system called the *gray code,* which is not the

Figure 7.10

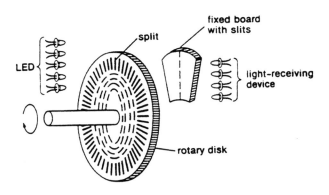

same as the normal binary system. Using the gray code helps reduce errors in reading the shaft position. The disadvantage of the absolute encoder is that multiple lights and detectors are needed. Table 7.1 shows a four-bit gray code.

Incremental Encoders. In this type of encoder, a metal disk with two tracks of slotted holes is mounted on the motor shaft. An LED is mounted on one side of each track, while a phototransistor is mounted on the other side. This means that we have a total of two LEDs on one side of the disk and two phototransistors on the other side. Each transistor produces a train of pulses as the disk rotates (Figure 7.11). The outputs of these transistors are *sinusoidal* (used to produce square waveforms). For every revolution, there are a certain number of pulses. Thus, the controller knows how many revolutions the motor has achieved. Besides giving the motor position, the encoder is also capable of knowing the direction of rotation.

As Figure 7.12 shows, two trains of square pulses come out of the detectors. Let's call them signal A and signal B. If signal A leads signal B, the motor runs in one direction, but if signal B leads signal A, the motor runs in the opposite direction. Channels A and B on the disk are arranged normally to be in quadrature (90° phase shift). A third channel—called the *index* channel (Z)—is also used. This gives a single output pulse per revolution and is used when establishing the *zero positive*.

Table 7.1

Decimal	Binary	4-bit Gray Code
0	0000	0000
1	0001	0001
2	0010	0011
3	0011	0010
4	0100	0100
5	0101	0111
6	0110	0101
—	.	—
—	.	—
15	1111	1000

Figure 7.11

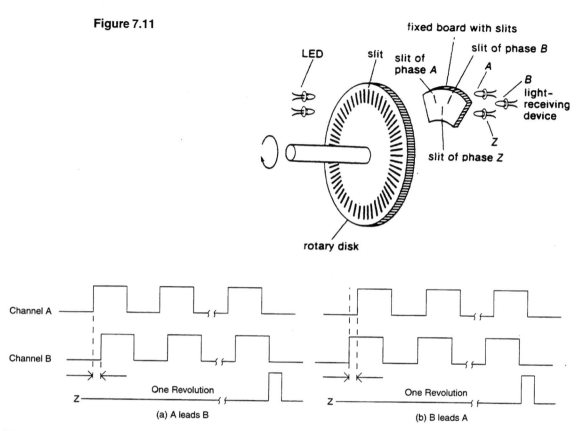

(a) A leads B   (b) B leads A

Figure 7.12

7.3   Digital Control of a DC Motor

Two methods are used in digital control: hybrid (analog and digital) system and digital systems. In the *hybrid system* as shown in Figure 7.13a, the analog output of the servomotor is fed back as a digital signal by the use of a shaft encoder—for example, an optical encoder. Figure 7.13b shows that the analog output of the servomotor is fed back using an analog transducer such as a potentiometer or a tachometer. The ADC converts this analog output to a digital signal to the computer. In both loops, the DAC is used to convert the digital error signal to an analog one.

The second method involves the *digital system,* where the DAC is not used to power the DC motor (Figure 7.13c). The controller sends its command to a pulse-width modulation (PWM) generator. This modulated pulse is fed to a PWM amplifier that drives the DC motor. In some microcontroller-based systems (such as the M68HC11), the MCU has an on-chip programmable timer that generates this PWM signal. This signal can be generated in two ways. The first is the PWM approach. The PWM switches the full voltage of the DC power supply on and off at a fixed frequency. By varying the duty cycle, we can control the average voltage applied to the motor (Figure 7.14a).

The second alternative is the pulse-length modulation approach. The PLM switches the full voltage of the power supply on and off at a fixed width but at a variable frequency

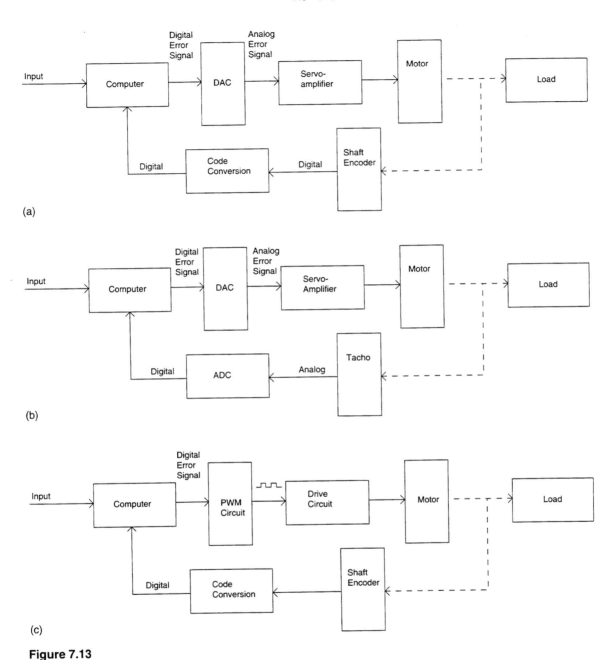

(a)

(b)

(c)

Figure 7.13

(Figure 7.14b). The MCU 68HC05b6, as an example, generates PLM pulses. PLM control is less popular than PWM but is used in some radio-controlled applications.

The duty cycle is calculated as follows:

$$DC = t/T \times 100$$

where t = length of the positive portion of the cycle

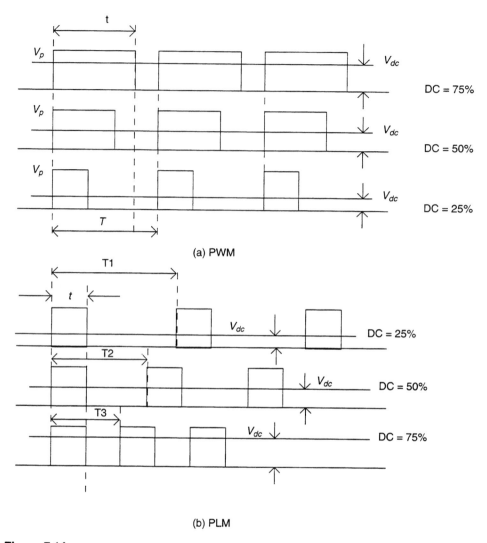

(a) PWM

(b) PLM

Figure 7.14

$$T = \text{cycle duration time}$$
$$V_{dc} = V_p \times t/T \text{ Volt}$$
$$V_p = \text{peak voltage value}$$

The higher the duty cycle, the higher the average voltage fed to the motor and the higher the resulting speed. In these systems, the computer makes the decision to reduce the error and produces a digital error signal that is sent to the motor driver.

DC motor control includes speed control, direction of rotation, and positioning. The speed of a DC motor is determined by the amount of control voltage (or error voltage) applied to the motor. The direction of rotation is determined by the polarity of this voltage. The position of an object is determined by comparing the command (control) position voltage with the feedback position voltage of the motor.

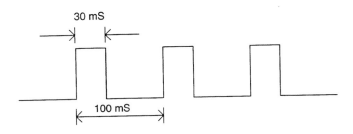

30 mS

100 mS

Figure 7.15

Exercises

1. Explain the terms servomechanism and servomotor.

2. What are the advantages and disadvantages of AC and DC servomotors?

3. For the square pulse-width wave in Figure 7.15, calculate the DC voltage V_{dc}.

4. What is the function of each compensator, P, I, and D?

5. Compare the hybrid and digital systems. Which do you prefer? Why?

6. Explain the differences between absolute and incremental encoders.

Program 1:   Speed Control of a Hybrid System with PI Compensator

The objective of this program is to implement a discrete-time (digital) version of the PI controller to maintain a desired rotational speed of a DC motor. Earlier, we discussed how to implement P, I, and D controllers. We explained in Chapter 6 how to interface the MC1408 to the 68HC11 (Figure 7.16). The velocity loop block diagram is shown in Figure 7.17.

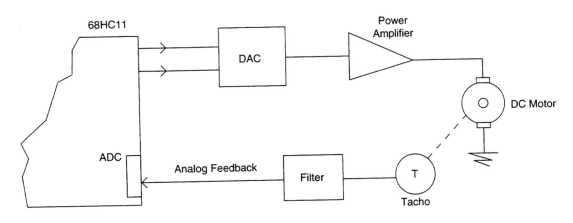

Figure 7.16

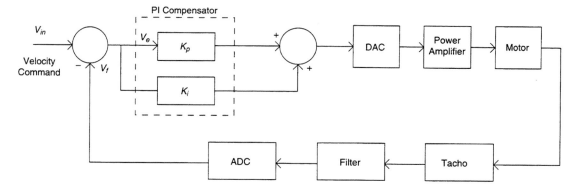

Figure 7.17

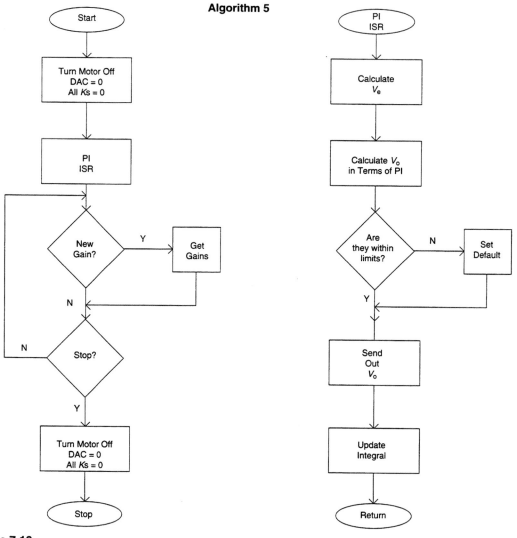

Algorithm 5

Figure 7.18

The PI term is calculated every time the MCU is interrupted. The sampling time t must be selected carefully.

$$t = \frac{\text{settling time}}{20}$$

The flowchart (Algorithm 5) for this routine is shown in Figure 7.18.

Program 2:   DC Motor Speed Measurement and Control in Digital Systems

We would like to use an optical encoder wheel as a feedback circuit from the motor shaft. The physical appearance of the opto-coupler and its circuit diagram are displayed in Figures 7.19 and 7.20.

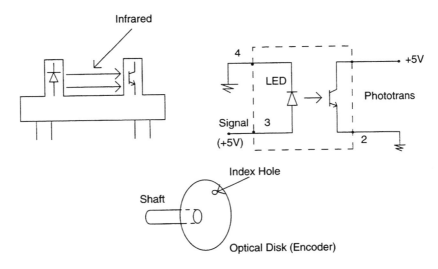

Figure 7.19

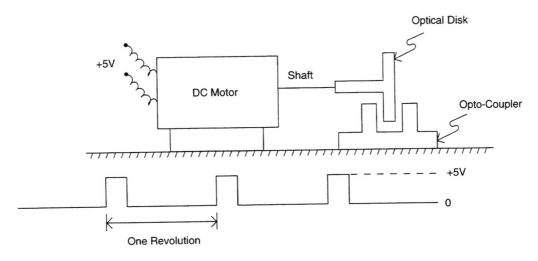

Figure 7.20

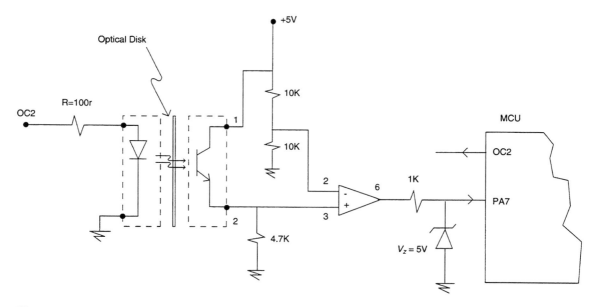

Figure 7.21

It consists of an LED, which generates infrared lights and a phototransistor mounted in one package. They are separated by a slot. When the LED is forward biased, it emits light that turns on the phototransistor. If an object is placed in the slot, the light beam (infrared) is broken and the phototransistor is turned off. As a result, a signal is generated that can be detected by the MCU.

The MCU sends a PWM signal on OC2 to the DC motor with a duty cycle of 50% initially. The motor turns and through its optical encoder (as feedback), the speed is calculated using the pulse accumulator and the equation below. (See Figure 7.21.) This speed is compared with the set speed (stored in memory) and is adjusted accordingly. The adjustment is done by varying the duty cycle of the PWM signal at frequent intervals (every 0.5 sec in our case). The motor speed can also be displayed on a digital display. The solution of this problem is divided into three stages.

Stage 1: PWM Signal Generation

The MCU generates a PWM signal on OC2. The user enters the duty cycle (DC) value manually in memory before executing the program, then the MCU generates the corresponding PWM signal.

Assume that the PWM frequency is 50 Hz with an initial DC of 50%.

$$T = 1/F = 1/50 \text{ Hz} = 20 \text{ mS}$$

Assume also the timer prescalar = 4; then

$$\text{Timer frequency} = \text{E-clock}/4 = 2 \text{ MHz}/4 = 0.5 \text{ MHz}$$
$$\text{Timer cycle period} = 2 \text{ μS}$$

Total number of timer cycles for a PWM of 20 mS = 20 mS/2 μS
$$= 10{,}000 \text{ cycles}$$
Number of timer cycles in the high portion of PWM $= 10{,}000 \times \%DC$
$$= DC \times 100$$
Number of timer cycles in the low portion of PWM $= (100 - DC) \times 100$

For example, assume we want to generate a PWM signal with DC = 60%. Then

Timer cycles in hi portion $= 100 \times 60 = 6000$ cycles
Timer cycles in lo portion $= 100 \times 40 = 4000$ cycles

The user stores the initial %DC value in a byte called DC_VALUE. This value is copied to another memory byte called SPDCOMND. The program keeps incrementing or decrementing this value to adjust the motor speed to the set speed. Figure 7.22 presents the flowchart.

```
DC_VALUE    RMB  1              ;initial %DC value
SPDCOMND    RMB  1              ;varying %DC value
```

Stage 2: RPM Calculations

The rpm of a motor can be calculated as follows:

$$\text{rpm} = \frac{n(\text{no. of counted holes in } T)}{T \text{ (sec)}} \times \frac{1 \text{ rev}}{m(\text{no. of holes per rev.})} \times \frac{60 \text{ sec}}{1 \text{ min}}$$

$$= \frac{n \times 60}{m \times T} \text{ rev/min}$$

In our special case here, $m = 1$ because there is one index hole only in the optical disk, which represents one tooth per one revolution. Now rpm becomes:

$$\text{rpm} = \frac{n \times 60}{T} \text{ rev/min}$$

The program calculates the number of pulses (n) fed to the pulse accumulator (PA7 pin) as shown in Figure 7.21, in an interval time T. Multiply n times 60 and divide the result by T. Assume $T = 100$ mS = 0.1 sec. Then

$$\text{rpm} = n \times 600 \text{ rev/min}$$

The program generates a square wave of predetermined duty cycle on OC2. As the shaft rotates, another square wave is generated as an output of the disk encoder mounted on the motor shaft, as shown in Figure 7.20. The disk has one hole—that is, each detected pulse represents one revolution.

The number of these pulses (n) in a certain time T are stored in a 2-byte storage called REVOL_NO. The high byte has the number of the pulse accumulator overflows, and the low byte has the remainder. The program calculates the rpm every 10 mS as explained above in the equation and stores this value in a 3-byte storage. We use the high 2 bytes and store them in memory as RPM_VLUE. This rpm value can be displayed on any of the display types (LED or LCD) mentioned in Chapter 8.

Figure 7.23 displays the flowchart.

```
REVOL_NO   RMB  2              ;number of revolutions
RPM_VLUE   RMB  2              ;motor rpm value
```

Figure 7.22

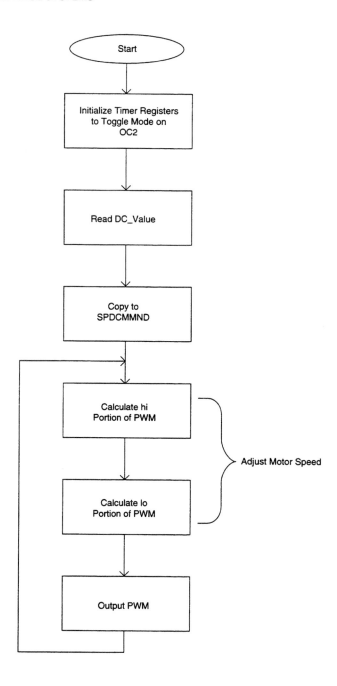

Stage 3: Scheduling

The calculated rpm value is checked every 500 mS (or less) and compared with the stored set speed. If the motor is running faster, the program slows it down by decreasing the PWM duty cycle and vice versa. The flowchart is shown in Figure 7.24.

```
SET_SPD  RMB 2                    ;set speed of motor
```

Figure 7.23

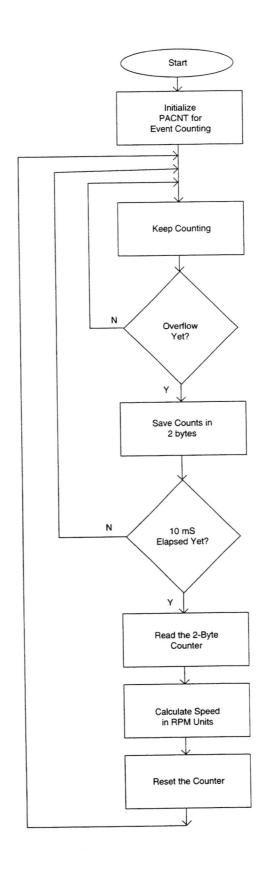

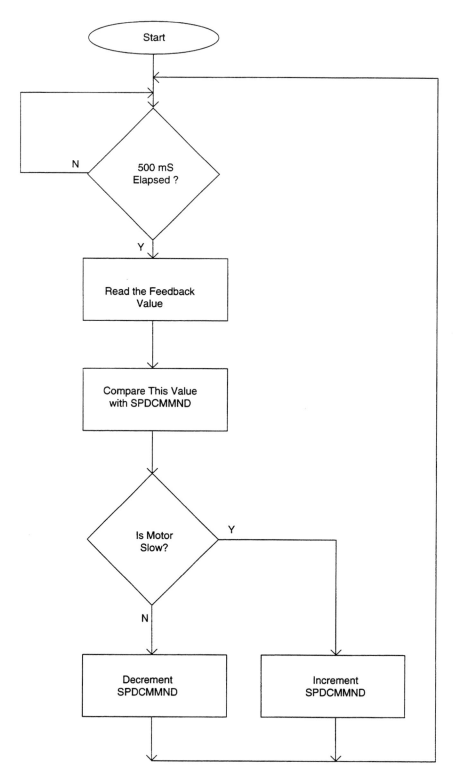

Figure 7.24

Exercises

1. We can use the same technique to measure and display the rpm of an engine gear using "Hall-effect" switches, as shown in Figure 7.25. These switches close when sensing a magnetic field. The circuit diagram uses the Hall effect as a counting device (instead of the opto-coupler). This method is the way to measure the rpm of an engine in real life. Modify the above program to implement this technique if the gear has 36 teeth.

2. Use a 4×4 keypad to enter the command word DC_VALUE to the MCU instead of doing it manually. Draw a flowchart to accomplish that.

3. Use a two-hole disk with a 90° out of phase to determine the direction of the motor using a pattern look-up table.

4. Assume that an absolute encoder is connected to the motor shaft. The absolute encoder uses the gray code, as explained earlier. Since there are 16 gray codes, the change occurs every 360°/16 = 22.5°. Table 7.2 shows the relation between the shaft angle and the gray code. Write a routine to move the motor 225°, then stop.

5. Repeat the above problem, if a gearbox is connected to the motor. The gearbox has a ratio of 5:1 (speed reduction), and the disk is mounted on the gearbox.

Figure 7.25

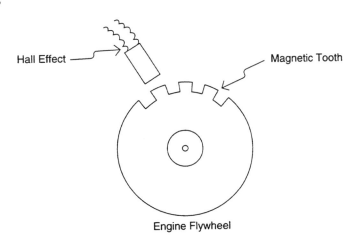

Engine Flywheel

Table 7.2

Motor Shaft Angle (deg)	Gray Code
0	0000
22.5	0001
45	0011
67.5	0010

6. Write routines to drive a DC motor if it is connected to one of the following motor-driver-power integrated circuits (ICs):

 a. Motorola MPC1710A, which includes an H-bridge composed of *n*-channel MOSFET

 b. SGS Thompson L293D, which includes two H-bridges composed of bipolar transistors

 c. Allegro Microsystems UDN2993B, which includes two H-bridges

7.4   Brushless DC Motor Control Circuits

The brushless servosystem can operate at higher speeds and provide higher peak torques than conventional DC motors. The physical structure of this motor provides better cooling, which allows it to produce much higher power in relation to its size. The commutation is achieved by solid-state switches and not by brushes. Because of these advantages, the brushless servomotors are very popular in industry applications.

A commutation technique for a three-phase brushless motor is shown in Figure 7.26. The three phases are connected as a Y-connection, which means that all windings are 120° apart. Six transistors are connected to the winding to control the commutation. Two transistors are on at any time to pass the current through two coils (phases). The conduction is always continuous in one leg while the other is being commutated. For example, Q1 is on between 0° and 60°, and Q2 is on too. Therefore, the current is flowing in coils AB. Between 60° and 120° Q1 is still on, but Q5 is off and Q6 is on. Thus the current is flowing in coils AC.

Exercises

1. Explain the reasons brushless motors are more popular than the other types of DC motors.

2. Explain the operation of the three-phase brushless motors.

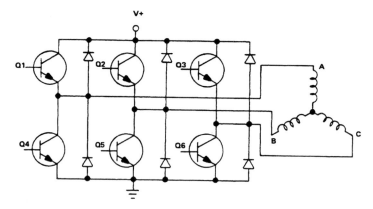

Figure 7.26

7.5   Stepper Motor Control

AC and DC motors rotate constantly, while stepper motors "step" from one position to another when they rotate. They move, then stop, then move again and so on. Stepper motors are popular in computer and instrumentation applications. For example, they are used to position the read/write head over the desired track of a floppy disk or hard drive in a personal computer. They are also used in printers; one stepper motor advances the paper and another moves the printhead to the next character. As mentioned earlier, a stepper motor moves in steps. Each step has an angle. These angles are called *step angles*. One stepper motor could have a step angle of 1.5° while another could have a step angle of 15° and so on.

Types of Stepper Motors

Three basic types of stepper motors are in common use:

1. Permanent magnet (PM)

2. Variable reluctance (VR)

3. Hybrid

Figure 7.27 shows several wiring methods for two-, three-, and four-phase stepper motors. To speed up the motor response, an external resistance (R_{ext}) must be connected to the phase coils, as Figure 7.28 indicates. R_{ext} must be three times the phase resistance.

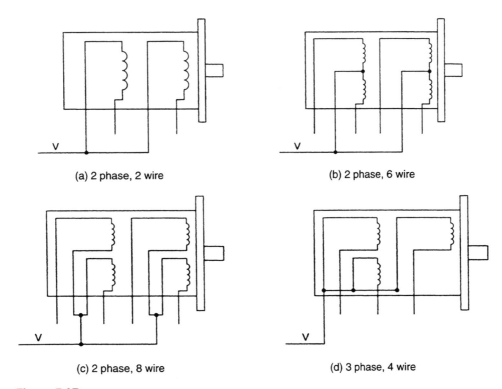

(a) 2 phase, 2 wire

(b) 2 phase, 6 wire

(c) 2 phase, 8 wire

(d) 3 phase, 4 wire

Figure 7.27
From A. K. Stiffler, *Design with Microprocessors for Mechanical Engineers*, McGraw-Hill, New York. Copyright 1992 by McGraw-Hill. Reprinted with permission.

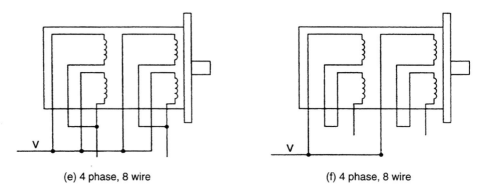

(e) 4 phase, 8 wire   (f) 4 phase, 8 wire

Figure 7.27 *(continued)*

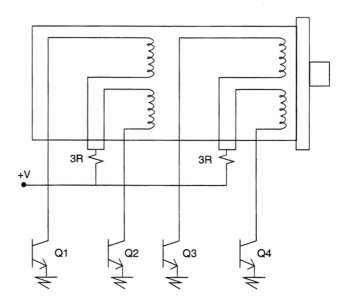

Figure 7.28

Stepper Motor Controllers

There are three types of controllers, as shown in Figure 7.29.

Direct Control

The MCU can be programmed to directly control the electronic switches (transistors) a, b, c, and d—included in the driver circuit—as shown in Figure 7.29a. The program sends pulses in phases A, B, C, and D in a certain order to turn the motor on in one direction and sends these pulses in the opposite order to move the motor in the opposite direction. The pulse duration determines the pulse rate and the latter determines the motor speed.

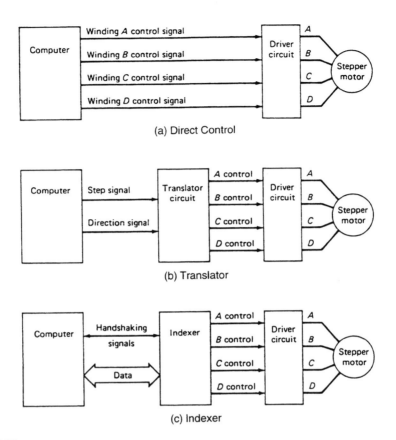

Figure 7.29
From P. Lawrence, *Real-Time Microcomputer System Design*, McGraw-Hill, New York. Copyright 1987 by McGraw-Hill. Reprinted with permission.

Translator

A translator circuit (shown in Figure 7.29b) can be used to generate the necessary control signals for the motor. It generates the proper winding excitation sequence. Translators can be configured to generate both full-step and half-step sequences. These translators contain internally a drive circuit, flywheel diodes, and a mechanism to reverse the motor direction if needed. See Appendix B (p. 370) for more details on this translator.

Indexer

The indexer (shown in Figure 7.29c) accepts commands from the MCU specifying the start/stop rate, accel and decel rate, slew rate, and the final position. It then generates the required stepping rate profile and informs the MCU when the final position has been reached. This may be done by using an interruptor or by setting a bit in a status register.

Types of Drivers

Unipolar Drive

The simple type of switch set is the unipolar arrangement depicted in Figure 7.30. It is called a *unipolar drive* because current can only flow in one direction through any particular motor

Figure 7.30

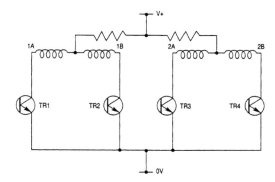

terminal. A bifilar-wound motor must be used since reversal of the stator field is achieved by transferring current to the second coil. The current is determined by the winding resistance and the applied voltage. The disadvantage of this drive is its inability to utilize all the coils on the motor. At any time, the current flows only in one half of each winding. If we could utilize both sections at the same time, we could get a 40% increase in amp turns for the same power dissipation in the motor.

Bipolar Drive

This configuration is arranged as an H-bridge circuit similar to the DC motor H-bridge. Therefore, the motor can step in two directions.

Because switched inductive loads generate a large rush current that can damage the transistors and also the MCU port, the MCU must be protected using different techniques as follows:

1. *Opto-couplers.* There is no direct coupling between the motor and the MCU.
2. *Diodes.* The diodes across the coils (as shown in Figure 7.31) prevent current in the reverse direction and protect the MCU.
3. *Resistance.* Resistance limits the rush current. However, it must be sized correctly to pass an adequate current to drive the transistors.

Chopper Voltage Drive

The drive applies higher voltages than the stepper's rated voltage. High voltage causes the current to rise quickly, which makes the motor respond quickly too.

Program 3: Interfacing a Stepper Motor Using the "Full-Step" Configuration

The unipolar driver circuit has four open-collector transistors. The technique is to generate pulses from the MCU on the output lines PB0 to PB3 as shown in Figure 7.31 in a certain order, using the look-up table (Table 7.3). These pulses are applied to coils A1, A2, B1, and B2.

How we can control the motor speed? We control it by controlling the pulse rate. If we send fast pulses, the motor moves fast and vice versa. We can control this rate by adjusting the pulse width using a time-delay subroutine. Here is the program.

```
Stepper1.asm


            ;Registers equates

        ORG $0000
```

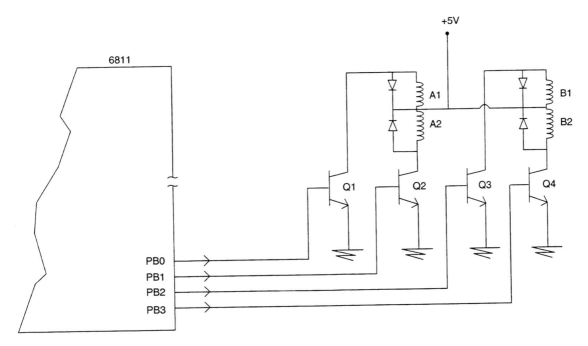

Figure 7.31

Table 7.3

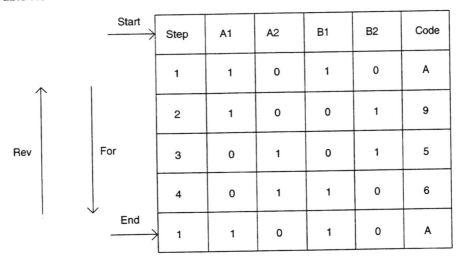

Step	A1	A2	B1	B2	Code
1	1	0	1	0	A
2	1	0	0	1	9
3	0	1	0	1	5
4	0	1	1	0	6
1	1	0	1	0	A

```
                        ; This is the look-up table
STTBL      FCB   $A
           FCB   $9
           FCB   $5
           FCB   $6
ENDTBL     FCB   $A

                  ORG  $C000
```

```
                 LDX   #BASE
                 LDAA  #$80
                 STAA  TFLG1,X      ;clear flg
START            LDY   #STTBL
BEG              LDAA  0,Y           ;first row in table
                 STAA  PORTB,X
                 JSR   DELAY
                 INY
                 CPY   #ENDTBL       ;end of table yet?
                 BNE   BEG           ;no, branch
                 BRA   START         ;yes, start all over again

DELAY            LDD   TCNT,X
                 ADDD  #4000         ;20ms delay
                 STD   TOC2,X
HERE             BRCLR TFLG1,X,$80,HERE ;wait till time delay elapsed
                 LDAA  #$80
                 STAA  TFLG1,X      ;clear flag
                 RTS
```

Exercises

1. Modify the above program to reverse the motor direction.

2. With the "half-step" technique, we use the same driving circuit used in the above program. In the full-step configuration, the technique is to send two pulses at the same time to the motor phases. But in the half-step configuration, the technique is to send two pulses the first time to the coils, then a single pulse the next time, and so on. Obviously, we use more steps here than in the full-step configuration, but the resolution is better. Modify the stepper1.asm to utilize the "half-step" configuration. The look-up table is included as Table 7.4.

3. Develop a routine for the translator/driver UCN5804B BiMOS II made by Allegro Microsystems to drive a stepper motor. The pinout is shown in Appendix B.

   In some applications, the MCU is too busy to be bothered with controlling the stepper motor directly. Some smart stepper controllers called *translators*, as mentioned earlier, can be used to drive the stepper motor.

4. The output compare OC2 to OC5 can be directly connected to the stepper motor driver circuit used previously. Write a routine to generate four signals through these channels using the OCI multicontrol feature.

5. Repeat problem 3 for the driver MC3479P, made by Motorola.

7.6   Process Control Systems

We mentioned at the beginning of this chapter that control systems can be divided into two major groups: servo control and process control. In process control systems, the variables manipulated are those used most often in manufacturing. The primary object of process

Table 7.4

Step	A1	A2	B1	B2	Code
1	1	0	1	0	A
2	1	0	0	0	8
3	1	0	0	1	9
4	0	0	0	1	1
5	0	1	0	1	5
6	0	1	0	0	4
7	0	1	1	0	6
8	0	0	1	0	2
1	1	0	1	0	A

Start → (at Step 1)

End → (at last Step 1)

Rev. ↑ / For. ↓

control is to regulate one or more of these variables, keeping them at a constant value called a *set point*. The set-point changes occur only occasionally and are usually less than 10% of the full load. Output responses are slow, often on the order of minutes or hours. This is much longer than usual for servosystems. Process control systems may be subdivided into two categories: batch (discrete) and continuous processes.

Discrete-State Systems

Two types of control strategies are associated with discrete state systems. One type is used to control the *value* of one or more variables in the system. The second type is *sequential*, referring to the progress of the system through a defined set of discrete states. We will explain the first type only.

Value Control

The value control discrete system has two states only: on/off. The following example explains this type of control.

Program 4: Discrete-State Process Control

The process control shown in Figure 7.32 is mainly controlled by a microcomputer (in our case by the M68HC11) to perform the following process:

1. Assume that there is a substance in the tank that must be mixed with a solution *(a)*.
2. The tank is filled with solution *a* through a solenoid valve (Va).
3. When the solution reaches level *La*, the mixer motor starts agitation and the heater starts heating for about a half minute.
4. The drainage valve *(Ve)* opens to empty the tank.

The tank has two level sensors that send signals through comparators to the input lines PC0 and PC1. The output lines PB0 to PB3 provide signals to the solenoid valves *Va*, and *Ve*, the heater, and the motor through the signal drivers. Let's write a routine to perform this process.

```
                        ;Equates are responsibility of the user
                        ;init routine here

                           ORG $C000


                           LDX #BASE
        START              LDAA #01              ;sol. Va is on
                           STAA PORTB,X
        LOOP               BRCLR 0,PORTC,X,LOOP  ;loop till La is reached
```

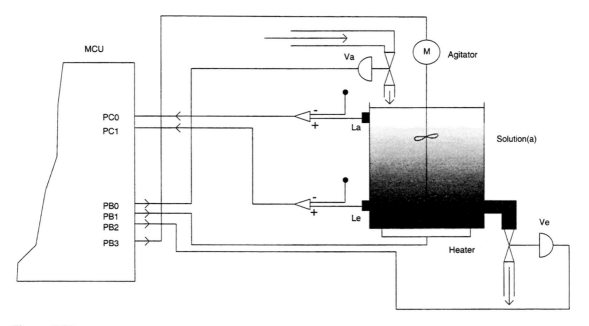

Figure 7.32

```
                    LDAA #$A
                    STAA PORTB,X            ;turn on motor and heater
                    JSR  DLY1               ;delay for 1 min.
                    LDAA #4
                    STAA PORTB,X            ; turn on Ve to empty tank
        LOOP1       BRCLR 1,PORTC,X,LOOP1 ;loop till Le is reached
                    BRA  START

        DLY1        ;Time delay of 1 min.
                    RTS
```

Note that we assumed that level sensors send positive signals.

Exercise

Modify the above program to perform the same process except that solution *b* replaces the substance in the tank.

7.7   Machine Control in Industrial Systems

The technology applied to the control of machines has changed significantly in the last 15 years. Numerical control (NC), computer numerical control (CNC), programmable logic controller (PLC), and robotics are expanding technologies that operate in a magnificent coordinated system. They provide programmability, expandibility, reliability, maintainability, and—best of all—productivity.

Computer-Integrated Manufacturing

The Computer and Automation Systems Association (CASA) of the Society of Manufacturing Engineers (SME) defines computer-integrated manufacturing (CIM) as "the integration of the total manufacturing enterprise through the use of integrated systems and data communications coupled with new managerial philosophies that improve organizational and personnel efficiency."

CIM may include CNC, computer-aided design (CAD), computer-aided manufacturing (CAM), and robots.

Robot Controller

The block diagram of the robot controller is basically the same as the standard block diagram of any computer: CPU, memory, and I/O ports (modules). The controller has a network of CPUs (MPUs/MCUs), each having a different responsibility within the system.

The controller also receives feedback from the robot's axes on joint position and velocity and responds accordingly with outputs to the servo drives to change the current position or velocity.

The subject of the control of industrial machines is beyond the scope of this book. The following case study provides a general idea by explaining some of their control aspects—for example, solenoids, directional valves, cylinders, and limit switches.

At the end of this case study, a program was created to control a three-axis hypothetical robot performing a simple pick-and-place task.

Solenoids

The solenoid is an electromechanical device. In this device, electrical energy is used to cause mechanical movement. Figure 7.33 shows a flat-armature solenoid. It consists of an electromagnetic coil wound on a yoke and a flat armature made of iron.

When the coil is energized, magnetic lines circulate from the yoke, via an air gap to the armature. An electromagnetic force (F) is generated at each pole. This force is a pulling force that attracts the armature to the yoke. Two forces act on the armature, one at each pole.

When the solenoid coil is energized, the plunger is in an output position. Due to the open gap in the magnetic path, the initial current in the coil is high. As the plunger moves into the coil, closing the air gap, the current drops to a lower value.

Directional Control Valves

The function of directional control is to direct oil or air to various places in the system so that it can operate. Directional control is accomplished by valves. These valves are operated by solenoids. Three different classes of valves are used in pneumatic or hydraulic systems:

1. Two-way (two ports and two internal passages) valves
2. Three-way (three ports and three internal passages) valves
3. Four-way (four ports and four internal passages) valves

Our discussion here considers only the pneumatic system. Figure 7.34 shows a four-way valve consisting of four ports: two ports let the air in (inlet ports) and the other two let the air out (outlet ports). The valve also contains a movable part called a *spool* that moves against a spring by the action of a solenoid plunger when the solenoid is energized. The ANSI graphic symbol of a four-way valve is shown in Figure 7.34a.

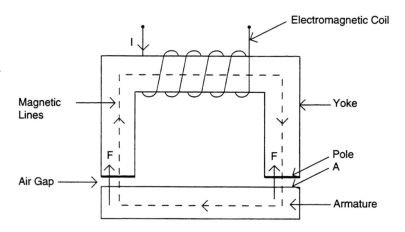

Figure 7.33

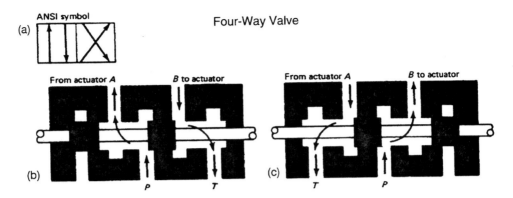

Figure 7.34

When the solenoid is deenergized, the valve is normally closed—that is, the spool is in its normal position and inlet port A is disconnected from outlet port B, as shown in Figure 7.34b. When the solenoid is energized, the plunger pushes the spool against the spring, the inlet port is connected to the outlet port as shown in Figure 7.34c, and the air passes from A to B. If the solenoid is deenergized again, the spring returns the spool to its original position and closes the passage between A and B. This type of valve is called a *single-solenoid, spring-return valve.*

Usually the control valves are used to move a piston inside a cylinder. A simple pneumatic system consists of an air compressor, control valve, and piston. The cylinder is called a *double-acting cylinder* because it has two ports at its ends. At reset condition, the solenoid is deenergized, the air passes from port A, and the piston moves from right to left (retracts). The exhaust air behind the piston leaves through port B to the atmosphere.

When the solenoid is energized, the valve changes its status. Compressed air passes through port B to extend the piston and the exhaust air leaves through the port.

7.8   Case Study 1

Robot Systems

Assume a three-axis robot (pick-and-place), as in Figure 7.35. These axes and their motions are defined as follows:

Base: CW/CCW
Arm: Extend/Retract
            Up/Down
Gripper: Open/Close

The extend/retract cylinder extends and retracts the arm. The arm has two internal limit switches, as the figure shows, to tell the control unit when the arm is fully extended or fully retracted, and so on, with the base and gripper as explained on the next page.

There are two ways to control high-voltage devices connected to the MCU. First, we can use DC electromechanical relays with high current-rating contacts. Second, we can also use solid-state relays with opto-couplers.

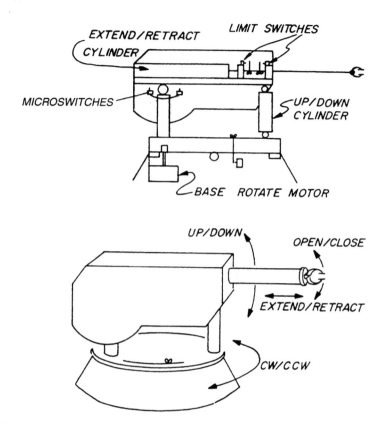

Figure 7.35
From W. Burns, *Practical Robotics*. Copyright 1986 by Prentice Hall. Reprinted by permission of Prentice Hall, Upper Saddle River, New Jersey.

The robot control circuit is shown in Figure 7.36. Port B is used to control the solenoid on/off modes, which operate with 110 Vac. We are going to use the solid-state relays for the reasons mentioned in Chapter 6.

If any bit of port B is high, the LED of the opto-coupler is on, which turns the opto-triac on and passes 110 Vac to the solenoid. The solenoid in this case is energized, which shifts the spool of the pneumatic valve and changes its status. The valves are four-way, double-solenoid, center position.

For example, if solenoid A of the gripper (end effector) valve is energized, the piston extends and the gripper closes. If solenoid B now is energized (while A is deenergized), the piston retracts and the gripper opens. If both solenoids are deenergized, the air passing in this valve is blocked and the piston keeps its status without change.

There are also eight microswitches for the cylinders, two microswitches for each. Each pair of switches senses whether the particular piston is extended or retracted. This information is fed back to the MCU via port C, as shown in Figure 7.37.

A *kinetic* diagram is used to illustrate the different motions of a robot. The diagram has a home position followed by a series of positions to perform the task required of the robot. These positions are defined in Program 5.

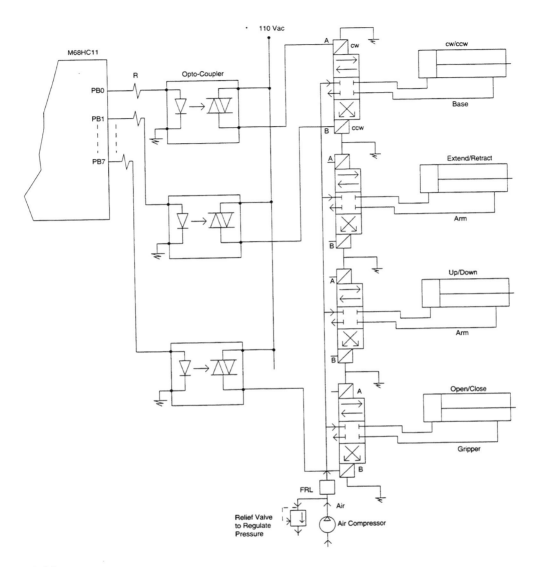

Figure 7.36

Program 5:   Moving an Industrial Pneumatic Robot to Perform a Pick-and-Place Task

Assume the kinetic diagram for this robot shown in Figure 7.38. The robot picks the workpiece from position (b) and places it at (d). According to this diagram, the robot motion is defined as follows:

1. At home position (a), the gripper is closed, the arm is retracted and fully down, and the base is pivoted CW.
2. When the start switch is pressed, the robot moves to position (b), opens its gripper, rotates the base CCW, and picks the workpiece.

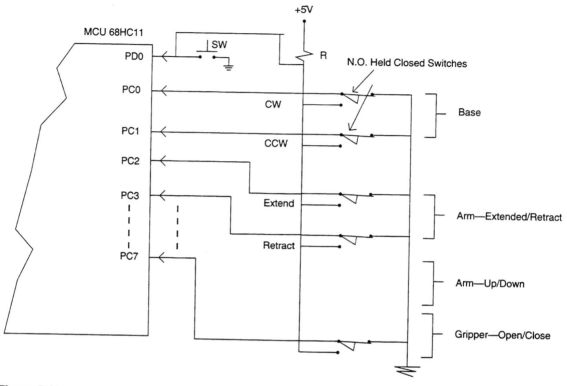

Figure 7.37

3. The arm is lifted up to position (c).
4. At position (d), the arm is extended and the gripper is opened to place the workpiece.
5. The base swings CW to position (e).
6. The arm is down to position (f).
7. The arm is retracted to position (a), home position.

Figure 7.38

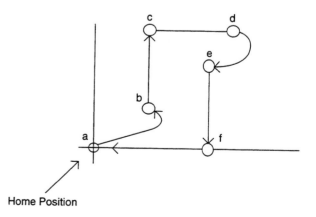

Home Position

```
                    ; Equates here

                            ORG $C000

                    LDX     #BASE
                    LDDA    #00
                    STAA    DDRC,X      ;PortC is input
                    STAA    DDRD,X      ;PortD is input

BEGIN               BRSET PORTD,X,$0,BEGIN ;is sw pressed?
                    LDAA    #01
                    STAA    PORTB,X     ;yes, turn base CW at (b)
LOOP_1              BRCLR PORTC X,$01,LOOP_1 ;is base fully CW?
                    LDAA    #$10        ;yes
                    STAA    PORTB,X     ;arm is up at (c)
LOOP_2              BRCLR PORTC,X,$40,LOOP_2 ;is gripper fully open?

LOOP_3              BRCLR PORTC,X,$10,LOOP_3 ;is arm fully up?
                    LDAA    #$40        ;yes
                    STAA    PORTB,X     ;gripper opens

                    LDAA    #$80
                    STAA    PORTB,X     ;gripper closed
LOOP_4              BRCLR   PORTC,X,$80,LOOP_4 ;is gripper fully closed?

                    LDAA    #$04        ;yes
                    STAA    PORTB,X     ;arm is extended at (d)
LOOP_5              BRCLR   PORTC,X,$04,LOOP_5 ;is arm fully extended?

                    LDAA    #$02        ;yes
                    STAA    PORTB,X     ;base swings CCW at (c)
LOOP_6              BRCLR   PORTC,X,$02,LOOP_6 ; is base fully CCW?
                    LDAA    #20         ;yes
                    STAA    PORTB,X     ;arm is down to (F)
LOOP_7              BRCLR   PORTC,X,$20,LOOP_7 ;is arm fully down?
                    LDAA    #$08        ;yes
                    STAA    PORTB,X     ;Home position
LOOP_8              BRCLR   PORTC,x,$08,LOOP_8 ;is arm fully retracted?
                    BRA     BEGIN       ;repeat the cycle

                    END
```

7.9   Case Study 2

Electronic control systems in vehicles using microcontrollers are numerous; they may include modules for

- Powertrain control
- Vehicle control
- Body control

- Instrumentation control
- Entertainment control

In this case study, we will focus on the engine control system that is part of the powertrain module. The engine control system consists of sensing devices, an electronic control unit (ECU) (MCU in automotive applications is called ECU), and actuating devices. (See the block diagram for such a system in Figure 7.39.) The input devices sense the engine operating conditions such as engine temperature, mass air flow, oil pressure, and so on. The ECU evaluates these conditions using data tables and algorithms to determine the output to the actuating devices.

The actuating devices are commanded by the ECU to take certain actions in response to the input devices. Examples of actuating devices are fuel injector drivers, solenoid drivers, and so forth.

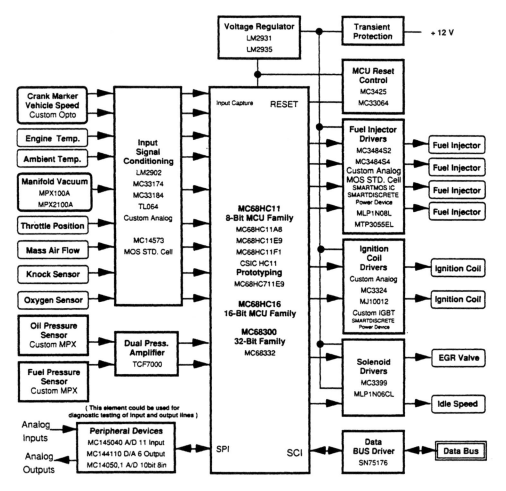

Figure 7.39
Copyright of Motorola. Used by permission.

In the following discussion, we are going to see how the ECU controls the amount of fuel going to the cylinders via fuel injectors.

Air/Fuel Ratio

The air/fuel (A/F) ratio is defined as stoichiometric when the engine is supplied with the exact quantity of air required from complete combustion and is equal to 14.7:1 by weight.

By definition, the air/fuel lambda (λ)

$$= \frac{\text{current air/fuel ratio}}{\text{stoichiometric air/fuel ratio (14.7)}}$$

A number of products of incomplete combustion occur as exhaust gases, such as C_O, H_2, HC, NO, NO_2, and N_2O.

If $\lambda < 1$, this is called a *rich mixture* because there is excess fuel. The exhaust gases are C_O, H_2, and HC.

If $\lambda > 1$, this is called a *lean mixture* because there is an excess of air (oxygen). The exhaust gases are NO_x.

The exhaust gases entering the catalytic converter are treated to comply with the U.S. emission standards. In the catalytic process, CO, H_2, and HC are oxidized to form CO_2 and H_2O, while NO_x are reduced to N_2 and O_2.

The engine control system regulates upstream from the catalytic converter with the aid of exhaust gas oxygen sensors. The control system adjusts the lambda value depending on these sensors' readings.

Open-Loop Control

Open-loop control involves the direct calculations of lambda without knowing the current A/F combustion status. The exhaust sensor feedback information is not used.

We must mention that for a cold engine, an increase in the air/fuel ratio is desired due to poor fuel vaporization, which decreases the amount of usable fuel. The amount of fuel increase depends on engine temperature and is a correction factor to the injection PWM. Within the ECU ROM, there are data tables to establish cold start fuel based on engine coolant temperature.

Closed-Loop Control

When the exhaust gas sensor is warmed up and its signal exceeds the specified threshold value (about 450 mV) of the comparator, the system responds first by leaning out the mixture, as shown in Figure 7.40. This is accomplished by sending a PWM signal with low DC to the injector drivers to reduce the amount of fuel delivered to the combustion chamber.

When the exhaust sensor reads lean ($\lambda > 1$), the ECU sends a PWM signal with higher DC to increase the delivered fuel, and lambda ramps up in the rich direction ($\lambda < 1$), as illustrated in the figure. The process of lean, rich, lean, . . . continues during the closed-loop control mode.

When an increase in engine load and throttle angle occurs, a corresponding increase in fuel mixture richness is required because of the sudden increase in air that results in a lean mixture.

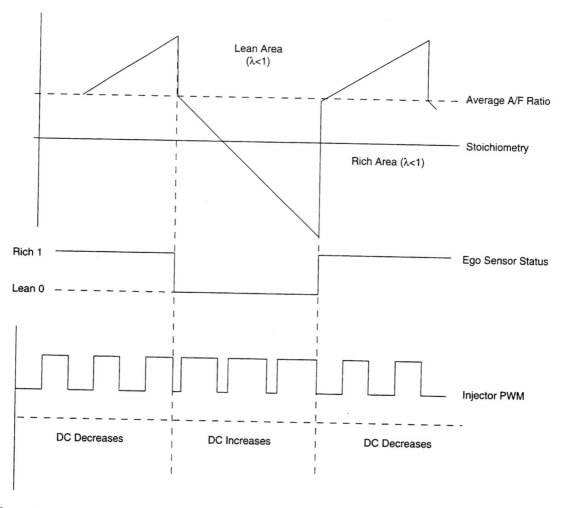

Figure 7.40

The following are some factors that affect fuel consumption and that should be taken into consideration when calculating the injector's pulse width:

1. Battery voltage
2. Temperature sensors
3. Vehicle speed

The ECU determines when the fuel shutoff can occur by evaluating the throttle position, engine rpm, and vehicle speed.

Program 6:   Controlling the Injectors of a Gasoline Engine

Using the above data, Figure 7.41 presents the flowchart for controlling the amount of gasoline injected (using PWM) into the combustion chambers via the injectors.

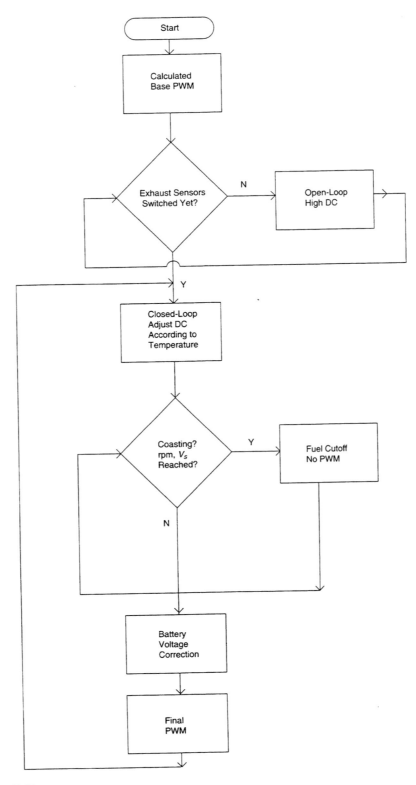

Figure 7.41

8 Microcontroller-Based Instrumentation Systems

This chapter covers the operation of some popular displays and transducers and explains how they are interfaced to a microcontroller (in our case, the M68HC11). Programs and exercises will help you master these topics. In addition, a case study will show you how to interface a transducer (a pressure transducer) to its signal conditioning circuit, to display the pressure on an LCD.

8.1 Displays

The most popular displays used with MCUs are light-emitting diodes (LEDs), light crystal displays (LCDs), vacuum fluorescent displays (VFDs), and DC plasma displays.

We will approach each type to learn how it functions and how it is interfaced to the MCU using several experiments. (The only exception are LEDs, which were covered in Chapter 6.)

The incoming signal is sent to a driver/decoder to amplify and convert the data to the display format, and then send the ready signal to the display hardware.

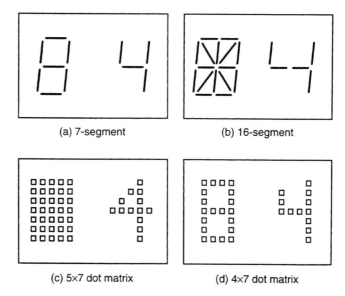

(a) 7-segment   (b) 16-segment

(c) 5×7 dot matrix   (d) 4×7 dot matrix

Figure 8.1
Adapted from A. K. Stiffler, *Design with Microprocessors for Mechanical Engineers,* McGraw-Hill, New York.
Copyright 1992 by McGraw-Hill. Reprinted with permission.

Display Formats (Fonts)

There are several display formats for all displays (LEDs, LCDs, and VFDs), as Figure 8.1 shows.

The seven-segment display is the most popular because it is cheap and easy to interface. However, this display is limited to numeric and a small number of alphabetic characters. The 16-segment display has full alphanumeric characters but has found only limited acceptance. The 5×7 dot matrix can display a wide range of alphanumeric characters but requires more complex circuitry or software. The modified 4×7 dot matrix is a modified version of the 5×7 dot matrix and is used for displaying hexadecimal information.

Bar Graph Drives

LED bar graph displays consist of a linear array of LEDs driven by a device that decodes an analog or digital signal into a bar graph indicator display code. There are two methods of displaying the information on a bar graph array, as Figure 8.2 shows. The first is called a *moving point* display, appropriate when one LED is used, at a time. If more than one LED is used, forming a lighted string, that is called a *bar graph.*

Seven-Segment Display

As explained in Chapter 6, there are two versions of a seven-segment display: common anode and common cathode. Some digits are left-hand decimal, right-hand decimal, or ±1 overflow. Some modules contain two or more digits with their decimal points (Figure 8.3). In large displays, one segment could have four LEDs connected in series.

Dot Matrix Techniques

In dot matrix displays, you cannot find a separate pin for each segment as in a seven-segment display. For example, a 5×7 matrix display has 12 pins for the array and one pin for

Figure 8.2

One LED on

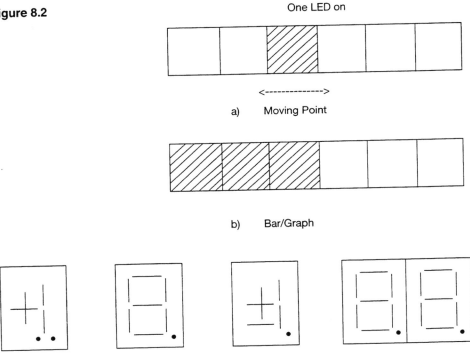

a)   Moving Point

b)   Bar/Graph

Figure 8.3

the decimal point, as indicated in Figure 8.4. The dot matrix is arranged in columns and rows and the driver circuits must be employed to generate characters either by column or by row strobing methods. Assume we want to display the letter P on a dot matrix display, as shown in Figure 8.5. Each column must be turned on one at a time from left to right while the remaining columns are off. For column 1, all rows must be off (logic 0) to turn

Figure 8.4
From A. K. Stiffler, *Design with Microprocessors for Mechanical Engineers,* McGraw-Hill, New York. Copyright 1992 by McGraw-Hill. Reprinted with permission.

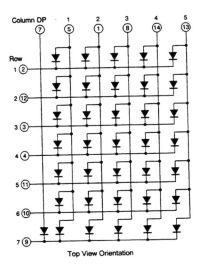

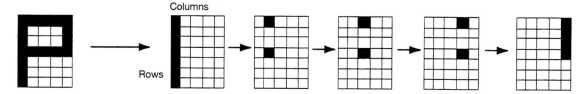

Figure 8.5

Table 8.1

Character	Column	Row Input
P	1	00
	2	F6
	3	F6
	4	F6
	5	70

column 1 on. On the next refresh cycle, for column 2, rows 1 and 4 are off, and so on, until column 5 is reached. This sequence is repeated continuously at the refresh rate.

Each character is stored in memory as a block of 5 bytes. See Figure 8.5 for P character bytes; also see Table 8.1.

Multiplexing

A technique used with multiple digits is called *multiplexing*. Multiplexing involves sending data to all display digits at the same time but only the digit that must be displayed will be enabled. Each display is turned on or refreshed at a frequency called the *refresh rate*. A minimum refresh rate is 100 Hz. If the refresh rate is very low, flickering occurs. The maximum "on time" *(t)* for each character is

$$t = \frac{1}{Nf}$$

where
N = number of display digits
f = refreshing frequency

The duty cycle is calculated as follows:

$$\% \, DC = \frac{1}{N} \times 100$$

For example, if we want to display the number 5 on the second display (D2), the MCU sends the data serially or in parallel to the driver/decoder. The driver/decoder will amplify, latch, and convert the coming signal to its seven-segment format. This format is sent to all four digits at the same time through lines *a, b, . . . , h*. To display the data on D2 only, this display *only* is enabled through enable lines.

Multiplexing Advantages

The major advantages of multiplexing include the following:

1. It reduces the amount of code (program) used for the driver/decoder.
2. Segments appear brighter with less current than without multiplexing.
3. It reduces components and wiring.

Program 1: Interfacing Two-Digit, Seven-Segment LEDs Using Discrete Drivers and Multiplexing

In this experiment, the MCU sends data to both common anode digits through port b at the same time, but if data are to be displayed on a specific display, this display only will be enabled (Figure 8.6a). For example, if we want to display the character *HI* on these digits, the technique is to send the code for *I* to both digits, but D1 is enabled only by sending a logic 1 to Q_1 to turn it on and supply 5 V to this digit for a short time. Similarly, the code for *H* is sent again for both digits and D2 is enabled only by sending a logic 1 to Q_2 to turn it on. (The seven-segmented code for HI is shown in Figure 8.6b.) This process is repeated again and again. The human eye will see that the word *HI* is displayed at the same time on the displays. For example, if $f = 100$ Hz, the minimum refresh rate (f) is:

$$f = 1/2t_{min} = 100 \text{ Hz}$$
$$t_{min} = 5 \text{ mS}$$

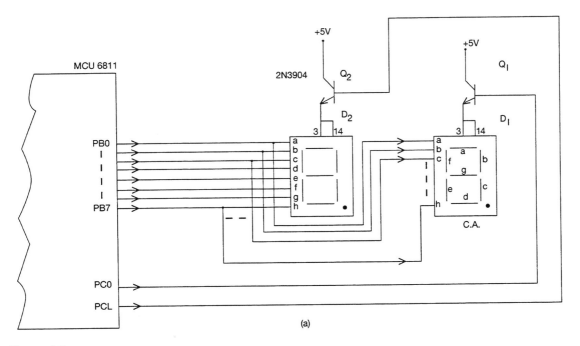

(a)

Figure 8.6

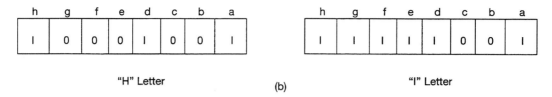

(b)

Figure 8.6   *(continued)*

This means that the minimum time delay used without flickering is about 5 mS.
If we choose $t_{min} = 2.5$ mS, then $f = 200$ Hz. Here is the program.

```
;DISPLAYING THE MESSAGE 'HI' ON A TWO MULTIPLEXED DISPLAYS USING
;DIRECT INTERFACING AND 2 TRANSISTORS TO SWITCH THE 5 V IF A
;DISPLAY IS ENABLED.


BASE  EQU $1000
PORTC EQU $3
PORTB EQU $4
DDRC  EQU $7
                          ORG $C000

              LDX  #BASE
              LDAA #$FF
              STAA DDRC,X       ;portc is output
START         LDAA #$F9         ;s.s. code for "I"
              STAA PORTB,X
              LDAB #1
              STAB PORTC,X      ;turn on Q1
              JSR  DELAY_25

              LDAA #$89         ;s.s. code for "H"
              STAA PORTB,X
              LDAB #2
              STAB PORTC,X      ;turn on Q2
              JSR  DELAY_25
              BRA  START
              END

DELAY_25      LDY  #$2CA        ;2.5 mS delay
AGN           DEY
              BNE  AGN
              RTS
```

Exercises

1. Redo the above program if the displays are common cathode.
2. Display a counter that counts from 00 to 99 on the above display. Use the seven-segment display format look-up table (Table 8.2).

Program 2:   ## Using a Decoder/Driver IC for Seven-Segment Display

In this experiment, we are going to use the 7447 common anode decoder/driver IC shown in Figure 8.7 to display counts 0 to 9. The chip converts the BCD inputs to seven-segment display code. For example, if we send 0001, it will display 1; if we send 0101, it will display 5.

We do not need a look-up table to convert the hex code to seven-segment display code if we use this chip. The inputs of this chip are A, B, C, and D and the outputs are a, b, c, d, e, f, g, and h. The A, B, C, and D are the BCD data needed to be displayed, and the a, b, c, . . . , h are the converted code of the seven-segment display and are connected to the analogous pins (also a, b, c, . . . , h) of the display chip itself. *Ripple blanking input* (RBI) is used to blank out the leading zeros. For example, if we would like to display the number 10 on a three-digit display as shown in Figure 8.8, the number should appear as 10 and not as 010. If the RBI is low and the BCD inputs ABCD to the driver are also low, the display is blanked.

If an input is not zero, the RBO *Ripple Blanking Output* is high regardless of the RBI status. The RBO of the first digit could be connected to the RBI of the second one and RBO of the second could be connected to the RBI of the third and so on.

Table 8.2

Display	Common Anode	Common Cathode
0	C0	3F
1	F9	06
2	A4	5B
3	B0	4F
4	99	66
5	92	6D
6	82	7D
7	F8	07
8	80	7F
9	98	67
A	88	77
C	C6	39
E	86	79
F	8E	71
G	82	70
H	89	76
⋮	7F	80
Blank	FF	00

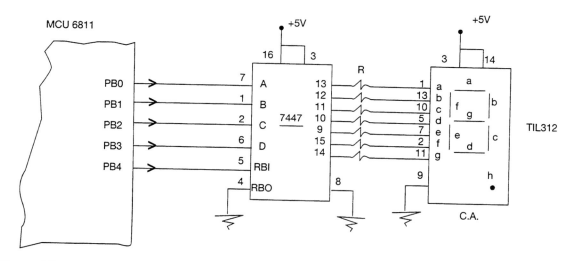

Figure 8.7

We would like to display, for example, a counter starting at 0 and ending at 9 and keep repeating this counting over and over again. Each count takes one second to be displayed. Here is the program.

```
;COUNTING FROM 0 TO 9 ON a 7-SEGMENT DISPLAY COMMON ANODE
;USING 7447 DECODER DRIVER AS SHOWN IN FIGURE 8.7.
;RBI PIN IS HIGH ALL THE TIME TO AVOID BLANKING AND CONNECTED TO PB4
;TO AVOID BLANKING

BASE  EQU $1000
PORTB EQU $4
```

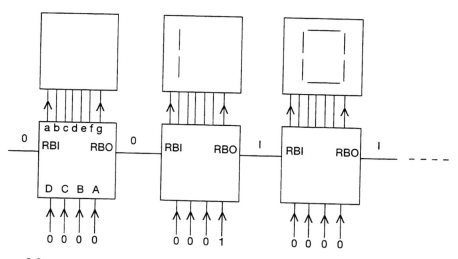

Figure 8.8

```
                        ORG  $C000

                LDX   #BASE
START           LDAA  #$10        ;"0" number
REPEAT          STAA  PORTB,X
                JSR   DELAY
                INCA
                CMPA  #$1A         ;is it "10" yet?
                BEQ   START        ;yes, start all over again
                BRA   REPEAT       ;no, keep incrementing
                END

DELAY           LDAB  #3           ;time delay for 1 sec
NEXT            LDY   #$FFFF
AGN             DEY
                BNE   AGN
AGN1            DECB
                BNE   NEXT
                RTS
```

Exercises

1. Using three digits (units, tens, and hundreds) with the same hardware as above, write a program to count from 000 to 999.

2. Display the number 7 in the above program in a flashing manner. Use the feature of RBI to blank the display and adjust the times for the following cases:
   a. On-time is the same as off-time.
   b. On-time is longer than off-time.
   c. On-time is shorter than off-time.

3. Rewrite Program 3 if the common cathode decoder/driver 7448 chip is used. This driver has the same pin-out as its sister 7447.

Program 3: **Counting from 0 to 9 on the Seven-Segment Display on a High-to-Low Pulse using Interrupt**

See the diagram in Figure 8.9.

```
;COUNTING FROM 0 TO 9 ON a 7-SEGMENT DISPLAY COMMON ANODE USING
;INTERRUPT
;USING 7447 DECODER DRIVER
;RBI PIN IS HIGH ALL THE TIME AND CONNECTED TO PB4

BASE  EQU $1000
PORTB EQU $4


                        ORG  $C100

                LDX   #BASE
START           LDAA  #$F0         ;"0" number
```

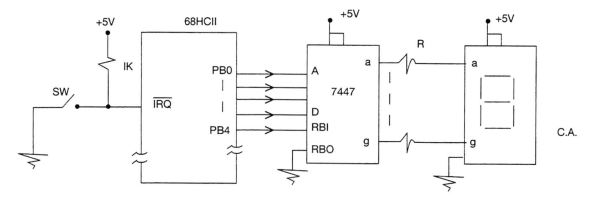

Figure 8.9

```
REPEAT          STAA  PORTB,X
                CLI
WAIT            BRA   WAIT         ;wait for the interrupt

ISR             INCA
                JSR   DLY
                CMPA  #$FA         ;is it "10" yet?
                BEQ   START        ;yes, start all over again
                BRA   REPEAT       ;no, keep incrementing
                RTI

DLY             LDAB  #2           ;time delay
DLY1            LDY   #$FFFF
AGN             DEY
                BNE   AGN
                DECB
                BNE   DLY1
                RTS

                ORG   $FFF2

                FDB   ISR
```

Exercise

A photocell circuit, depicted in Figure 8.10, is used to detect and count the number of bottles moving on a conveyor belt. The circuit is connected to three-digit LED displays. Show that there is more than one way to solve this problem. Write different routines to accomplish this.

Program 4:   Interfacing a 5×7 Dot Matrix Display

Figure 8.11 shows the circuit for multiplexing two 5×7 dot matrix displays. The columns are driven by MPS6531 Darlington transistors. The rows are driven by

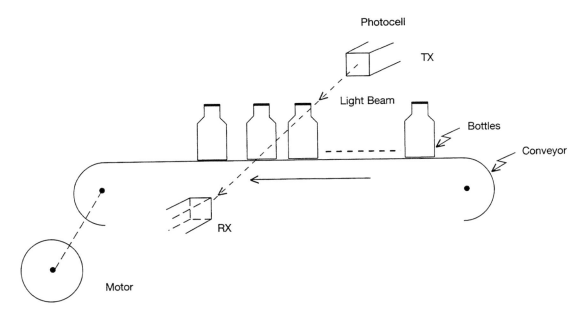

Figure 8.10

2N3904 general-purpose transistors. A digit is selected via PB7. If it is set, the unit's digit is selected while if it is cleared, the ten's digit is selected.

To display two characters, 10 bytes are stored in RAM—5 for each character, as explained earlier. Here is the program.

```
BASE   EQU $1000
PORTC  EQU $3
PORTB  EQU $4
DDRC   EQU $7


              ORG $C000


              LDX  #BASE
              LDAA #$0F
              STAA DDRC,X       ;pc0-pc3 output
          ;unit's digit
CHRCTR_1      LDY  #CHTR_1
START         LDAB #$10         ;turn on col1
              STAB PORTC,X
              LDAA 0,Y          ;get char 1

LOOP1         STAA PORTB,X      ;send character byte to rows
              JSR  DELAY
              INY
              LSRB              ;next column
              CMPB #0           ;5 columns yet?
              BNE  LOOP1        ;no, repeat
          ;ten's digit
```

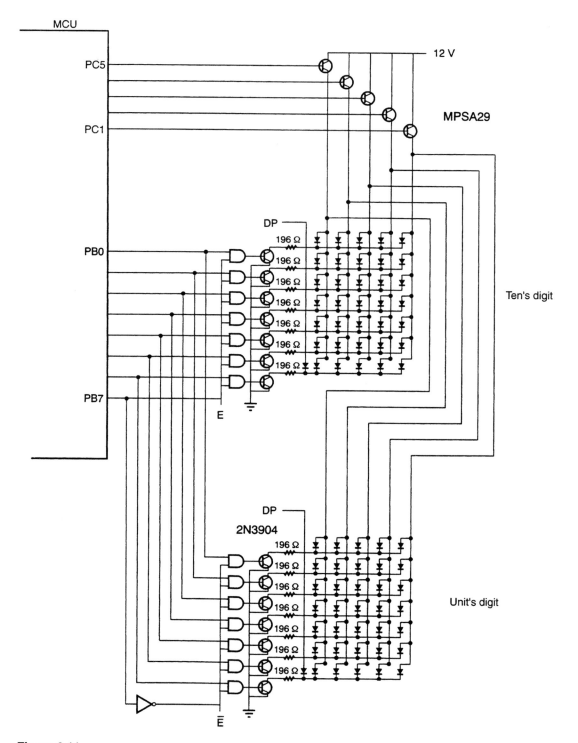

Figure 8.11

Adapted from A. K. Stiffler, *Design with Microprocessors for Mechanical Engineers,* McGraw-Hill, New York. Copyright 1992 by McGraw-Hill. Reprinted with permission.

```
                LDY   #CHTR_2
                LDAB  #$10        ;turn on col1
                STAB  PORTC,X
                LDAA  0,Y          ;get char 2,right hand
                ORAA  #$80         ;enable ten's digit

LOOP2           STAA  PORTB,X      ;send character byte to rows
                JSR   DELAY        ;2.5 mS delay
                INY
                LSRB               ;next column
                CMPB  #0           ;5 columns yet?
                BNE   LOOP2        ;no, repeat
                BRA   CHRCTR_1     ;start all over
                END

DELAY           PSHY
                LDY   #$2CA        ;2.5 mS delay
AGN             DEY
                BNE   AGN
                PULLY
                RTS

                ORG $C100

CHTR_1          FCB $00,$7F,$7F,$7F,$7F ;"I" character
CHTR_2          FCB $00,$77,$77,$77,$00 ;"H" character
```

Program 5: Interfacing an Alphanumeric LED 16-Segment Display

As mentioned earlier, the LED 16-segment display is one of many available fonts for alphanumeric display. It displays more characters than its counterpart, seven-segment display.

We choose the smart HPDL-1414 display made by Hewlett Packard. The display has four digits, built-in RAM, ASCII decoder, and LED drive circuitry. See Appendix B for more technical detail about this display. The circuit connections are shown in Figure 8.12.

The ASCII character is sent to pins D0 to D6 (7 bits). Address input A0, A1 are used to select the digit. See the truth table (Appendix B.)

The write ($\overline{\text{WR}}$) pin is connected to PC2 on the MCU. PC2 is used to strobe the corresponding digit low, then high. The chip contains a driver to drive each segment and convert the ASCII character to a 16-segment format. To blank any digit, D5 must equal D6, in the ASCII character set. Assume we would like to display the message "GOOD". Here is the program.

```
                ORG $C000

;Equates for Portb, PortC, DDRC, BASE

                LDX   #BASE
                LDAA  #$FF
                STAA  DDRC,X       ;Port C is output
```

```
START           LDAA    #'D'
                STAA    PORTB,X
                LDAB    #04
                STAB    PORTC,X       ;enable DIG0
                BCLR    PORTC,X,$4
                BSET    PORTC,X,$4    ;strobe the digit

                JSR     DELAY_25
                LDAA    #'O'
                STAA    PORTB,X
                LDAB    #05
                STAB    PORTC,X       ;enable DIG1
                BCLR    PORTC,X,$4
                BSET    PORTC,X,$4    ;strobe the digit

                JSR     DELAY_25
                LDAB    #06
                STAB    PORTC,X       ;enable DIG2
                BCLR    PORTC,X,$4
                BSET    PORTC,X,$4    ;strobe the digit

                JSR     DELAY_25
                LDAA    #'G'
                STAA    PORTB,X
                LDAB    #07
                STAB    PORTC,X       ;enable DIG3
                BCLR    PORTC,X,$4
                BSET    PORTC,X,$4    ;strobe the digit
                JSR     DELAY_25
                BRA     START
                END
DELAY_25         |
                RTS
```

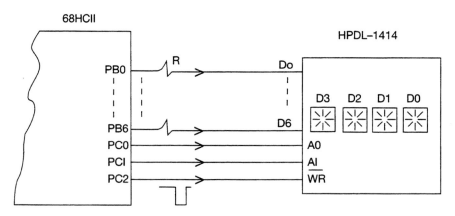

Figure 8.12

Exercise

Develop an algorithm to test an IC automatically. The hardware is shown in Figure 8.13. For simplicity, we chose the 7400 quad two input NAND gate as the test element. The circuit uses the HPDL-1414 as a result display. The operator places the test element and pushes the start button "SW".

If the element is good, then the display shows the message GOOD. If the element is bad, the display shows the message "BAD". When the system is ON but not actively testing, the display shows the message "OK".

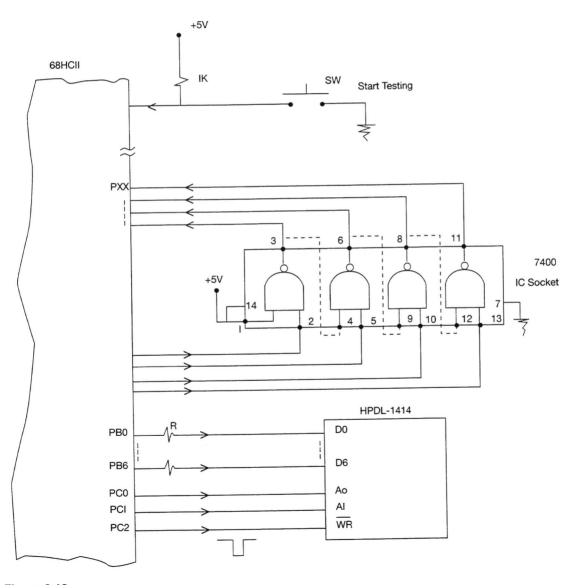

Figure 8.13

Liquid Crystal Displays

LCDs are becoming more popular these days. They are optically passive. This means that they form characters and symbols not by emitting light but by manipulating it. The heart of an LCD is a liquid crystal fluid sealed between two glass plates. Liquid crystal molecules are organized in distinct planes. The molecules in each plane align themselves so they all point in the same direction. This unique molecular structure lets the fluid manipulate light.

In the most common type of LCD, the glass plates have a coating that aligns the LC molecules in the two outermost planes, such that the molecules in one plane are perpendicular to the molecules in the other. Between these extremes, the molecular orientation of successive planes increase gradually from 0° to 90°, forming a spiral as shown in Figure 8.14a. The effect of this twist is to change the orientation of light as it travels through the display. To accomplish that twist, the display should have two polarizers: one vertical and the other horizontal.

There is another way to change the orientation of liquid crystal molecules by applying an electric field, as shown in Figure 8.14b. The field is generated across a pair of transparent electrodes, one deposited on each glass plate. When an electric field is applied to an LCD, the liquid crystal molecules realign themselves so that they are parallel to the field. As a result, polarized light does not get twisted as it passes through the liquid crystal material. If the two polarizers are perpendicular as before, light getting into the display will be blocked

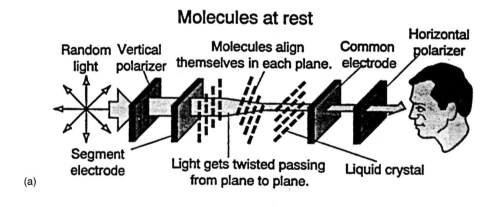

(a)

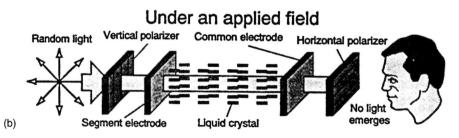

(b)

Figure 8.14

in the front polarizer. Thus the LCD produces dark characters on a light background. Characters are created where front and rear electrodes overlap, forming capacitors. Charging a particular capacitor activates the corresponding LCD segment.

Viewing Modes

Common viewing modes are reflective, transmissive, and transflective.

Reflective displays use reflection to illuminate the display. A reflector is either mounted behind the display or incorporated into the rear polarizer, reflecting light back toward the surface. They are the least expensive and consume little power.

Transmissive displays use backlighting for illumination. They are used in indoor instruments.

Transflective displays use reflection and backlighting. They are used in outdoor applications such as gasoline pumps.

LCDs are activated by applying voltage between the segment and the common electrodes. Typical driving voltage is 5 V and the allowable AC frequency range of the driving voltage is 30 to 100 Hz. The typical driving signal is a symmetrical square wave. Flicker may be seen by using a drive frequency below 30 Hz. If the frequency used is higher than 100 Hz, power consumption increases and may cause *ghosting*. Ghosting occurs when voltage from an energized element is partially on.

Usually the LCD driver circuit contains the oscillator and XOR circuit. Several multiplex formats have become popular. See the waveforms shown in Figure 8.15. Multiplexing

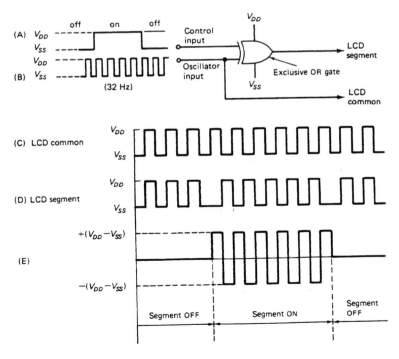

Figure 8.15
From J. Peatman, *Design with Microcontrollers*, McGraw-Hill, New York. Copyright 1988 by McGraw-Hill. Reprinted with permission.

in LCD circuits is very common. Some drivers employ 1/3 or 1/4 multiplexing. The term *1/3 multiplexing* means that the backplane (BP) of an LCD display is divided into three daisy-chained conductors. The term *1/4 multiplexing* means that the backplane is divided into four daisy-chained conductors.

As shown in Figure 8.16, there is a 2×4 multiplexed format for seven-segment display, showing the backplane connections. Since it is 1/4 multiplexing, it uses four daisy-chained backplanes: BP1, BP2, BP3 and BP4. The number of frontplanes used ordinarily equals $N/4$, where N is the total number of segments. In this example, the total number of segments is 8 per digit. The number of frontplanes = 8/4 = 2. The truth table for multiplexing is also shown.

As an example, if we wish to display segment a, the driver has to turn BP4 and FP1 on. In alphanumeric displays, we use the same multiplexing technique. This display has 16 segments per digit. For 1/4 multiplexing, the number of frontplanes = 16/4 = 4 per digit.

The Motorola MC14500 (master) LCD driver and MC145001 (slave) LCD driver are CMOS devices designed to drive LCD displays in a multiplexed-by-four configuration. The master unit generates both frontplane and backplane waveforms and is capable of independent operation. The slave unit generates only frontplane waveforms and is synchronized with the backplanes from the master unit. Several slave units may be cascaded from the master unit to increase the number of LCD segments driven in the system.

In the block diagram of the MC145000 (not shown), 48 bits of data are serially clocked into the shift register on the falling edges of the external data clock. The data are then latched into the 48-bit latch at the beginning of each frame period. The binary data present in the latch determines the appropriate waveform signal to be generated by the frontplane drive circuits, whereas the backplane waveforms are invariant.

Twelve frontplane and 4 backplane drivers are available from the master and 11 frontplanes from the slave.

When several such displays are used in a system, the 4 backplanes generated by the master are common to all segments. The 12 frontplanes are capable of controlling all 48 LCD segments (6 LCD digits) and the 11 frontplanes of each slave are capable of controlling 44 LCD segments (5 1/2 LCD digits).

As mentioned, the data are serially shifted first into the 48-bit locations of the shift register of the MC145000. After the 48 data bits have been shifted, the first bit to be entered has been shifted into bit location 1 in the latch, the second bit into bit location 2,

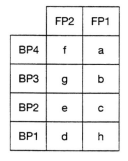

	FP2	FP1
BP4	f	a
BP3	g	b
BP2	e	c
BP1	d	h

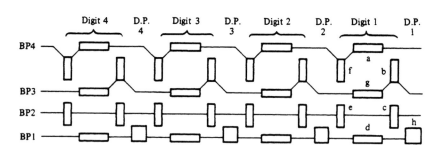

Figure 8.16
From J. Peatman, *Design with Microcontrollers,* McGraw-Hill, New York. Copyright 1988 by McGraw-Hill. Reprinted with permission.

Table 8.3

	Frontplanes											
Backplanes	FP1	FP2	FP3	FP4	FP5	FP6	FP7	FP8	FP9	FP10	FP11	FP12
BP1	4	8	12	16	20	24	28	32	36	40	44	48
BP2	3	7	11	15	19	23	27	31	35	39	43	47
BP3	2	6	10	14	18	22	26	30	34	38	42	46
BP4	1	5	9	13	17	21	25	29	33	37	41	45

and so on. Table 8.3 shows the bit location in the latch that controls the corresponding frontplane backplane intersection.

In addition to controlling seven-segment displays, the MC145000 and the MC145001 may be used to control displays using 5×7 dot-matrices.

Program 6: Interfacing an LCD Display

In this program, we wish to interface an LCD with five digits to the driver MC145000. The data from the MCU is clocked into the driver by the SCI.

The seven-segment table for the LCD is also provided and differs from the LED seven-segment display table shown earlier. Here's how the LCD table is generated.

To turn on any segment, a logic 1 is sent to this segment. For example, to display the numeral 0, segments a, b, c, d, e, and f are 1s and segments h and g are 0s, as shown in the truth table in Figure 8.17. To obtain the hex equivalent, you follow the arrow from a to f. Thus, the code for 0 is 11101101 = $EB, and so on for the other numbers.

The data to be displayed are stored in 5-byte RAM called BUFFER. These raw data must first be converted to seven-segment code before being sent to the driver. See the circuit in Figure 8.17.

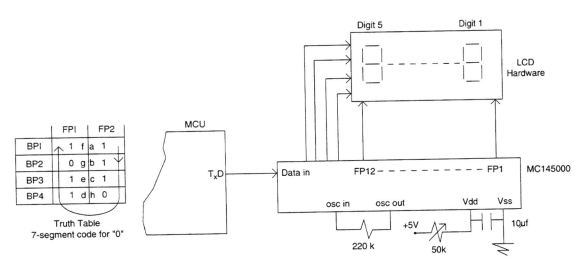

Figure 8.17

```
                         ORG $0000

BUFFER RMB 5                                ;Data to be displayed

                         ORG $C000

            ; Equates are responsibility of the user
              LDX   #BASE
              LDS   #$4A
              LDAA  #$30
              STAA  BAUD,X   ; 9600 baud
              LDAA  #00
              STAA  SCCR1,X  ;8 bits
              LDAA  #$08
              STAA  SCCR2,X  ;enable Tx,Rx

            ;Transmit routine

              LDY   #BUFFER
NEXT          LDAB  0,Y       ;get data from BUFFER
              PSHY            ;store BUFFER pointer
              LDY   #DTABLE
              ABY             ;get the right table address
              LDAA  0,Y       ;get the equivalent 7-seg
code
SEND          BRCLR SCSR,X,$80,SEND ;TD empty yet?
              STAA  SCDR,X    ;yes, transmit
              PULY            ;retrieve BUFFER pointer
              INY
              CPY   #BUFFER+5 ;end of BUFFER yet?
              BNE   NEXT      ;no, branch
HERE          BRA   HERE      ;yes, wait here

                         ORG $C100

DTABLE    FCB  $EB      ;0
          FCB  $60      ;1
          FCB  $C7      ;2
          FCB  $E5      ;3
          FCB  $6C      ;4
          FCB  $AD      ;5
          FCB  $AF      ;6
          FCB  $E0      ;7
          FCB  $EF      ;8
          FCB  $ED      ;9
```

Character-Type LCD Module

The LCD modules have all the controls and display built into one IC. They can operate under the control of either a 4-bit or 8-bit MCU or MPU to display alphanumeric characters, symbols, and other signs. The module provides the user with a character-type dot matrix display panel featuring a simple interfacing circuit.

The LCD module receives character codes (8-bits per character) from an MPU, latches the codes to its Display Data RAM (DD RAM), transforms each character code into a 5×7 dot-matrix character pattern, and displays the characters on its LCD screen.

The module incorporates a character-generating ROM, which produces more than 100 different 5×7 dot-matrix character patterns.

To display a character, a positional data is sent via the database from the MPU to the LCD module, where it is written into the data register. The module displays the corresponding character pattern in the specified position.

The display/cursor shift instruction allows the entry of characters in either the left-to-right or right-to-left direction.

Vacuum Fluorescent Displays

Basic Structure

A VFD is a troide consisting of the directly heated oxide cathode (called a *filament*), grid (mesh), and anode, which is a uphosphor coated on a conductive layer in the form of the desired display within a vacuum envelope. The electrons emitted from the filament strike the anode, which is coated with phosphor, and emit light. Reaching the anode, electrons are controlled by the voltage of the grid. If this voltage is positive relative to the filament voltage, electrons (which have negative charges) are attracted by the grid, accelerated, and diffused to the anode (emitting light). If voltage on the grid is negative relative to the voltage on the filament, electrons are repelled and never make it to the anode (no light). The higher the positive voltage on the grid, the faster electrons diffuse to the anode and the stronger the emitted light.

DC Plasma Displays

DC plasma display consists of two "electrodes" in the form of conductive traces on two sheets of glass. The cathode sheet is the bottom one, while the anode sheet is the top one. These two sheets of glass are fused together with a small gap filled with a "neon-argon" gas.

The DC plasma display operates by applying high voltage between the two electrodes. This high voltage will ionize the neon-argon gas and emit the "neon orange" color. When firing, a DC plasma display will maintain about 150 V across the electrodes as shown in Figure 8.18.

The dot inside the display symbol means that the filled gas is of the ionization type. The brightness of the display depends on the amount of current flowing in this tube. The amount of current also depends on the current-limiting resistor (R).

The total voltage between VS_p and VS_n is higher than 150 V, about 180 V. The difference between the two voltages is the drop voltage across R and the solid-state switches S1 and S2.

$$VS_p = 60 \text{ to } 70 \text{ VDC}$$
$$VS_n = -110 \text{ to } -120 \text{ VDC}$$

Figure 8.18

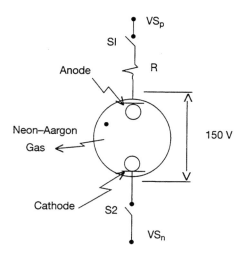

Advantages of DC Plasma Displays

The advantages are as follows:

1. Offer high-contrast viewing angle greater than $130°$
2. Long-distance readability
3. Low power consumption
4. The orange-red color of these displays is more readable and eye-pleasing than other display colors.

Applications

Because of these advantages, DC plasma displays are used on gas pumps, pinball games, and some machine tools and robots' control display panels.

Exercise

Develop a routine to identify a pressed key number in a 4×4 keypad. Display the key number on the previous LCD display.

8.2   Transducers

A transducer converts a certain parameter such as heat, force, stress, and so on into an electrical signal that can be measured. The transducer's output could be voltage, current, resistance, capacitance, or frequency.

    Terms relevant to the transducer output include the following:

1. *Accuracy* is specified in terms of a percent error. Error is the difference between the correct output and the actual output of the transducer.
2. *Offset* is the minimum analog value.
3. *Span* is the maximum analog value minus the minimum analog value.
4. *Step size* is the smallest change at the input of the transducer that will result in change in the output and is equal to:

$$Span/2^n$$

or

$$\frac{Rated\ maximum\ output}{2^n}$$

5. *Resolution* has the same definition as *step size* and can be expressed in units or percentage.

6. *Repeatability* is a measure of how the output will return to a given value when the same precise input is applied several times:

$$Repeatability = \frac{(max - min) \times 100}{full\ scale}$$

7. *Dead time* is the length of time from the application of a step change at the input of transducer until the output begins to change.

EXAMPLE 1

Consider an analog range of -2 V to +10 V. The output digital range is 8 bits. Calculate:

a. Offset

b. Span

c. Step size

d. Percent resolution

Solution

$$Offset = minimum\ value = -2\ V$$
$$Span = 10 - (-2) = 12\ V$$
$$Step\ size = 12\ V/2^8 = 46\ mV$$
$$Percent\ resolution = \frac{1}{2^8} \times 100 = 0.391\%$$

EXAMPLE 2

A temperature between 0°C and 100°C is converted into a 0 to 5V signal. This signal is fed to an 8-bit ADC with a 5.0 V reference. Calculate:

a. Resolution

b. Actual measurement range of the system

c. Hex output of 70°C

Solution

$$Step\ size = \frac{5V - 0V}{2^8} = 0.019\ V$$

$$Full\text{-}scale\ output\ (FSO) = full\text{-}scale\ input\ (FSI) - step\ size$$
$$= 5.0\ V - 0.019\ V = 4.98\ V$$

a. $Resolution = \dfrac{100°C}{2^8} = 0.39°C/bit$

That means if the temperature is $0°C$, the output is 00H. It will stay at this value until the temperature reaches $0.39°C$, then the output will change to 01H. By the same token, the output reaches FFH at $99.61°C$ and stays at this value until the temperature reaches $100°C$. The actual measurement range is:

b. $0.39°C$ to $99.61°C$

c. For $100°C$, the output is 100H (256D); the hex value of $70°C$ is

$$\frac{70°C \times 256}{100°C} = 179D = B3H$$

Selected Transducers

Speed Transducers

Speed sensing can be divided into rotational and linear applications. Rotational speed sensing has application areas such as engine speed control and antilock braking and traction control systems. Linear sensing is used in ground-speed monitoring and crash avoidance.

Acceleration sensors are used in air bag deployment, ride control, antilock braking, and traction systems. In automotive applications, the sensing must be accurate, the devices must be rugged and reliable and must function in the presence of oil, grease, dirt, and so on. These requirements have limited the use of other sensors such as optical sensors.

In the area of rotational speed applications, the most popular sensors are variable reluctance sensors (VRS), Hall sensors, and magnetic resistance element (MRE) devices.

The VRS and Hall sensors have been used widely in automotive applications, while the MRE devices have recently come out with a new technology.

In the area of linear speed applications, optical, infrared, laser, ultrasonic, and radar devices are used.

Variable Reluctance Sensors (VRS). By monitoring the crankshaft-mounted trigger wheel, the VRS is the primary sensor to provide the necessary ignition information to the control module. The trigger wheel is a ferromagnetic toothed timing wheel mounted on the crankshaft. The wheel is designed to have 36 teeth spaced at $10°$ increments for the $360°$, with an empty space for one "missing tooth."

This means that the trigger wheel has a total of 35 teeth. The empty space provides a fixed point of reference for piston travel identification.

The VRS is a magnetic transducer with a pole piece wrapped with fine wire. The fine wire, when exposed to a change in flux lines, will induce a differential voltage (Figure 8.19). This output voltage is a function of sensor-to-tooth air gap, tooth and sensor pole piece width, and the engine RPM.

When the trigger wheel is rotating, the passing ferromagnetic teeth cause a change in VRS reluctance. The varying reluctance alters the number of magnetic flux lines passing through the wire windings and induces Vac signal proportional to both the rate of change in magnetic flux and the number of coil windings.

As a tooth approaches the pole piece of the sensor, a positive differential voltage is induced and the sensor has low reluctance. This voltage is positive as long as the flux charge is increasing. When the pole piece is in the center of the tooth, there is no change in

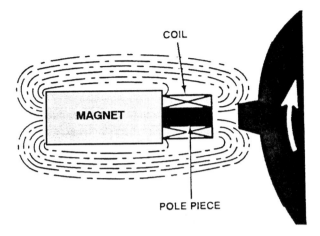

COIL

MAGNET

POLE PIECE

Figure 8.19

flux and the output voltage is 0 V. As the tooth moves away from the pole piece, the output voltage becomes negative.

At normal engine running speeds, the VRS signal is a sine waveform with an increased amplitude adjacent to the missing tooth region.

Hall-Effect Sensors. The basic Hall sensor is simply a small sheet of semiconductor material. A constant voltage source forces a constant bias current to flow in the semiconductor sheet. If the biased Hall sensor is placed in a magnetic field oriented at right angles to the Hall current, an output voltage is generated and is directly proportional to the strength of the magnetic field as shown in Figure 8.20a. This is the Hall effect, discovered by E. H. Hall in 1879.

The basic Hall sensor is a transducer that will generate an output voltage if the applied magnetic field changes in any manner. This magnetic field is created by either electromagnets or permanent magnets. The schematic symbol of a Hall sensor is shown in Figure 8.20b.

The output voltage of the basic Hall sensor is very small and must be conditioned (amplified) using transistors or op-amps. This output is linear and proportional to the applied magnetic flux density.

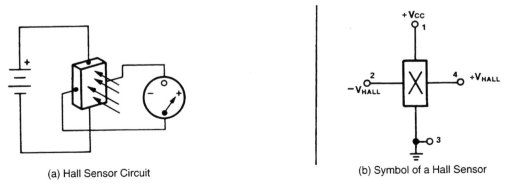

(a) Hall Sensor Circuit

(b) Symbol of a Hall Sensor

Figure 8.20
Adapted from *Integrated and Discrete Semiconductors Manual.* Courtesy of Allegro MicroSystems, Inc., Worcester, MA.

Some of these Hall devices have integrated drivers and temperature compensation elements. Examples of linear devices are UGN3501 and UGN3503 made by Allegro company.

The addition of a Schmitt trigger gives the Hall-effect circuit digital output capabilities. When the applied magnetic flux density exceeds a certain limit, the trigger provides a clean transition from off to on, without contact bounces. Built-in hysteresis eliminates oscillation.

An open-collector NPN output transistor added to the above circuit gives the switch digital logic compatibility. The transistor can sink enough current to drive many loads such as relays, triacs, SCRs and lamps.

The advantages of Hall sensors include the following:

1. The linear Hall-effect sensors detect motion, position, or changes in field strength. Energy consumption is low. The output is linear and the temperature stable. The sensor's frequency response is flat up to 25 KHz.

2. Hall-effect switches combine Hall voltage generators, signal amplifiers, Schmitt trigger circuits, and transistor output circuits on a single integrated circuit chip. The output is clean, fast, and without bounce. A Hall-effect switch typically operates up to a 100-KHz repetition rate and costs less than many electromechanical switches.

3. The Hall-effect sensor is immune to environmental contaminants and is suitable for use under severe service conditions.

EXAMPLE 3

Loads requiring sinking current up to 20 mA can be driven directly by the Hall switch. Assume a certain load requires 4 A to be switched when the Hall sensor is on. In this case, we must add another amplifier to the circuit as shown in Figure 8.21.

Explain how the circuit works and find out the values of $R1$ and $R2$.

Solution

We select $R1 = 1$ K. When the Hall switch is off (insufficient magnetic flux), I1 flows through $R1$ to the base of the transistor 2N5812.

$$I1 = 12 \text{ V}/1 \text{ K} = 12 \text{ mA} \quad (\text{ignore } V_{be})$$

This current is sufficient to turn on $Q1$ and saturate it. When $Q1$ is saturated, it shorts the base of $Q2$ to ground and turns it off. No current can flow through the load.

Figure 8.21

Adapted from *Integrated and Discrete Semiconductors Manual.* Courtesy of Allegro MicroSystems, Inc., Worcester, MA.

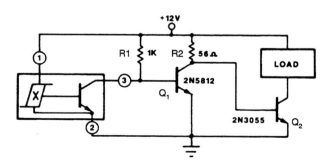

When the Hall switch turns on (sufficient magnetic flux), it shorts the base of $Q1$ to ground allowing $I2$ to flow in the base of $Q2$.

$$Ib2 = 200 \text{ mA for transistor 2N3055 to be saturated.}$$
$$R2 = 12 \text{ V}/200 \text{ mA} = 60 \text{ ohms}$$
$$Ic2 = \text{hfe} \times Ib2 = 20 \times 200 \text{ mA} = 4 \text{ A}$$

Stress and Strain

Strain is the deformation of an object per unit length when a force is applied on the object as shown in Figure 8.22.

The relation between stress and strain is as follows:

$$E = \frac{\sigma}{\epsilon} = \frac{\text{stress}}{\text{strain}}$$

where

E = Modulus of elasticity

σ = stress

ϵ = strain

Strain Gauges. The bonded resistance strain gauge is by far the most popular strain measured tool. It consists of a grid of a very fine wire or a metal foil bonded to this insulating backing. The resistance of the wire or foil varies linearly with strain.

The force will act on the sensitive areas, as shown Figure 8.23a. Some factors must be taken into consideration when designing the strain gauges: the high resolution and the effect of the temperature.

The wheatstone bridge technique (explained in Chapter 6) is the answer to both requirements. By using a dummy gauge as shown in Figure 8.23b (R4) and 8.23c, we can provide temperature compensation. Both gauges (active and dummy) change in resistance from temperature effect, but these effects will cancel each other. When a force *(F)* acts on strain gauge R3, its resistance changes that unbalance the prebalanced circuit. An output V is detected that is amplified through an op-amp and V_o is measured or calculated. Now V_o is proportional to the change in resistance, thus proportional with the acting force. There is another factor called the *gauge factor,* which is defined as:

$$G = \text{gauge factor} = \frac{\text{change in resistance}}{\text{strain}} = \frac{\Delta R/R}{\epsilon}$$

and it is close to 2.

Figure 8.22
From M. Jacob, *Industrial Control Electronics: Applications and Design.* Copyright 1988 by Prentice Hall. Reprinted by permission of Prentice Hall, Upper Saddle River, New Jersey.

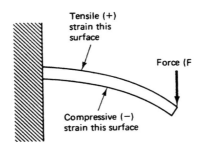

EXAMPLE 4

In Figure 8.23b, if a strain gauge with a gauge factor of 2 and $R_1 = R_2 = 2000$, both active and dummy resistance = 1000 ohms, Vs = 10 V. When acted on by R_3, its resistance changes to 1002 ohms. If the bridge output is connected to an op-amp with a gain of 1000, find V_0.

Solution

$$V = V_2 - V_1$$

$$= \frac{10 \times 1002}{3002} - 10 \times \frac{1000}{3000}$$

$$= 10(\frac{1002}{3002} - \frac{1000}{3000})$$

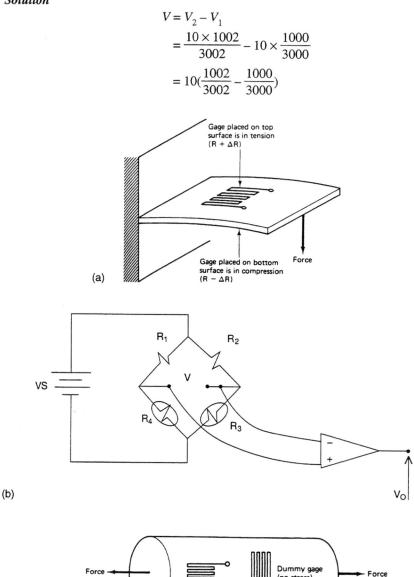

Figure 8.23

Parts (a) and (c) from M. Jacob, *Industrial Control Electronics: Applications and Design.* Copyright 1988 by Prentice Hall. Reprinted by permission of Prentice Hall, Upper Saddle River, New Jersey.

$$= 4.44 \text{ mV}$$
$$V_o = 1000 \times 4.44 \text{ mV} = 4.44 \text{ V}$$

Temperature Transducers

The temperature is one of the most measured dynamic variables in industry today. Temperature transducers can be classified into three major types: thermoresistive, thermocouple, and semiconductor.

Thermoresistive Transducers. Examples of this kind are resistive temperature detectors (RTDs) and thermoresistors.

RTDs have a positive temperature coefficient. A temperature *coefficient* of a resistance is defined as the percent change in its resistance with the change in temperature. The term *positive* means that the resistance increases proportionally with temperature. If the resistance of an RTD is known at a certain temperature, it can be calculated at any another temperature as follows:

where

$$R_2 = R_1 \{1 + \alpha (T_2 - T_1)\}$$
$R_1 =$ resistance at T_1
$R_2 =$ resistance at T_2
$\alpha =$ temperature coefficient

RTD contains a wire made of nickel, copper, or platinum.

A thermistor is another type of thermoresistor and is made of a semiconductor material. Thermoresistors have a *negative* temperature coefficient—that is, if the temperature increases, their resistance decreases.

In any application using a thermoresistor, only the linear portion should be used.

One of the thermoresistor's advantages is its sensitivity to temperature.

Thermocouple. A thermocouple is made by joining two different metals at one end, as shown in Figure 8.24. The fixed end is called the *hot end*. The free end is called the *cold end*. When the joined end is heated up, a voltage difference at the cold end can be measured. This voltage difference could be in μV; therefore it must be amplified before being used by means of an op-amp.

Semiconductor Temperature Transducers. Temperature transducers are available in many forms of ICs. These devices are accurate and inexpensive; they include LM34, LM35, and LM399 made by National Semi-Conductor, AD590 made by Analog Devices,

Figure 8.24

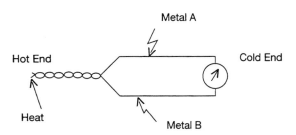

or the MTS10X series made by Motorola. To interface these devices to an MPU or MCU, they are usually connected to the ADC first.

The LM34 does not require any external calibration or trimming to provide typical accuracies of ±1/2° F at room temperature. The LM34's low output impedance, linear output, and precise inherent calibration make interfacing to readout or control circuitry especially easy.

The MTS10X series are designed for use in temperature-sensing applications in auto-motive, consumer and industrial products requiring low cost and high accuracy. The pre-cise temperature accuracy is ±2°C in the range -40°C to +150°C. They have a linear Vbe versus temperature. The higher the temperature, the lower the Vbe. (See Appendix B.)

Program 7: Data Acquisition—Interfacing a Thermometer to MCU 6811

In many applications, analog data have to be digitized and transferred into a computer's memory. The process by which the computer acquires these digitized analog data is referred to as *data acquisition*.

In this exciting experiment, we are going to interface a "thermometer" to our MCU. To interface any analog device to MCU, we must take advantage of the on-board ADC port, which you are familiar with from Chapter 5, to convert these analog values to digital values.

In the following program, we will use an IC temperature transducer "LM34" made by National Semi-Conductor, which has three leads, to measure the temperature. See appendix B for more detail. The output voltage of this chip is linearly proportional to the input temperature. Its output voltage sensitivity is +10 mV/1°F. No trimming or calibration circuits are required. The supply voltage is between +5 and +20 V. The output voltage is calculated as follows:

$$V_o = (10 \text{ mV}/1°F) \times (\text{temperature in Fahrenheit})$$

Assume that we are going to measure a temperature in the range of 0°F to 100°F. The output voltage range is 0 V to 5 V. Since the 100°F temperature will produce 1 V only out of the IC LM34, we need a signal conditioning circuit (op-amp) with a gain of 5.

The op-amp is a noninverting amplifier with $R_1 = 4K$ and $R_2 = 1K$ to produce a gain of 5 as shown in Figure 8.25. For simplicity, we used a look-up table with only 16 readings.

Let's see how we should design a look-up table such as Table 8.4, which contains all the information we need for the program.

$$\text{Resolution} = 100°F/16 = 6.25°F$$

which is very low and generates inaccurate readings. The highest resolution is obtained if we use all 256 readings. The resolution in this case is 0.39°F.

$$\text{Step size} = \frac{5V}{2^4} = 0.312 \text{ V}$$

$$\text{The full-scale output (FSO)} = \text{full-scale input (FSI)} - \text{step size}$$
$$= 5.0 \text{ V} - 0.312 \text{ V} = 4.7 \text{ V}$$

The hex value of FSO = 0FH.

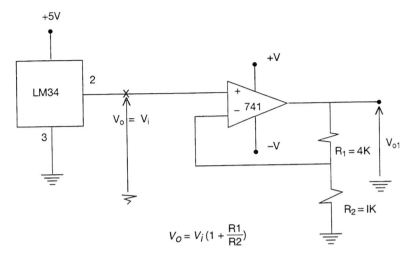

Figure 8.25

```
Thermo.asm

BASE    EQU  $1000
ADCTL   EQU  $30
OPTION  EQU  $39
ADR4    EQU  $31
TEMP0   EQU  $10


              ORG $C100

   ; This is the look-up table for the thermometer

TABL    FCB $00,$06,$0D,$13,$19,$1F,$26,$2C
        FCB $32,$38,$3F,$45,$4B,$51,$58,$5E


              ORG $C000

             LDX  #BASE
             LDAA #$20
             STAA ADCTL,X      ;continuous scanning
             LDAA #$80
             STAA OPTION,X     ;start conversion

STRT_CNV     LDY  #TABL
             LDAA #26
DELAY        DECA              ;time delay for
             BNE  DELAY        ;conversion

             LDAB ADR4,X
             LSRB
```

```
LSRB
LSRB
LSRB                  ;div reading by 16
ABY                   ;form address of temperature
LDAA 0,Y
STAA TEMP0            ;store temperature here
BRA  STRT_CNV         ;get another reading
END
```

The LM35 is similar to the LM34, except it has a sensitivity of 10 mV/1°C.

Exercises

1. Modify the above routine so that you can display the temperature on two 7-segment display hardware digits.

2. If we intend to use the LM35 to display temperature in Centigrade, what changes to the program must be made?

3. You can use the same technique to interface "a photometer" using a different look-up table. A phototransistor (shown in Figure 8.26) is used as a light transducer to measure the light intensity. Use your own look-up table to develop an algorithm to accomplish that.

4. The look-up table method is used in special cases as illustrated in the above program. Another method is to convert the hex value stored in ADC registers directly to Fahrenheit. The Motorola temperature sensor MTS105 (shown in Appendix B) can be used to accomplish that.

$$\text{The ADC resolution} = \frac{5\,\text{V}}{256} = 0.0195\,\text{V}$$
$$\text{The sensor sensitivity} = 12.6\,\text{mV/1°F}$$

Table 8.4

ACCB	ACCA	°F	V_o volt	V_1 volt
00	0	0	0	0
01	6	6	0.06	0.3
02	0D	13	0.13	0.65
03	13	19	0.19	0.95
04	19	25	0.25	1.25
05	1F	31	0.31	1.55
06	26	38	0.38	1.90
07	2C	44	0.44	2.20
08	32	50	0.50	2.50
09	38	56	0.56	2.80
0A	3F	63	0.63	3.15
0B	45	69	0.69	3.45
0C	4B	75	0.75	3.75
0D	51	81	0.81	4.05
0E	58	88	0.88	4.40
0F	5E	94	0.94	4.70

Figure 8.26

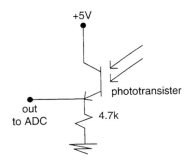

The temperature reading is given as follows

$$T = \text{AD\_Reading} \times \frac{19.5\,\text{mV}}{12.6\,\text{mV}}\,{}^{\circ}\text{F}$$

Develop a routine to read the temperature directly.

Solid-State Pressure Transducers

The solid-state pressure sensor is designed utilizing a monolithic silicon piezoresistor, which generates a change in output voltage depending on variations in applied pressure. The resistive element (the strain gauge) is ion-implanted on a thin silicon diaphragm. Applying pressure to the diaphragm results in a resistance change in the strain gauge, which in turn causes a change in the output voltage.

Motorola has introduced the MPX series of pressure sensors. Besides the on-chip temperature compensation and calibration offered on some series, amplifier signal conditioning has been integrated in other series such as MPX5000 to allow interface directly to any MCU. These pressure sensors also offer extremely linear output proportional to the applied pressure. High sensitivity and excellent long-term repeatability make these sensors suitable for almost every application.

The schematic of a pressure sensor is shown in Figure 8.27. The typical output voltage is in millivolts.

Types of Pressure Measurements. Motorola pressure sensors support three types of pressure measurements: absolute pressure, differential pressure, and gauge pressure.

1. **Absolute Pressure Sensors.** Measure an external pressure relative to a zero-pressure reference (vacuum) sealed in the cavity under the diaphragm. The numbering system of these sensors ends with A, AP, AS, or ASX—for example, MPX100A, MPX2100AP, MPX4100AS, or MPX2200ASX. The more common absolute applications include altimeters, barometers, and vacuum measurements.

2. **Differential Pressure Sensors.** Measure the difference between pressures applied simultaneously to the opposite sides of the diaphragm. The numbering system ends with D or DP. Typical applications are air-flow and filter status.

3. **Gauge Pressure Sensors.** Any measurement referenced to barometric pressure is called a *gauge measurement*. The sensor will measure positive and negative pressures. The gauge sensor refers to a chip that has the reference cavity vented to the atmosphere. The numbering system ends with GP, GVP, GS, GVS, GSX, or GVSX. Typical applications are blood pressure, engine vacuum, and tire pressure.

Figure 8.27
Copyright of Motorola. Used by permission.

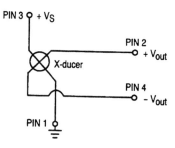

An example of one of Motorola's pressure sensors is the MPX4115. It is an altimeter/barometer/absolute (BAP) sensor, designed to sense absolute air pressure. This sensor integrates on-chip bipolar op-amp circuitry and thin film resistor networks to provide a high output signal and temperature compensation.

8.3   Case Study 3

Figure 8.28 shows the overall system architecture chosen for an application involving a barometric pressure gauge using pressure sensor MPX2100AP. This system serves as a building block, from which more advanced systems can be developed. Enhanced accuracy, resolution, and additional features can be integrated in a more complex design.

There are some preliminary concerns regarding the measurement of barometric pressure that directly affect the design considerations for this system. Barometric pressure is the air pressure existing at any point within the earth's atmosphere. This pressure can be measured as an absolute pressure (with reference to absolute vacuum) or can be referenced to some other value or scale. The meteorology and avionics industries traditionally measure the absolute pressure, and then reference it to a sealevel pressure value. This complicated process is used in generating maps of weather systems. The atmospheric pressure at any altitude varies due to changing weather conditions over time. Therefore, it can be difficult to determine the significance of a particular pressure measurement without additional information. However, once the pressure at a particular location and elevation is determined, the pressure can be calculated at any other altitude. Mathematically, atmospheric pressure is

Figure 8.28
Copyright of Motorola. Used by permission.

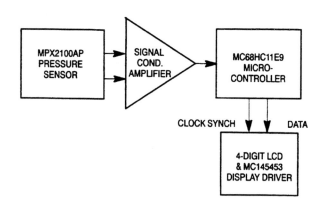

Table 8.5

Altitude Versus Pressure Data	
Altitude (ft)	Pressure (in-Hg)
0	29.92
500	29.38
1000	28.85
6000	23.97
10,000	20.57
15,000	16.86

exponentially related to altitude. This particular system is designed to track variations in barometric pressure once it is calibrated to a known pressure reference at a given altitude.

For simplification, the standard atmospheric pressure at sealevel is assumed to be 29.9 in-Hg. "Standard" barometric pressure is measured at a particular altitude at the average weather conditions for that altitude over time. The system described in this text is specified to accurately measure barometric pressure variations up to altitudes of 15,000 ft. This altitude corresponds to a standard pressure of approximately 15.0 in-Hg. As a result of changing weather conditions, the standard pressure at a given altitude can fluctuate approximately ± 1 in-Hg in either direction. Table 8.5 indicates standard barometric pressures at several altitudes of interest.

System Overview

To measure and display the correct barometric pressure, this system must perform several tasks. The measurement strategy is outlined in Figure 8.28. First, pressure is applied to the sensor. This produces a proportional differential output voltage in the millivolt range. This signal must then be amplified and level-shifted to a single-ended, MCU compatible level (0.5–4.5 V) by a signal conditioning circuit. The MCU will then sample the voltage at the analog-to-digital converter (ADC) channel input, convert the digital measurement value to inches of mercury, and then display the correct pressure via the LCD interface. This process is repeated continuously.

There are several significant performance features implemented into this system design. First, the system will digitally display barometric pressure in inches of mercury, with a resolution of approximately one-tenth of an inch of mercury. To allow for operation over a wide altitude range (0–15,000 ft), the system is designed to display barometric pressures ranging from 30.5 in-Hg to a minimum of 15.0 in-Hg. The display will read "lo" if the pressure measured is below 30.5 in-Hg. These pressures allow for the system to operate with the desired resolution in the range from sealevel to approximately 15,000 ft. An overview of these features is shown in Table 8.6.

Design Overview

The following sections are included to detail the system design. The overall system will be described by considering the subsystems depicted in the system block diagram, Figure 8.28. The design of each subsystem and its function in the overall system will be presented.

Table 8.6

System Features Overview	
Display Units	in-Hg
Resolution	0.1 in-Hg
System Range	15.0–30.5 in-Hg
Altitude Range	0–15,000 ft

Pressure Sensor

The first and most important subsystem is the pressure transducer. This device converts the applied pressure into a proportional, differential voltage signal. This output signal will vary linearly with pressure. Since the applied pressure in this application will approach a maximum level of 30.5 in-Hg (100 kPa) at sealevel, the sensor output must have a linear output response over this pressure range. Also, the applied pressure must be measured with respect to a known reference pressure, preferably absolute zero pressure (vacuum). The device should also produce a stable output over the entire operating temperature range.

The desired sensor for this application is a temperature compensated and calibrated, semiconductor pressure transducer, such as the Motorola MPX2100A series sensor family. The MPX2000 series sensors are available in full-scale pressure ranges from 10 kPa (1.5 psi) to 200 kPa (20 psi). Furthermore, they are available in a variety of pressure configurations (gauge, differential, and absolute) and porting options. Because of the pressure ranges involved with barometric pressure measurement, this system will employ an MPX2100AP (absolute with single port). This device will produce a linear voltage output in the pressure rate of 0 to 100 kPa. The ambient pressure applied to the single port will be measured with respect to an evacuated cavity (vacuum reference). The electrical characteristics for this device are summarized in Table 8.7.

As indicated in Table 8.7, the sensor can be operated at different supply voltages. The full-scale output of the sensor, which is specified at 40 mV nominally for a supply voltage of 10 Vdc, changes linearly with supply voltage. All nondigital circuitry is operated at a

Table 8.7

MPX210000AP Electrical Characteristics					
Characteristic	Symbol	Minimum	Typical	Max	Unit
Pressure Range	Pop	0		100	kPa
Supply Voltage	VS		10	16	Vdc
Full-Scale Span	V_{fss}	38.5	40	41.5	mV
Zero Pressure Offset	V_{off}				mV
Sensitivity	S		0.4		mv/kPa
Linearity			0.05		%FSS
Temperature Effect on Span			0.5		%FSS
Temperature Effect on Offset			0.2		%FSS

regulated supply voltage of 8 Vdc. Therefore, the full-scale sensor output (also the output of the sensor at sealevel) will be approximately 32 mV.

$$(\frac{8}{10} \times 40 \text{ mV})$$

The sensor output voltage at the system's minimum range (15 in-Hg) is approximately 16.2 mV. Thus, the sensor output over the intended range of operations is expected to vary from 32 to 16.2 mV. These values can vary slightly for each sensor, as the offset voltage and full-scale span tolerances indicate.

Signal Conditioning Circuitry

To convert the small-signal differential output signal of the sensor to MCU-compatible levels, the next subsystem includes signal conditioning circuitry. The operational amplifier circuit is designed to amplify, level-shift, and ground reference the output signal. The signal is converted to a single-ended, 0.5–4.5 Vdc range. The schematic for this amplifier is shown in Figure 8.29.

This particular circuit is based on classic instrumentation amplifier design criteria. The differential output signal of the sensor is inverted, amplified, and then level-shifted by an adjustable offset voltage (through Roffset1). The offset voltage is adjusted to produce 0.5 V at the maximum barometric pressure (30.5 in-Hg). The output voltage will increase for decreasing pressure. If the output exceeds 5.1 V, a zener protection diode will clamp the output. This feature is included to protect the A/D channel input of the MCU. Using the transfer function for this circuit, the offset voltage and gain can be determined to provide 0.1 in-Hg of system resolution and the desired output voltage level. The calculation of these parameters is illustrated below.

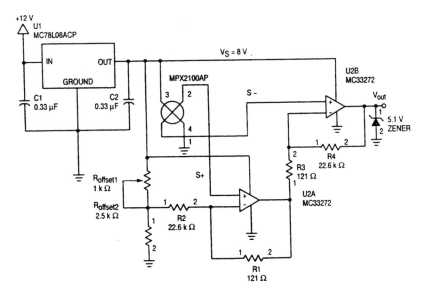

Figure 8.29
Copyright of Motorola. Used by permission.

In determining the amplifier gain and range of the trimmable offset voltage, it is necessary to calculate the number of steps used in the analog-to-digital conversion process to resolve 0.1 in-Hg.

$$(30.5 - 15.0)\text{in-Hg} \times \frac{10 \text{ steps}}{\text{Hg}} = 155 \text{ steps}$$

The span voltage can now be determined. The resolution provided by an 8-bit A/D converter with low- and high-voltage references of 0 and 5 V, respectively, will detect 19.5 mV of change per step.

$$\text{VRH} = 5 \text{ V}, \text{VRL} = 0\text{V}$$
$$\text{Sensor output at 30.5 in-Hg} = 32.55 \text{ mV}$$
$$\text{Sensor output at 15.0 in-Hg} = 16.26 \text{ mV}$$
$$\text{Sensor output} = \text{SO} = 16.18 \text{ mV}$$
$$\text{Gain} = \frac{3.04 \text{ V}}{\text{SO}} = 187$$

Note: 30.5 in-Hg and 15.0 in-Hg are the assumed maximum and minimum absolute pressures, respectively.

This gain is then used to determine the appropriate resistor values and offset voltage for the amplifier circuit defined by the transfer function shown below.

$$V_{\text{out}} = \frac{R_2 + 1}{R_1} \times V + V_{\text{off}}$$

V is the differential output of the sensor.

The gain of 187 can be implemented with:

$$R_1 = R_3 = 121 \text{ ohms}$$
$$R_2 = R_4 = 22.6 \text{ k}$$

Choosing Roffset1 to be 1 k and Roffset2 to be 2.5 k, V_{out} is 0.5 V at the presumed maximum barometric pressure of 30.5 in-Hg. The maximum pressure output voltage can be trimmed to a value other than 0.5 V, if desired via Roffset1. In addition, the trimmable offset resistor is incorporated to provide offset calibration if significant offset drift results from large weather fluctuations.

Microcontroller Interface

The low cost of MCU devices has allowed for their use as a signal processing tool in many applications. The MCU used in this application, the MC68HC11, demonstrates the power of incorporating intelligence into such systems. The on-chip resources of the MC68HC11 include an 8-channel, 8-bit A/D, a 16-bit timer, and SPI (serial peripheral interface—synchronous), and SCI (serial peripheral interface—asynchronous), and a maximum of 40 I/O lines. This device is available in several package configurations and product variations that include additional RAM, EEPROM, and/or I/O capability. The software used in this application was developed using the MC68HC11 EVB development system.

The following software algorithm outlines the steps used to perform the desired digital processing. This system will convert the voltage at the A/D input into a digital value, convert this measurement into inches of mercury, and output this data serially to an LCD display interface (through the on-board SPI). This process is outlined in greater detail below.

Algorithm Description

The algorithm is as follows:

1. Set up and enable A/D converter and SPI interface.
2. Initialize memory locations, initialize variables.
3. Make A/D conversion, store result.
4. Convert digital value to inches of mercury.
5. Determine if conversion is in system range.
6a. Convert pressure into decimal display digits.
6b. Otherwise, display range error message.
7. Output result via SPI to LCD driver device.

The signal-conditioned sensor output signal is connected to pin PE5. The MCU communicates to the LCD display interface via the SPI protocol.

LCD Interface

To digitally display the barometric pressure conversion, a serial LCD interface was developed to communicate with the MCU. This system includes an MC145453 CMOS serial interface/LCD driver, and a four-digit, nonmultiplexed LCD. In order for the MCU to communicate correctly with the interface, it must serially transmit 6 bytes for each conversion. This includes a start byte, a byte for each of the four decimal display digits, and a stop byte. For formatting purposes, decimal points and blank digits can be displayed through appropriate bit patterns. The control of display digits and data transmission is executed in the source code through subroutines BCDCONV, LOOKUP, SPI2LCD, and TRANSFER. A block diagram of this interface is shown in Figure 8.30.

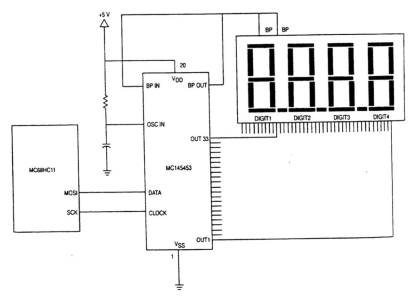

Figure 8.30
Copyright of Motorola. Used by permission.

- This code will be used to implement an MC68HC11 Microcontroller as a processing unit for a simple barometer system.
- The HC11 will interface with an MPX2100AP to monitor, store, and display measured barometric pressure via the 8-bit A/D channel.
- The sensor output (32 mV max) will be amplified to 0.5–2.5 Vdc.
- The processor will interface with a 4-digit LCD (FE202) via a Motorola LCD driver (MC145453) to display the pressure within ± one tenth of an inch of mercury.
- The system's range is 15.0–30.5 in-Hg.
- A/D & CPU register assignment.
- This code will use index addressing to access the important control registers. All addressing will be indexed off of BASE, the base address for these registers.

```
BASE      EQU $1000              ; register base of control register
ADCTL     EQU $30                ; offset of A/D control register
ADR2      EQU $32                ; offset of A/D results register
ADOPT     EQU $39                ; offset for A/D option register
PORTB     EQU $04                ; location of PORTB used for conversion
PORTD     EQU $08                ; PORTD Data Register Index
DDRD      EQU $09                ; offset of Data Direction Reg.
SPCR      EQU $28                ; offset of SPI Control Reg.
SPSR      EQU $29                ; offset of SPI Status Reg.
SPDR      EQU $2A                ; offset of SPI Data Reg.
;             User Variables
;         The following locations are used to store important
;         measurements and calculations used in determining
;         the altitude.  They are located in the lower 256 bytes
;         of user RAM

DIGIT1    EQU    $0001           ; BCD blank digit (not used)
DIGIT2    EQU    $0002           ; BCD tens digit for pressure
DIGIT3    EQU    $0003           ; BCD tenths digit for pressure
DIGIT4    EQU    $0004           ; BCD ones digit for pressure
COUNTER EQU      $0005           ; Variable to send 5 dummy bytes
POFFSET EQU      $0010           ; Storage location for max pressure
                                 ; offset
SENSOUT EQU      $0012           ; Storage location for previous
                                 ; conversion
RESULT  EQU      $0014           ; Storage of pressure (in Hg) in hex
                                 ; format
FLAG    EQU      $0016           ; Determine if measurement is within
                                 ; range

;         **MAIN PROGRAM**
;
;         The conversion process involves the following steps:
;
;
;         1.    Set-Up API device-        SPI_CNFG
;         2.    Set-Up A/D, Constants     SET_UP
;         3.    Read A/D, store sample    ADCONV
```

```
;       4.   Convert into in-Hg          IN_HG
;       5.   Determine FLAG condition     IN_HG
;       a.   Display error                ERROR
;       b.   Continue Conversion          INRANGE
;       6.   Convert hex to BCD format    BCDCONV
;       7.   Convert LCD display digits   LOOKUP
;       8.   Output via SPI to LCD        SPI2LCD
```

```
;       This process is continually repeated as the loop CONVERT
;       runs unconditionally through BRA (the BRANCH ALWAYS
;       statement)
;       Repeats to step 3 indefinitely.
```

```
              ORG     $C000      ; DESIGNATES START OF MEMORY MAP
                                 ; FOR USER CODE
              LDX     #BASE      ; Location of base register for
                                 ; indirect adr
              BSR     SPI_CNFG   ; Set-up SPI Module for data x-
                                 ; mit to LCD
              BSR     SET_UP     ; Power-Up A/D, initialize
                                 ; constants
CONVERT       BSR     ADCONV     ; Calls subroutine to make an
                                 ; A/D conversion
              BSR     DELAY      ; Delay routine to prevent LCD
                                 ; flickering
              BSR     IN_G       ; Converts hex format to in of
                                 ;     hg
```

```
;       The value of FLAG passed from IN_HG is used to determine
;       if a range error has occurred.  The following logical
;       statements are used to either allow further conversion or
;       jump to a routine to display a range error message.
```

```
              LDAB    FLAG       ; Determines if a range Error
                                 ; has occurred

              CMPB    #$80       ; If No Error detected
                                 ; then (FLAG<>80),

              BEQ     INRANGE    ; system will continue
                                 ; conversion  process
              BSR     ERROR      ; If error occurs (FLAG<>80), branch
                                 ; to ERROR
              BRA     OUTPUT;  Branches to output ERROR code to
                                 ; display
;
;              No Error Detected, Conversion Process Continues
;
INRANGE       JSR     BCDCONV    ; Converts Hex Result to BCD
              JSR     LOOKUP     ; Uses Look-Up Table for BCD-
                                 ; Decimal
```

```
OUTPUT      JSR     SPI2LCD   ; Output transmission to LCD
            BRA     CONVERT   ; Continually converts using
                             ; Branch Always

;           Subroutine SPI_CNFG
;               Purpose is to initialize SPI for transmission
;               and clear the display before conversion.
SPI_CNFG    BSET    PORTD, #$20; Set SPI SS Line High to
                             ;prevent glitch;

            LDAA    #$38      ; Initializing Data Direction
                             ; for Port D
            STAA    DDRD,X    ; Selecting SS, MOSI, SCK as
                             ; outputs only
            LDAA    #$5D      ; Initialize SPI-Control
                             ; Register
            STAA    SPCR,X    ; selecting SPE, MSTR, CPOL,
                             ; CPHA, CPRO
            LDAA    #$5       ; sets counter to x-mit 5 blank
                             ; bytes
            STAA    COUNTER
            LDAA    SPSR,X    ; Must read SPSR to clear SPIF
                             ; Flag
            CLRA              ; Transmission of Blank Bytes to
                             ; LCD

ERASELCD    JSR     TRANSFER  ; Calls subroutine to transmit
            DEC     COUNTER
            BNE     ERASELCD

            RTS

;           Subroutine SET_UP

;           Purpose is to initialize constants and to power-up
            A/D and to initialize POFFSET used in conversion
            purposes.

SET_UP      LDAA    #$90    ; selects ADPU bit in OPTION register
            STAA    ADOPT.X     ; Power-UP of A/D complete
            LDD     #$0131+$001A ; Initialize POFFSET
            STD     POFFSET     ; or Pmax + offset voltage (5 V)
            RTS

;           Subroutine DELAY
;           Purpose is to delay the conversion process
;           to minimize LCD flickering.
DELAY       LDA     #$FF        ; Loop for delay of display
OUTLOOP     LDB     #$FF        ; Delay = clk/255;255
INLOOP      DECB
            BNE     INLOOP
```

```
              DECA
              BNE      OUTLOOP
              RTS

;          Subroutine ADCONV
;          Purpose is to read the A/D input, store the
;          conversion into SENSOUT.  For conversion purposes
;          later.

ADCONV        LDX      #BASE       ;loads base register for
                                   ;indirect addressing

              LDAA     #$25
              STAA     ADCTL,X     ;initialize A/D cont. register
                                   ;SCAN=1,MULT=0

WTCONV        BRCLR ADCTL,X  #$80,WTCONV; Wait for completion
                                   ; of conversion flag

              LDAB     ADR2,X      ; Loads conversion result into
                                   ; Accumulator

              CLRA
              STD      SENSOUT     ; Stores conversion as SENSOUT

              RTS

;          Subroutine IN_HG
;          Purpose is to convert the measured pressure SENSOUT,
;          into units of in-Hg, represented by a hex value of
;          305-150 This represents the range 30.5 - 15.0 in-Hg

IN_HG             LDD  POFFSET;   Loads maximum offset for
                                   ; subtraction
                  SUB  SENSOUT; RESULT = POFFSET-SENSOUT in hex
                                   ; format
                  STD  RESULT ; Stores hex result for P, in Hg
                  CMPD     #305
                  BHI      TOHIGH

                  CMPD     #150
                  BLO      TOLOW

                  LDAB     #$80
                  STAB     FLAG
                  BRA      END_CONV
TOHIGH            LDAB     #$FF
                  STAB     FLAG
                  BRA      END_CONV

TOLOW             LDAB     #$00
                  STAB     FLAG
END_CONV RTS
```

```
;          Subroutine ERROR
;                    This subroutine sets the display digits to
;                    output an effort message having detected an out
;                    of range measurement in the main program from
;                    FLAG

ERROR              LDAB       #$00    ; Initialize digits 1,4 to
                                      ; blanks
                   STAB       DIGIT1
                   STAB       DIGIT4

                   LDAB       FLAG    ; FLAG is used to determine
                   CMPB       #$00    ; if above or below range.
           .       BNE        SET_HI  ; If above range GOTO SET_HI

                   LDAB       #$0E    ; ELSE display LO on display
                   STAB       DIGIT2  ; Set DIGIT2+L,DIGIT3=0
                   LDAB       #$7E
                   STAB       DIGIT3
                   BRA        END_ERR ; GOTO exit of subroutine

SET_HI             LDAB       #$37    ; Set DIGIT2=H,DIGIT3=1
                   STAB       DIGIT2
                   LDAB       #$30
                   STAB       DIGIT3

END_ERR            RTS

;          Subroutine BCDCONV
;                    Purpose is to convert ALTITUDE from hex to BCD
;                    uses standard HEX-BCD conversion scheme
;                    Divide HEX/10 store Remainder, swap Q & R,
;                    repeat process until remainder = 0.

BCDCONV            LDAA       #$00    ; Default Digits 2,3,4 to 0
                   STAA       DIGIT2
                   STAA       DIGIT3
                   STAA       DIGIT4
                   LDY        #DIGIT4 ; Conversion starts with
                                      ; lowest digit
                   LDD        RESULT  ; Load voltage to be converted
CONVLP             LDX        #$A     ; Divide hex digit by 10
                   IDIV               ; Quotient in X, Remainder in D
                   STAB       0,Y     ; stores 8 LSB's of remainder as
                                      ; BCD digit

                   DEY
                   CPX        #$0     ; Determines if last digit
                                      ; stored
                   XGDX               ; Exchanges remainder &
                                      ;quotient
```

```
                    BNE        CONVLP
                    LDX        #BASE    ; Reloads BASE into main program

;         Subroutine LOOKUP
;              Purpose is to implement a Look-Up conversion
;              The BCD is used to index off of TABLE
;              where the appropriate hex code to display
;              that decimal digit is contained.
;              DIGIT4,3,2 are converted only.

LOOKUP              LDX        #DIGIT+4 ; Counter starts at 5
TABLOOP             DEX                 ; Start with Digit4
                    LDY        #TABLE   ; Loads table base into Y-
                                        ; pointer
                    LDAB       0,X      ; Loads current digit into B
                    ABY                 ; Adds to base to index off
                                        ; TABLE
                    LDAA       0,Y      ; Stores HEX segment result in
                                        ; acc A
                    STAA       0,X
                    CPX        #DIGIT2 ; Loop condition complete,
DIGIT2
                                        ; Converted
                    BNE        TABLOOP

                    RTS

;         Subroutine SPI2LCD
;              Purpose is to output digits to LCD via SPI
;              The format for this is to end a start byte.
;              four digits, and a stop byte.  This system
;              will have 3 significant digits:  blank digit
;              and three decimal digits.

;                           Sending LCD Start Byte
SPI2LCD             LDX        #REGBASE
                    LDAA       SPSR,X   ; Reads to clear SPIF flag
                    LDAA       #$02     ; Byte, no colon, start bit
                    BSR        TRANSFER ; Transmit byte
;                   Initializing decimal point
;                              & blank digit

                    LDAA       DIGIT3; Sets MSB for decimal pt.
                    ORA        #$80 ; after digit 3
                    STAA       DIGIT3

                    LDAA       #$00     ; Set 1st digit as blank
                    STAA       DIGIT1
;
;         Sending four decimal digits
;
```

```
                    LDY      #DIGIT1  ; Pointer set to send 4 bytes
DLOOP               LDAA     0,Y      ; Loads digit to be x-mitted
                    BSR      TRANSFER ; Transmit byte
                    INY               ; Branch until both bytes sent
                    CPY      #DIGIT4+1
                    BNE      DLOOP
;
;         Sending LCD Stop Byte
;
                    LDAA #$00          ; end byte requires all 0's
                    BSR  TRANSFER; Transmit byte

                    RTS


;         Subroutine TRANSFER
;         Purpose is to send data bits to SPI
;         and wait for conversion complete flag bit to be set.

TRANSFER LDX        #BASE
                    BCLR     PORTD,X #$20; Assert SS Line to start x-
                                            mission
                    STAA  SPDR,X; Load Data into Data Reg.,X-mit
XMIT                BRCLR SPSR,X #$80 XMIT ; Wait for flag
                    BSETPORTD,X #$20   ; DISASSERT SS Line
                    LDAB  SPSR,X       ;  Read to Clear SPI Flag

                    RTS


;         Location for FCB memory for look-up table
;         There are 11 possible digits:  blank, 0-9

TABLE    FCB  $7E,$30,$6D,$79,$33,$5B,$5F,$70,$7F,$73,$00


                    END
```

9 Microcontroller-Based Systems in Communications

The electronic communications industry has been one of the fastest-growing technological industries. Modern digital communications systems are gradually replacing the traditional analog communication systems. Of course, one chapter is not adequate to cover all aspects of digital communications. This chapter touches on some commonly used systems and techniques, such as standards, protocols, and modems. Especially useful features of the chapter include the programming examples and exercises on the communications and protocols between multiple microcontrollers.

9.1   Introduction to Communications Systems

Parallel transmission is called *parallel-by-bit* transmission, and it takes a single pulse *(t)* to be transmitted. This transmission is usually used for a short distance and within a computer.

Serial transmission is called *serial-by-bit* transmission. Each bit takes one pulse to be transmitted. For 8 bits, it takes 8*t*. This method is generally used for long distances.

Configuration

Data communications circuits can be generally categorized as either two-point or multipoint. A *two-point* configuration involves the transfer of information between two locations or stations only. The transfer could occur between a mainframe and a terminal, two mainframes, or two terminals.

A *multipoint* involves transfer of information between three or more stations. The transfer could occur between a mainframe and several terminals or between any combinations of mixed mainframes and terminals.

Topologies

The *topology* of a data communications circuit identifies how these locations (stations) are interconnected. The most common topologies are point-to-point, star bus or multidrop, ring or loop, and mesh. Figure 9.1 shows these topologies.

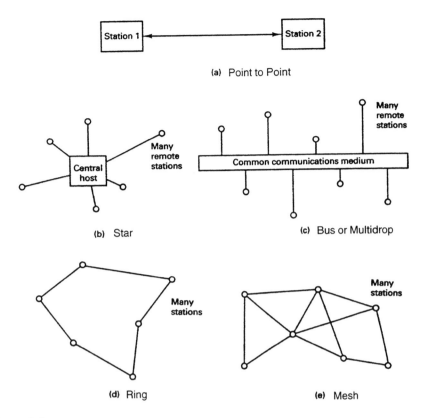

Figure 9.1

9.2 Standards Organizations for Data Communications

These organizations have been established to ensure that all data communications users comply with uniform standards. The more important organizations are:

International Standards Organization (ISO)

The ISO creates rules and sets standards for graphics, document exchange, and related technologies. It is also responsible for endorsing and coordinating the work of the other standards organizations.

Consultative Committee for International Telephony and Telegraphy (CCITT)

This organization includes U.S. government representatives and representatives from other countries. It has developed three sets of specifications: the V Series for modem interfacing, the X Series for data communications, and the I and Q Series for Integrated Services Digital Network (ISDN).

American National Standards Institute (ANSI)

This is the official standards agency for the United States. It is the U.S. voting representative for the ISO.

Institute of Electrical and Electronics Engineers (IEEE)

The IEEE is a U.S. professional organization of electronics, computer, and communications engineers.

Electronic Industries Association (EIA)

The EIA is a U.S. organization that establishes and recommends industrial standards and standards for data and telecommunications.

Standard Council of Canada (SCC)

The SCC is the official standards agency for Canada.

9.3   Communications Modes

There are three modes of data communications circuits, as Figure 9.2 shows.

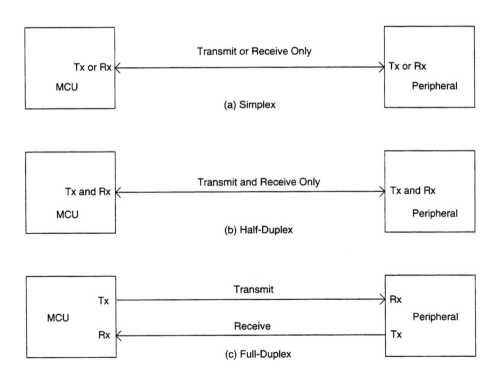

Figure 9.2

Simplex

In this case the data transmission is unidirectional; information can be sent in one direction only. Simplex lines are called *receive only, transmit only,* or *one-way-only* lines. Commercial television and radio systems are examples of simplex transmission.

Half-Duplex (HDX)

With this approach, transmission is possible in both directions but not at the same time. Citizens Band (CB) radio is an example of HDX.

Full-Duplex (FDX)

Here transmission is possible in both directions and at the same time. A standard telephone system is an example of FDX.

9.4   Data Communications Protocols

A *protocol* is a set of customs or regulations dealing with formality or precedence. A data communications network protocol is a set of rules governing the orderly exchange of data.

In a data communications circuit, the station that is presently transmitting is called the *master* and the receiving station is called the *slave.* In a centralized network, the primary station controls when each secondary station can transmit. When a secondary station is transmitting, it is the master and the primary station becomes the slave. Initially, the primary is the master. The primary station solicits each secondary by "polling" it. A poll is an invitation from the primary to a secondary to transmit a message. Secondaries cannot poll a primary. Secondary stations cannot select the primary. Transmissions from the primary go to all the secondaries and it is up to the secondary stations to individually decode each transmission and determine if it is intended for them. When a secondary transmits, it sends only to the primary (the master).

Data link protocols are generally categorized as either asynchronous or synchronous. As a rule, asynchronous protocols use an asynchronous data format and asynchronous modems, whereas synchronous protocols use a synchronous data format and synchronous modems.

Asynchronous Serial Protocol

The serial format for asynchronous protocols is shown in Figure 9.3. It always starts with the start bit, the transmitted ASCII data, then ends with a stop bit. Logic 1 is called *mark,* while logic 0 is called *space.* The start bit is always low and the stop bit(s) always high.

The total bits including start to stop is called a *frame.* The term *asynchronous* means not clocked—that is, there is no common clock between the transmitting and receiving devices. In digital systems, a clock is used to synchronize the transfer of data between different parts of the system.

In asynchronous systems, transmission and reception are synchronized by frames. Figure 9.4 shows transmitted/received frames. Refer to Chapter 5 for more information on asynchronous systems.

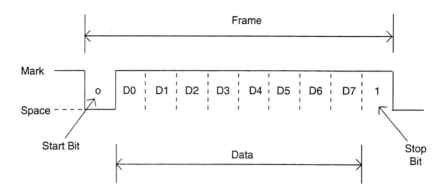

Figure 9.3

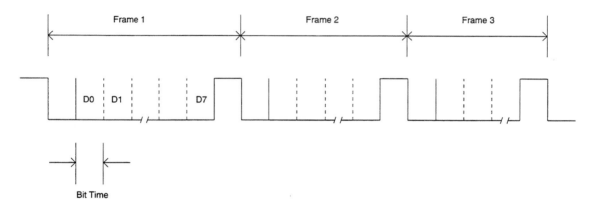

Figure 9.4

Synchronous Protocol

In synchronous communications, data are not transmitted on a character-by-character basis. They are transmitted in blocks. Each block consists of a sequence of characters. Each block is synchronized as an entity rather than character by character. This type of communication can support high data rates by keeping the receiver in sync with the transmitter. Most synchronous networks use one of the protocols developed by IBM. These protocols are binary synchronous communication (BISYNC) or synchronous data-link control (SDLC). If a synchronous modem is used, it generates a receiver clock signal that is automatically synchronized with the received bit stream.

9.5   Modems

If the peripheral is located at a great distance (100 ft) from the computer (data terminal equiptment [DTE]), it cannot be connected by simply using a longer cable. As the connecting cable gets longer, its resistance combines with capacitances along the line to form a natural integrator that distorts square waves until they are unacceptable as digital signals. To overcome this signal distortion, a device called a *modem* (MOdulator/DEModulator) is used to convert the parallel digital signals used by the DTE into serial analog signals that are better suited for transmission over wires.

Three types of modulation are used by data links: *amplitude shift keying* (ASK), *frequency shift keying* (FSK), and *phase shift keying* (PSK). ASK is used for low speeds of less than 100 bps. FSK is used for medium speed up to 1200 bps. PSK is used for speeds up to 4800 bps. For 9600 bps, a combination of PSK and ASK is used and called *quadrative amplitude modulation* (QAM). (See Figure 9.5.)

The modem will provide the FSK service. When the modem converts the digital pulses (0,1) into a sine wave (FSK) for transmission, this is called *modulation*. When the modem converts the sine wave (FSK) received back into digital pulses, this is called *demodulation,* as shown in Figure 9.6.

The modem can be a simplex-mode, half-duplex-mode, or full-duplex mode modem. Figure 9.7 shows a half-duplex and full-duplex modem. Modems (data communication equiptment [DCE]) are used to enable computers (DTEs) to use public telephone lines to exchange information. The telephone system accommodates a range of frequencies between 300 and 3300 Hz—that is, a bandwidth of 3000 Hz.

This is adequate to transmit voice signals but it distorts digital data. To use the audio characteristics of the telephone lines to best advantage, a modem encodes the digital "1s" and "0s" into frequencies within this bandwidth.

Before communication can take place, the parties must agree on who will transmit and receive on which band (high or low). The two modems at each end of telephone lines are called *originate* and *answer* modems.

The originate modem transmits on a low band and receives on a high band, while the answer modem transmits on a high band and receives on a low band, as shown in Figure 9.7.

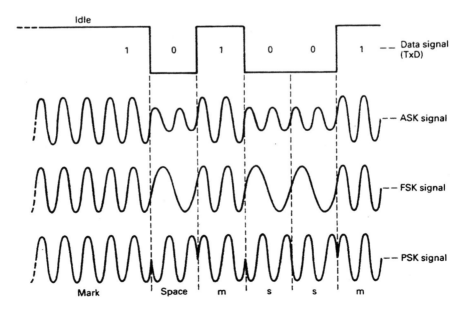

Figure 9.5
From P. Bates, *Practical Digital and Data Communications.* Copyright 1987 by Prentice Hall. Reprinted by permission of Prentice Hall, Upper Saddle River, New Jersey.

Figure 9.6

From R. Tocci, *Microprocessors and Microcomputers: The 6800 Family.* Copyright 1986 by Prentice Hall. Reprinted by permission of Prentice Hall, Upper Saddle River, New Jersey.

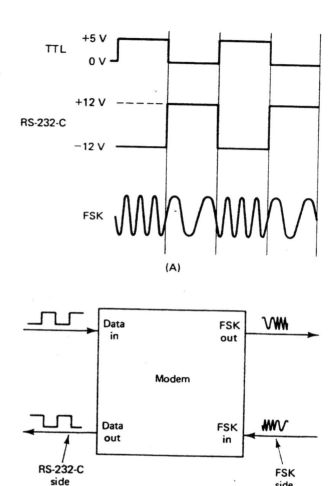

In FSK, the mark-space frequencies for the low and high bands are 1270 Hz, 1070 Hz, 2225 Hz, and 2025 Hz respectively.

Data Transfer

To facilitate the transfer of data from the parallel bus of the computer to the serial bit stream used by the modem (and vice versa), a device called a *Universal Asynchronous Receiver/Transmitter* (UART) is used. The UART converts data from serial to parallel form and also converts parallel data to serial form. In addition, the UART also adds start and stop bits and error-checking bits to the data for transmission. As a receiver, it interprets and strips these extra bits from the data.

When a computer sends a character to the UART, the character will be loaded into a character buffer register. The UART shifts a start bit, followed by the character bits into a larger serial-in shift register. Finally, the proper error-checking bit and a specified number of stop bits are also shifted into the register. The character data block is now ready for transmission. At the receiver, the start bit of the incoming character block is detected,

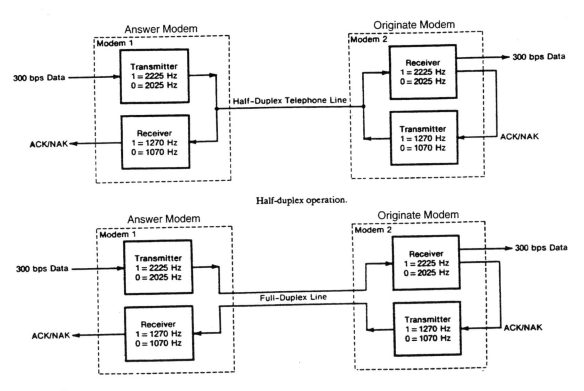

Half-duplex operation.

Figure 9.7

causing the bits that follow to be shifted into the serial shift register. The error-checking bit is compared to an error bit generated by the receiver as the character block is shifted into the register. The start and stop bits are stripped from the character frame.

The UART and a number of other asynchronous support chips make up the computer/modem interface. The physical location of the interface and its complexity is dependent on the particular computer system and the type of modem being used.

Stand-alone modems do not require an on-board UART because there must be one present in the computer's asynchronous adapter board or its serial interface connector circuit.

LSI Modem Example

NMOS and CMOS chips are available that contain modulator, demodulator, and control logic for various low to medium-speed modems. Motorola has introduced several modem chips: the NMOS such as the MC6860 and the CMOS MC14412 modems.

The M6860 Digital Modem

This modem uses NMOS technology in a 24-pin package and includes the modulator, demodulator, and control logic to interface directly with a UART chip and data coupler. Figure 9.8 shows a block diagram for a Bell 103 originate modem using the MC6860 interfaced to an 8251A USART.

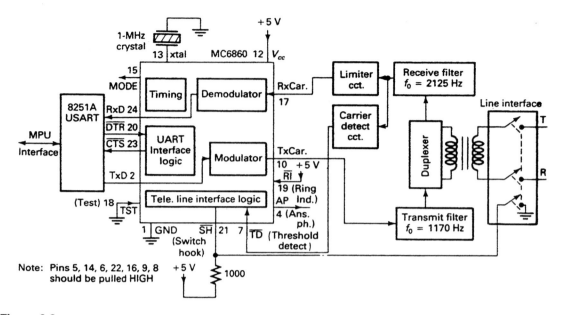

Figure 9.8

From P. Bates, *Practical Digital and Data Communications.* Copyright 1987 by Prentice Hall. Reprinted by permission of Prentice Hall, Upper Saddle River, New Jersey.

Here is how it works:

1. Line $\overline{SH}$ (switch hook) goes low and places the MC6860 in originate mode and is ready to receive the 2225-Hz (mark) tone from the remote answering modem.

2. At 450 mS after receiving the mark signal, the MC6860 begins transmitting its mark tone (1270 Hz).

3. At about 750 mS after first receiving the 2225-Hz tone, $\overline{CTS}$ (clear-to-send) signal goes low and data can be transmitted as well as received. After $\overline{SH}$ goes low, the mark carrier from the remote modem must be received within 17 mS maximum.

4. Insufficient carrier level results in the carrier detect circuit causing $\overline{TD}$ (threshold detect) high which causes the 6860 to make $\overline{CTS}$ high and turns off the TX Car (transmit carrier) signal.

5. The FSK carrier output is synthesized by a D/A converter within the MC6860 and its wave shape is a "stair-step" approximation of a sine wave full of undesirable harmonics. The transmit filter must alternate these harmonics to prevent their interfacing with the received FSK carrier.

6. The $\overline{RI}$ (ring indicate) input is meant to be driven from a ring-detector circuit when used. A low on $\overline{RI}$ places the MC6860 in the answer mode and causes the answer pho (AP) output to go high, which is normally used to cause a switch-hook relay to close (answering the phone).

7. The mode output indicates the modem state: HIGH indicates originate mode and LOW indicates answer mode. Mode signal can be used to switch (transpose) the receive and transmit filters in auto-originate-answer modems.

9.6 Motorola 6850 UART (the ACIA)

ACIA stands for *Asynchronous Communications Interface Adapter*. The block diagram in Figure 9.9 shows the basic structure of this chip. The major components include the following:

1. Transmit data register (TxDR) and transmit shift register (TSR)

2. Receive data register (RxDR) and receive shift register (RSR)

3. Control register

4. Status register

5. Data bus buffers

6. Control lines

Control Register

The functions of the control register bits are shown in Figure 9.10. See Table 9.1 for bit functions.

If CR7 is set, the receive interrupt is enabled, but if it is reset, the interrupt is disabled.

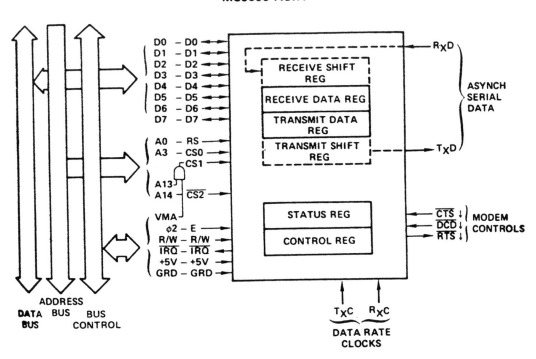

Figure 9.9

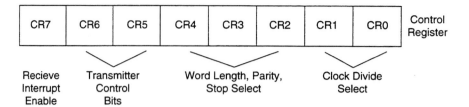

Figure 9.10

Table 9.1

CR1	CR2	Clock Divide Ratio
0	0	divided by 1
0	0	divided by 16
1	0	divided by 64
1	1	Master Reset

CR4	CR3	CR2	Data Bits	Parity	Stop Bits
0	0	0	7	Even	2
0	0	1	7	Odd	2
0	1	0	7	Even	1
0	1	1	7	Odd	1
1	0	0	8	None	2
1	0	1	8	None	1
1	1	0	8	Even	1
1	1	1	8	Odd	1

CR6	CR5	Function
0	0	$\overline{RTS}$ = 0 and transmit interrupt disabled
0	1	$\overline{RTS}$ = 0 and transmit interrupt enabled
1	0	$\overline{RTS}$ = 1 and transmit interrupt disabled

Status Register

The status register contains 8 bits (read-only). The functions of these bits are described below; also see Figure 9.11.

Receive Data Register Full (RDRF)

This bit is set when the receive data register is full (data are received).

Transmit Data Register Empty (TDRE)

This bit is set when the transmit data register is empty (data are transmitted).

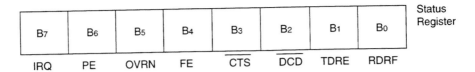

B7	B6	B5	B4	B3	B2	B1	B0	Status Register
IRQ	PE	OVRN	FE	$\overline{CTS}$	$\overline{DCD}$	TDRE	RDRF	

Figure 9.11

Data Carrier Detect ($\overline{DCD}$)

The DCD is set when a modem detects a carrier modulating signal.

Clear to Send ($\overline{CTS}$)

This active low signal is generated by the modem and sent to UART in response to a request-to-send ($\overline{RTS}$) signal (this is called a *handshake*). This bit reflects the status of the UART's $\overline{CTS}$ input line. When the modem is ready to transmit data, the $\overline{CTS}$ line goes low. Thus the microcomputer can determine if the modem is ready by reading the B3 of the status register.

Framing Error (FE)

This bit is set if any of the received stop bits are 0. The MCU/MPU is constantly checking this bit in the receiving mode.

Overrun (OVRN)

This bit is set if a data word is shifted into the receiver shift register before the previous one was transferred to the receiver data register. This previous data could be lost. If this happened, the OVRN bit is set.

Parity Error (PE)

If a particular parity (odd or even) is used, the receiver keeps checking for this parity when the UART receives data. If the received parity is different from the parity used, this flag is set.

Interrupt Request ($\overline{IRQ}$)

When the $\overline{IRQ}$ output is low, this bit is set. It can be cleared again when the microcomputer performs a read to RxDR or a write to TxDR.

In addition to the UART chip, the interface also contains a controller to supervise the operation of the UART and various status registers that it uses to monitor the input line, the modem, and data flow status. The interface may also contain a bigger memory, an on-board clock, and other support circuits to provide functional flexibility for the interface.

The ACIA M6850 can be used in different applications in communications systems. Typical applications are in Figure 9.12.

The first application shows how a terminal (printer) is connected to a nearby microcomputer. The RS-232c is used to interface the ACIA to the terminal.

The second application shows how the above terminal is connected to a remote microcomputer through telephone lines. As mentioned before, data transmission across telephone lines always require modems to communicate.

The third application illustrates how two remote microcomputers communicate with each other via modems.

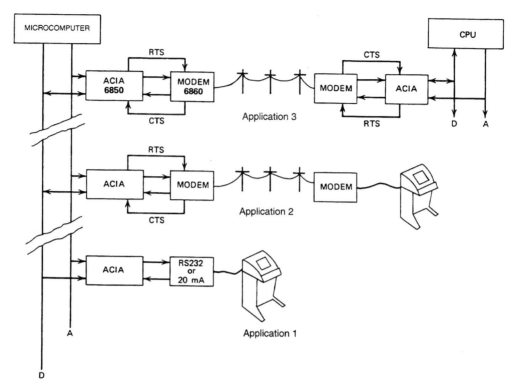

Figure 9.12
From J. Greenfield and W. Wray, *Using Microprocessors and Microcomputers.* Copyright 1988 by Prentice
Hall. Reprinted by permission of Prentice Hall, Upper Saddle River, New Jersey.

Program 1:   Communication Between the ACIA and a Modem

This program illustrates the communications between an ACIA and a modem. Assume the
control/status register (ACIASC) at address $8000 and the data register (ACIADA) at
address $8001. In the software master reset, CR0 and CR1 are set to 1, allowing the control
register to be programmed with the control word.

```
ACIASC   EQU $8000                    ;status/control reg address
ACIADA   EQU $8001                    ;data reg address


                        ORG $C000


              LDAA  #3
              STAA  ACIASC            ;master reset
              LDAA  #5
              STAA  ACIASC            ;no interrupt,7 data,odd parity,
                                      ;RTS low
         ;transmit routine


TRNS_CHR      LDAB  ACIASC
              ANDB  #2                ;trans buffer empty yet?
              BEQ   TRNS_CHR          ;no, wait
```

```
                 STAA  ACIADA        ;yes, transmit data

                 RTS
            ;receive routine

RCV_CHR          LDAB  ACIASC
                 ANDB  #1            ;rec buffer ready yet?
                 BEQ   RCV_CHR       ;no, wait
                 LDAA  ACIADA        ;yes, receive data
                 RTS
```

Exercises

1. Using the above program, write a routine to detect any errors received.

2. After studying the modem and the UART, respond to the following:
   a. What are the types and functions of modems?
   b. Draw the programming model of the ACIA.

3. A system is equipped with two synchronous serial communication ports operating in full-duplex mode. The transmission rate for both bidirectional channels is 9600 bps. Each frame of information is 1064 bits long and carries 128 bytes of data.
   a. The system is programmed such that it interrupts the CPU every time it is ready to accept or provide another byte of data. At what rate will the system interrupt the CPU while receiving a frame?
   b. It is estimated that the interrupt service routing (ISR) takes about 50 µS for each data byte and that the composite traffic amounts to about 25 frames/second. What fraction of its time does the CPU spend servicing the system?

9.7   Serial Interface Standards

EIA RS-232C Standard

The RS-232C standard was adopted in 1969 by the Electronic Industries Association (EIA). The standard refers to the modem as a DCE (data communications equipment) and to the computer or a terminal as a DTE (data terminal equipment). The specification limits the baud rate to 19,200 with a 50-ft cable. In practice, much longer cables can be accommodated, but with lower data rates.

The RS-232C standard requires that a logic 0 level be presented by voltage in the range +3 V to +25 V. Most microcomputer interfaces use +12 V or +15 V to represent a logic 0 and –12 V or –15 V to represent a logic 1.

The RS-232C standard specifies the pin assignments for a 25-pin connector, as shown in Table 9.2. There are four data signals, twelve control signals, three timing signals, and two ground signals. The other four pins are not used. A typical 25-pin connector and an RS-232C microcomputer interface using eight signal lines are shown in Figure 9.13.

The additional control signals are used for handshaking between the microcomputer system and peripheral devices. The RS-232C "transmitted data line" is used for data traveling from "peripheral to MCU" and must be connected to the MCU receive line.

Table 9.2

		Data		Control	
Pin	Signal name	From DTE to DCE	To DTE from DCE	From DTE to DCE	To DTE from DCE
1	Protective ground				
2	Transmitted data	x			
3	Received data		x		
4	Request to send ($\overline{RTS}$)			x	
5	Clear to send ($\overline{CTS}$)				x
6	Data set ready ($\overline{DSR}$)				x
7	Signal Ground				
8	Data carrier detect ($\overline{DCD}$)				x
9/10	Reserved for data set testing				
11	Unassigned				
12	Secondary data carrier detect				x
13	Secondary clear to send				x
14	Secondary transmitted data	x			
15	Transmit signal element timing				x
16	Secondary received data		x		
17	Receive signal element timing				x
18	Unassigned				
19	Secondary request to send			x	
20	Data terminal ready ($\overline{DTR}$)			x	
21	Signal-quality detector (indicates probability of error)				x
22	Ring indicator				x
23	Data signal rate select (allows selection of two different baud rates)				x
24	Transmit signal element timing			x	
25	Unassigned				

From J. Uffenbeck, *The 8086/8088 Family: Design, Programming, and Interfacing.* Copyright 1988 by Prentice Hall. Reprinted by permission of Prentice Hall, Upper Saddle River, New Jersey.

The control pins $\overline{DTR}$, $\overline{DSR}$, $\overline{CTS}$, $\overline{RTS}$, and $\overline{DCD}$ establish a protocol between the modem (DCE) and terminal (DTE). These five signals are also included in the modem control signals. These signals are explained below.

1. *Data Carrier Detect* ($\overline{DCD}$). This signal is output by the DCE and indicates that the modem has detected a valid carrier.
2. *Data Terminal Ready* ($\overline{DTR}$). This signal is output by the DTE to indicate that it is ready for communications. It can be used to switch on a modem.
3. *Data Set Ready* ($\overline{DSR}$). This signal is output by the DCE in response to DTR and indicates that the DCE is on and connected to the communications channel.
4. *Request to Send* ($\overline{RTS}$). This signal is output by the DTE to indicate that it is ready to transmit data.
5. *Clear to Send* ($\overline{CTS}$). This signal is output by the DCE and acknowledges RTS. It indicates that the DCE is ready for transmission.

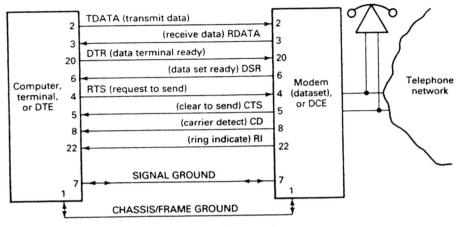

RS232-C data and control lines

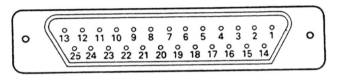

DB-25 connector (female, front view)

Figure 9.13
From P. Bates, *Practical Digital and Data Communications.* Copyright 1987 by Prentice Hall. Reprinted by permission of Prentice Hall, Upper Saddle River, New Jersey.

Nonstandard 9-Pin RS-232C Connectors

Because only 9 of 25 pins of the RS-232C interface are used for the majority of applications, many computer and peripheral manufacturers have developed 9-pin RS-232C connectors. Figure 9.14 illustrates this connector used with PC microcomputers. The 9-pin connector is used to save connector space on the back panels of computers and peripherals.

Figure 9.14
From J. Uffenbeck, *The 8086/8088 Family: Design, Programming, and Interfacing.* Copyright 1988 by Prentice Hall. Reprinted by permission of Prentice Hall, Upper Saddle River, New Jersey.

Front view

Pin		
1	DCD	(8)
2	RD	(3)
3	TD	(2)
4	DTR	(20)
5	SG	(7)
6	DSR	(6)
7	RTS	(4)
8	RTS	(5)
9	RI	(22)

One of the main problems with the RS-232C is that the drivers and receivers are unbalanced or single-ended. The input and output signals are referenced to a common ground. If the ground potentials of the receiving and transmitting modes are not the same, a current will flow in the common ground-wire connection. Two new electrical interface standards—RS-422A and RS-423A—are used to overcome this problem.

RS-422A uses a differential transmitter and receiver. This eliminates the common ground wire. The receiver detects the difference between its two inputs as positive or negative. RS-423A is similar but uses a single-ended driver with a differential receiver.

The RS-449 is a new interface standard comparable with RS-232C. It specifies two data connectors, a 37-pin connector for the main interface signals and another 9-pin connector (for a total of 46 pins). The basic 37 lines include all the previously described RS-232C functions and introduce 10 new circuits.

Exercises

1. Which of the following electrical standards uses a balanced line driver and receiver?
   a. RS-232C
   b. RS-422A
   c. RS-423A
2. An RS-423A receiver can be used with an RS-232C.
   a. True
   b. False
3. List the sequence of logical transitions that a transmitted data line would undergo to transmit the word *RST*. Assume that the data protocol is as follows: start bit, 7 data bits, even parity, and 1 stop bit. How long does it takes to transmit a 2048-byte data buffer at a baud rate of 4096 bps?

9.8   Serial Standard Conversion

There are several serial communications standards in the microcomputer world today. These standards are:

1. TTL standard
2. RS-232C
3. 20-mA current loop
4. Kansas City (KCS)

CRTs and teletypewriters use the 20-mA loop and/or RS-232C standard. Modems operate on the TTL and/or RS-232C standard. Cassette recorders operate on the Kansas City standard. Microcontrollers operate on the TTL standard. For these MCUs to communicate with serial devices that use different standards, conversion between TTL and the particular standard must be provided.

TTL to/from RS-232C

To perform the conversion from the TTL to the RS-232C standard or vice versa, there are some popular integrated circuit devices on the market. The MC1488 is a TTL-to–RS-232C–level converter, and MC1489 is an RS-232C–to–TTL–level converter.

A minimum full-duplex RS-232C interface circuit requires two lines for transmitting and receiving. The MCU uses a TTL standard. To communicate with the RS-232C standard, these two integrated circuits must be used to do the conversion as shown in Figure 9.15. The RS-232C interface should not be used for data transmission over 50 feet, due to noise and loss problems.

Here is the conversion relation:

	TTL		RS-232C
Logic 0	0 V	⟷	+12 V
Logic 1	+5 V	⟷	−12 V

TTL to/from 20-mA Loop

This standard is used for CRT data terminals, teletypewriters, and printers. In this standard, zero current is a logic 0 and 20 mA is a logic 1. This conversion between the TTL and 20-mA standards is required. Opto-couplers are used.

	TTL		20-mA loop
Logic 0	0 V	⟷	0 mA
Logic 1	5 V	⟷	20 mA

The 20-mA current loop standard is used for transmission over 50 feet, up to 1 mile.

RS-232C to/from 20-mA Current Loop

In some applications conversions between the RS-232C and the 20-mA current loop are needed. The circuit in Figure 9.16 performs the conversion. In Figure 9.16a, if the input of the RS-232C is a logic 1 (−12 V), D1 is turned on and x has a 0 V. The LED now is forward biased and a 20-mA current loop will run in the 20-mA peripheral device. If this input is a logic 0 (+12V), D2 is turned on, and due to the voltage drop across the 1K, x has about 5 or 6 V. The LED is reverse biased and is turned off. Thus, the 20-mA loop is open (logic 0).

In Figure 9.16b, if a logic 1 (20 mA) turns on the LED, a −12 V (logic 1) is produced at y, the RS-232C output.

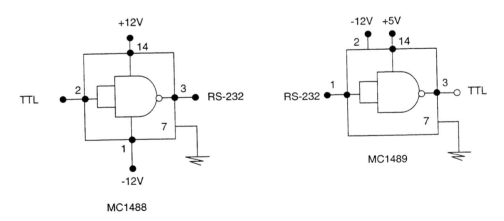

Figure 9.15

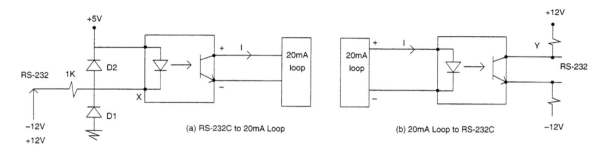

Figure 9.16

If there is no 20-mA current loop in the opto-coupler, the input phototransistor is turned off and y has +12 V (logic 0). Now the relation is:

	RS-232C	20-mA Current loop
Logic 0	+12 V ⟷	0 mA
Logic 1	−12 V ⟷	20 mA

TTL to/from KCS

KCS is used to interface cassette recorders to microcomputers. In KCS, a logic 0 is represented by four cycles of 1200 Hz and a logic 1 is represented by 8 cycles of 2400 Hz of audio sine wave frequencies.

The TTL/KCS converter circuit must make conversions between a 0V to 5V rectangular TTL waveform and a sine-wave form consisting of two separate sine-wave frequencies. Thus, a logic 0 TTL must be converted to a 1200-Hz sine wave and vice versa. A logic 1 TTL must be converted to a 2400-Hz sine wave and vice versa.

Example 1

Show the minimum connections required to interface a microcomputer serial port wired as a DCE to a video terminal serial port wired as a DTE.

Solution
Figure 9.17 shows that in this connection only three signals are required: transmitted data (TD), received data (RD), and signal ground (SG).

Program 2:   **Communication Protocol with a Remote Sensor Using the RS-232C Interface**

This program demonstrates how a remote transducer is interfaced to an MCU using a serial interface, RS-232c. The block diagram of such a circuit is shown in Figure 9.18. The digital information coming from a transducer consists of 8-bit data. These parallel data should be converted to serial data, then sent to the MCU through an RS-232c interface.

As mentioned earlier, hardware conversion between the TTL parts and the RS-232c must be used. For purposes of demonstration only, the transducer is simulated by eight switches, S0 to S7.

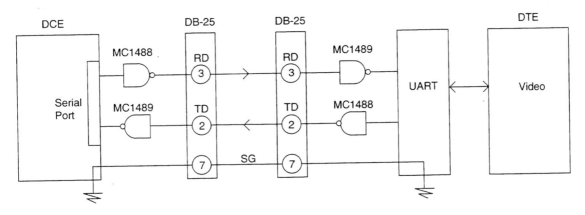

Figure 9.17

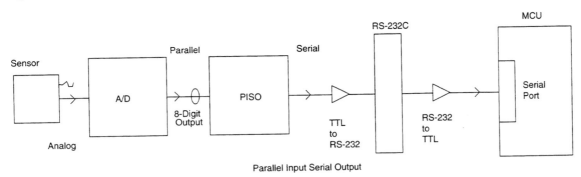

Parallel Input Serial Output

Figure 9.18

The chip MC74HC589, parallel-input serial-output (PISO), is used to convert the parallel data to serial data. The circuit diagram is shown in Figure 9.19. In a later exercise, these switches are replaced by the actual temperature transducer MTS102.

The pin assignments of the MC74HC589 are as follows:

$$SP = \text{serial shift/parallel load}$$
$$SC = \text{shift clock}$$
$$OE = \text{output enable}$$
$$QH = \text{serial output}$$
$$LC = \text{latch clock}$$

A rising edge on LC loads the parallel data into the data latch register. A falling edge on SP transfers the data latch to the shift register. A rising edge on both LC and SP will shift out the latched data serially. LC and SP are connected together. The MCU outputs 8 cycles for 8-bit data.

The circuit in Figure 9.19 works as follows:

1. When SW is pressed, the data set ready ($\overline{\text{DSR}}$) line goes low, indicating that data set is available from the transducer.

2. The MCU senses that and pulls the data terminal ready ($\overline{\text{DTR}}$) line high when it is ready to receive data.

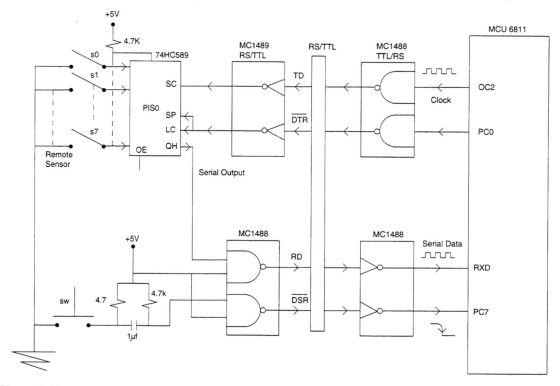

Figure 9.19

3. Eight clock pulses are generated by the MCU on the TD line to clock the 8-bit data out on the receiving line of the SCI via QH output.

4. Data is processed—that is, stored in memory, or displayed on a hardware display, and so on. In our program, it is stored in a buffer.

The pulses generated and their behavior are illustrated in Figure 9.20. Selected baud rate = 9600 bps. The clock rate of OC2 should have the same rate. Bit time = 1/baud rate = 1/9600 = 104 μS. Number of clock cycles/half period (as shown in Figure 9.21) = 52 μS/0.5 μS = 104 = 68 hex.

Here is the program.

```
              ; Equates are the user's responsibility

                      ORG $C000

              LDX   #BASE
              LDY   #BUFF
              LDAA  #$0F
              STAA  DDRC,X        ;pc0 in, pc7 out

         ;Initialization of SCI
```

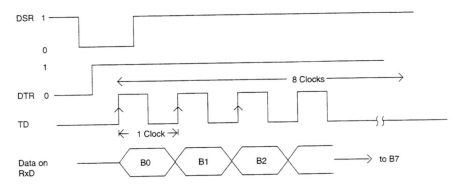

Figure 9.20

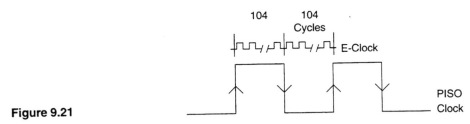

Figure 9.21

```
          LDD  #$3004
          STAA BAUD,X      ;baud 9600
          CLR  SCCR1,X     ;8 bit data
          STAB SCCR2,X     ;enable receive only

     ;Initialization of timer
          LDAA #$40
          STAA TFLG1,X     ;clear status flg
          LDAA #$40
          STAA TCTL1,X     ;oc2 toggles

START     LDAA #8
          STAA COUNTER
          BRSET PORTC,X,$7,START ;sw is pressed yet?
AGAIN     LDAA TOC2,X
          ADDD #$68
          STD  TOC2,X      ;generate clock
          DEC  COUNTER     ;8 cycles yet?
          BEQ  START
LOOP      BRCLR TFLG1,X,$40,LOOP ;wait till match occurs
          BRSET PORTA,X,$40,LATCH;clock it at begin of
                                 ;each cycle
          LDAA #$40
          STAA TFLG1,X     ;clear flg
          BSR  RECEIVE
          BRA  AGAIN
```

```
LATCH              BCLR PORTC,X,$0
                   BSET PORTC,X,$0
                   BCLR PORTC,X,$0
                   BSET PORTC,X,$0
                   RTS

                   ;Receiving routine

RECEIVE            BRCLR SCCR,X,20,RECEIVE   ;received data yet?
                   LDAA  SCDR,X              ;yes, clear flg
                   STAA  0,Y                 ;store data in buffer
                   INY
                   RTS
```

Exercises

1. Replace the simulating switches with the temperature transducer MTS102 (manufactured by Motorola). Add the necessary hardware components such as a sensor signal driver, A/D converter, and so on. Rewrite the above routine and display the temperature on a display (LED or LCD).

2. In energy management systems (EMS), remote temperature and humidity sensors are installed in different rooms and hallways in large buildings to monitor both the temperature and humidity. Every group of these sensors reports to a data gathering panel (DGP). The DGP contains signal drivers and A/D converters. Groups of DGPs report to a data control panel (DCP). The DCP contains an MCU that obtains information from DGPs and makes decisions about which ventilation system should be turned on or off.

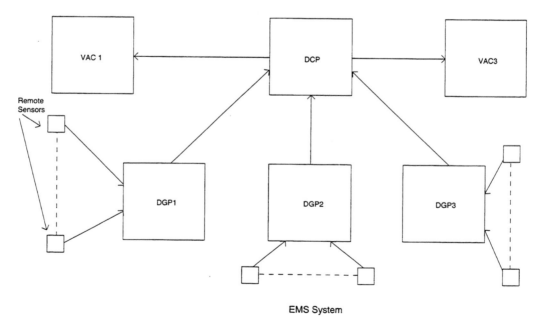

EMS System

Figure 9.22

The block diagram is shown in Figure 9.22. Design a flowchart or pseudocode for that system.

3. The letter W has the ASCII code 57 (hex). Draw the RS-232C, TTL, and 20-mA loop levels for this character. Configure the transmission for 7-bit, even parity and one stop bit, then draw the asynchronous frame to transmit this character.

9.9   IEEE 488 GPIB

The IEEE adopted the *Hewlett-Packard Interface Bus* (HPIB) and revised it as the *General Purpose Interface Bus* (GPIB) as a standard digital interface for programmable instrumentation. The standard consists of eight bidirectional data lines, supported by three handshake control signals and five general-purpose interface management lines. There are 24 in all, including 8 ground lines (Figure 9.23).

The bus is rated to handle data rates as high as 1 Mbps. The maximum cable length is restricted to 60 m at lower data rates. The standard allows up to 15 devices to be connected to the parallel bus at one time. It defines three types of devices that may be connected to the bus:

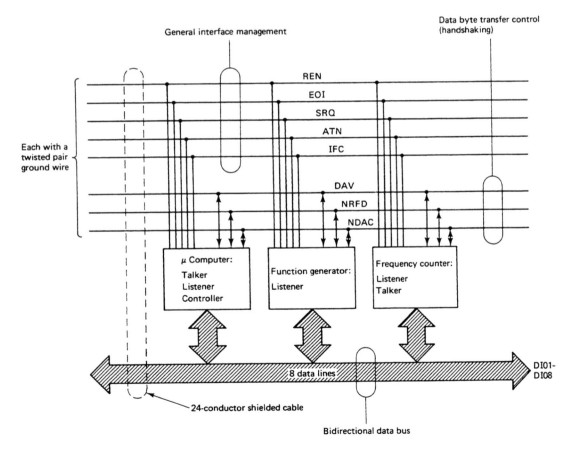

Figure 9.23
From J. Uffenbeck, *The 8086/8088 Family: Design, Programming, and Interfacing.* Copyright 1987 by Prentice Hall. Reprinted by permission of Prentice Hall, Upper Saddle River, New Jersey.

1. *Talkers* are devices capable of putting data onto the bus for transmission to listeners or the controller.

2. *Listeners* are devices capable of reading data on the bus output by talkers or the controller.

3. The *controller* programs devices on the bus. It decides which device will be allowed to talk or listen.

Electrical Characteristics

The voltage levels of this standard are TTL compatible and use a negative logic.
The requirement for a driver are:

$$Vol \le 0.5 \text{ V at 48 mA sink current}$$
$$Voh \ge 2.4 \text{ V at 5.2 mA source current}$$

And the requirements for a receiver are:

$$Vih \ge 2.0 \text{ V}$$
$$Vil \le 0.8 \text{ V}$$

Figure 9.24 shows the 24-pin GPIB connector and pin assignments. Each of the 8 control signals has a separate ground pin (EOT and REN share pin 24) and a shield.

The Intel 8293 chip is a GPIB transceiver that meets the full requirements of the standard and includes Schmitt trigger with 0.4 V of hysteresis. See Figure 9.25 for the chip schematic.

The maximum data rate is 1 Mbps, and the cable is restricted to 15 m. Normally, devices are connected in a "daisy chain" or star configuration with less than 4 m between connections.

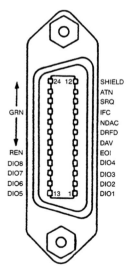

Contact	Signal Line	Contact	Signal Line
1	DIO 1	13	DIO 5
2	DIO 2	14	DIO 6
3	DIO 3	15	DIO 7
4	DIO 4	16	DIO 8
5	EOI (24)	17	REN (24)
6	DAV	18	Gnd. (6)
7	NRFD	19	Gnd. (7)
8	NDAC	20	Gnd. (8)
9	IFC	21	Gnd. (9)
10	SFQ	22	Gnd. (10)
11	ATN	23	Gnd. (11)
12	SHIELD	24	Gnd. LOGIC

Figure 9.24

Figure 9.25
From J. Uffenbeck, *The 8086/8088 Family: Design, Programming, and Interfacing*. Copyright 1987 by Prentice Hall. Reprinted by permission of Prentice Hall, Upper Saddle River, New Jersey.

GPIB Bus Structure

The eight *data lines* send data and command bytes from one device (the talker) to many devices on the bus (listeners).

Control lines can be divided into handshaking and management lines. The three main types of *handshaking lines* are as follows:

- *Data Valid* (DAV). A logical 1 on this line indicates that data are available on the data lines.
- *Not Ready for Data* (NRFD). When all listeners are ready for data, NRFD will be high. When they read the data byte from the data bus, NRFD goes low.
- *Not Data Accepted* (NDAC). The devices receiving data use NDAC to indicate to the transmitting device that the data have been read. When all devices have read the data, NDAC goes low.

These are the primary *management lines:*

- *Attention* (ATN). This line is used by the device controlling the bus to signal that a command is being sent. A logic 1 on this line indicates that the bytes being transmitted by the controller are commands.
- *Interface Clear* (IFC). The IFC line is used by the bus controller to initialize all devices on the bus. A logic 1 (low voltage) causes all devices connected to the bus to be reset to their initial states.
- *Service Request* (SRQ). It is used by the devices on the bus to signal the controller that they require service. When the controller detects a logic 1 on SRQ, it rolls the devices on the bus to determine which is requesting service.
- *Remote Enable* (REN). This line is used by the controller to enable devices on the bus to receive commands or data from the GPIB.
- *End or Identify* (EOI). The EOI has two functions. Any device transmitting data on the bus may assert the EOI (pull it to a low voltage) to signal the end of data transmission. The EOI is also used by the controller to initiate a parallel poll.

Program 3: **The M68HC11 MCU Reads Data from a GPIB Bus, Using a Handshake Procedure**

The MCU communicates with the bus through three lines: DAV, NFRD, and NDAC. The function of these lines have already been explained.

The GPIB bus uses a negative logic. The negative logic means that a logic 0 is high voltage and a logic 1 is low voltage. Some manufacturers use the positive logic because it is easier to use. We will also use the positive logic. These pins become $\overline{\text{DAV}}$, RFD, and DAC.

You recall from earlier chapters that our MCU can perform either a pulsed or an interlocked full I/O handshake. We choose the input handshake with the peripheral interlocked mode, as shown in Figure 9.26. In this mode, port C is used to read data from the GPIB bus. $\overline{\text{DAV}}$ is connected to the STRA and IRQ lines. DAC is connected to the STRB line. PD0 is connected to RFD. $\overline{\text{IRQ}}$ is configured as low edge-sensitive.

The protocol between the MCU and the GPIB bus follows these steps:

1. *Listener.* Asserts RFD high initially.
2. *Talker.* Sees RFD high, places the data on the bus, then asserts the $\overline{\text{DAV}}$ line low to notify the listener(s) that data are available. As Figure 9.26 shows, the $\overline{\text{DAV}}$ falling edge causes an interrupt on IRQ. This also will assert the RFD line low.
3. *Listener.* Reads the data and asserts DAC high to tell the talker that the data have been accepted.
4. *Talker.* When it detects that the DAC line is high, it deasserts the $\overline{\text{DAV}}$ line high. The MCU is programmed so that when it detects a rising edge on the STRA line, it deasserts DAC low.
5. *Listener.* Processes the data, then it asserts the RFD high again to be ready for a new handshake cycle. Here is the program.

```
; Equates are the user's responsibility

        ORG $C000

        LDX  #BASE
        LDAA #$20
```

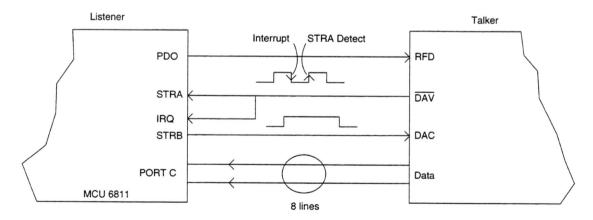

Figure 9.26

```
                STAA  OPTION,X    ;IRQ is falling edge
                LDY   #BUFF       ;data table
                LDAA  #0
                STAA  DDRC,X      ;port c is input
                LDAA  PIOC,X      ;clear staf flg
                LDAA  #$13
                STAA  PIOC,X      ;rising on stra,high on strb,
                                  ;handshake mode
                CLI
                BSET  DDRD,X,$0   ;pd0 is output
                BSET  PORTD,X,$1  ;pd0 is high initially
WAIT            WAI               ;wait for falling edge

                LDAA  PORTC,X     ;DAC is asserted high,read data
                STAA  0,Y         ;store in BUFF table
                INY
LOOP            BRCLR PIOC,X,$80,LOOP ;wait here till DAV
                                      ;asserts high,DAC
                                      ;deasserts low
                BSET  PORTD,X,$0  ;pd0 is high again
                BRA   WAIT

            ;INTERRUPT SERVICE ROUTINE

GPISR           BCLR  PORTD,X,$0
                RTI

                     ORG $FFF2

                FDB  GPISR
```

Exercise

Compare the RS-232C interface and the IEEE-488 GPIB bus with respect to the following:

1. The purpose of each standard
2. Maximum of data rate
3. Number of interfaces attached to the interconnecting lines
4. Types of information transmitted over the data lines
5. Total number of lines in each standard

9.10   VME Bus

The VME bus (or VMEbus) is a bus structure designed originally to allow the interconnection of the various printed-circuitboard modules that made up the Motorola's lines of VERSA modules. The VERSA modules are products designed to take full advantage of the capabilities of the 16- and 32-bit MC68000 family. A more compact version of the boards called VERSA module Euro cards was developed in Motorola's Munich facility and became a standard in Europe. In 1981, Motorola announced the VME bus as a nonproprietary standard in the public domain. What does *VME* stand for? *VM* stands for VERSA Module and *E* stands for Europe or Eurocard.

The VME bus system is designed to do the following:

1. Allow communication between devices without disturbing the internal activities of the other devices interfaced to the bus.
2. Specify protocols that define the interaction between devices.
3. Specify the electrical and mechanical system characteristics required to design devices that will reliably communicate with other devices interfaced to the VME bus.
4. Allow a broad range of design latitudes so that the designer can optimize cost and performance without affecting system compatibility.

Mechanical Structure

The mechanical structure consists of subracks, backplanes, and a plug-in board. Figure 9.27 shows a view of a 19-inch-wide subrack. This subrack includes I/O modules, the VMEbus module, the MVME820 mass storage module, chassis, backplane, and power supply.

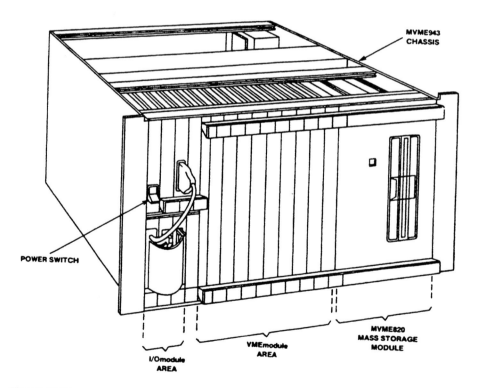

Figure 9.27

Backplanes may have up to 21 slots. Each backplane includes several 96-pin connectors and signal paths that bus their pins.

All VMEbus systems have a J1 backplane that provides all the signal paths needed for 24-bit addressing and 16-bit data. In systems that utilize a 32-bit address and data path, a J1/J2 backplane is provided.

The VMEbus boards could be either single or double height. A single-height board is 100 mm (high) × 160 mm (deep). A double-height board is 233.35 mm (high) × 160 mm (deep).

9.11 Communication Between MCUs

There are three possible communication configurations between several MCUs using SPI.

1. Hardware SPI at both master and slave
2. Software SPI at the master and hardware at the slave
3. Hardware SPI at the master and software at the slave

In the first alternative, data transfer can be handled full duplex. The configuration could be a single-master or a multiple-master system. For a single master, both master and slave write the data to be transferred to their respective serial peripheral data register. A data transfer is initiated when the master writes to its serial peripheral data register. A slave device can shift data at a maximum rate equals to the CPU clock.

In a multiple-master system, the master must pull the slave's $\overline{SS}$ line low prior to writing to its serial peripheral data register and initiating the transfer. (See Figure 9.28.)

In the second possibility, the system requires only two lines between the MCUs: data and clock lines. An SS line can be added in case of multiple slaves. The master provides the serial clock by toggling a different port pin.

In the third alternative, a situation similar to the one described in the preceding paragraph will occur. The following program illustrates this alternative.

Program 4: **Data Transfer Between Two MCUs**

In this program, the two MCUs communicating with each other are M68HC11 and M6805L3. Data are transferred between them on a single bidirectional line (half-duplex) with the clock supplied on an additional line.

A software handshake sequence is implemented on the same lines that provide the clock and data. This has the considerable advantage of minimizing the number of lines between the MCUs. The M68HC11 is the master and the M6805L3 is the slave. (See Figure 9.29.)

The M68HC11 has a hardware SPI while M6805L3 does not. The slave can use a software SPI. On the 68HC11, the MOSI and MISO pins must be connected, as shown in Figure 9.29. On the 6805L3, PD2 and PD3 are also connected. These lines are the software MOSI and MISO lines respectively. The slave select pins SS are configured output to prevent SPI faults and are not used by either chip.

MCU Protocol

Communication between two MCUs occurs through the handshake sequence. During the data transfer and during the handshake sequence, both SPIs (hardware and software) must operate in Wired-OR. The handshake sequence is shown in Figure 9.30a.

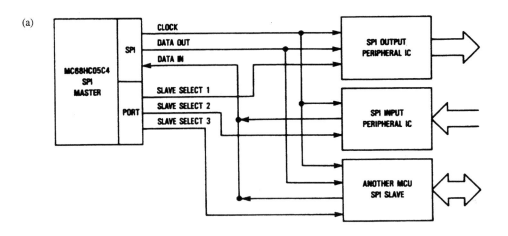

(a)

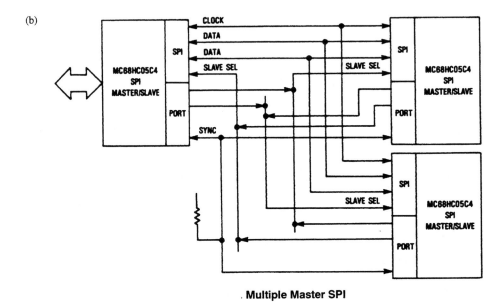

. **Multiple Master SPI**

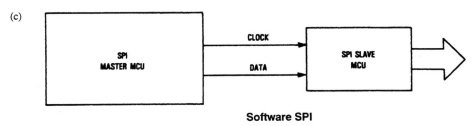

Software SPI

Figure 9.28
Copyright of Motorola. Used by permission.

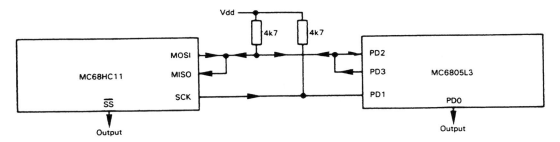

Figure 9.29
Copyright of Motorola. Used by permission.

The clock format is as follows: idle low, data output on positive edge and sampled on the negative edge. Both SPIs must be configured to operate with the same clock format.

The MSB of data is transferred first. Data are latched into both SPI shift registers on the falling edge of the clock. Once the last data bit is latched, the data line is released high. The protocol sequence is as follows:

1. The master releases the clock line by disabling the SPI.

2. The slave clears the data line by forcing the output clamp on PD3.

3. The master clears the clock by enabling the SPI.

4. When the clock goes low, the slave stores the data in the SPI data register, sets its data DDR to the correct state and enables SPI.

5. The slave releases data clamp.

6. The master starts the SPI clock by storing data in the SPI register.

7. Both master and slave detect end of transmission and read data from the SPI register.

8. The slave disables the SPI to ensure the data line is released.

The transfer protocol operates such that after initialization, the master MCU (68HC11) transmits a control byte to the slave MCU (6805L3). The control byte selects the subsequent data transfer direction and is either a slave listen address or a slave talk address.

If it is a slave listen address, the 6805L3 stays in receive mode. The next byte indicates the total number of bytes to be received, followed by the data stream, as Figure 9.30b shows. Once the last byte is transferred, the slave may wait for a new control byte or process a certain task. By the same token, the master may send a new control byte or return to another task.

If the control byte sent is a slave talk address, the slave will switch to transmit mode and send a byte count followed by the data stream to the master, as Figure 9.30c indicates.

Here is the program.

```
;M68HC11 Routine
;Equates are the user's responsibility
```

(a)
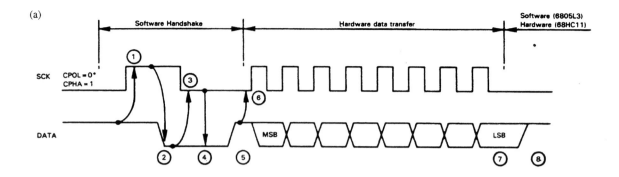

(b)
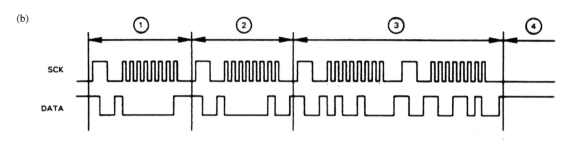

(1)   Master sends 'slave listen address' = 1

(2)   Master sends byte count   = 2

(3)   Master sends 2 bytes of data

(4)   SPI goes idle

(c)
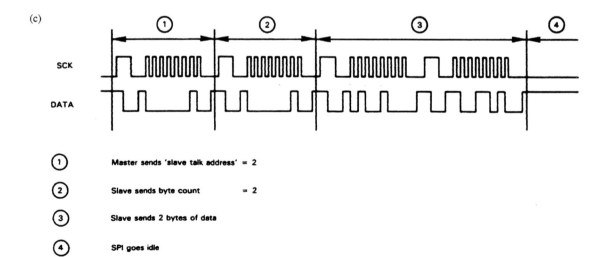

(1)   Master sends 'slave talk address' = 2

(2)   Slave sends byte count   = 2

(3)   Slave sends 2 bytes of data

(4)   SPI goes idle

Figure 9.30
Copyright of Motorola. Used by permission.

```
MSG FCB    24          ;number of bytes in data buffers

                       ORG $C000

               ;SPI initialization

               LDX  #BASE
               LDAA #$38
               STAA DDRD,X       ;SS,SCk,MOSI output,MISO input
               LDAA #$76         ;enable SPI,Wired-OR
               STAA SPCR,X
               LDAA #$18
               STAA PORTD,X      ;set MOSI and clock high first

MASTER         LDY  #MSG
               BSR  SND_MSTR     ;transmit data to slave
               LDY  #MSG
               BSR  RD_MSTR      ;read data back
               BRA  MASTER

SND_MSTR       LDAA #1           ;1st byte is a slave listen
                                 ;address
               BSR  XFR_MSTR     ;send to slave
               LDAA 0,Y
               TAB               ;ACCB holds the byte count
               BSR  XFR_MSTR
NXT_MSTR       INY               ;next byte
               LDAA 0,Y
               BSR  XFR_MSTR
               DECB              ;all bytes sent yet?
               BNE  NXT_MSTR     ;no,continue
               RTS

RD_MSTR        LDAA #2           ;1st byte is a slave talk
                                 ;address
               BSR  XFR_MSTR     ;send to slave
               LDAA #$FF
               BSR  XFR_MSTR
               TAB               ;ACCB holds the byte count
               STAA 0,Y
NXT_MST1       INY               ;next byte
               LDAA #$FF
               BSR  XFR_MSTR
               STAA 0,Y
               DECB              ;all bytes received yet?
               BNE  NXT_MST1     ;no,continue
               RTS

XFR_MSTR       BRCLR PORTD,X,#MISO,XFR_MSTR ;wait for slave to
                                            ;release data line
               BCLR SPCR,X,#SPE          ;SPI is disabled
```

```
          BRSET PORTD,X,#MISO,*   ;wait for slave to
                                  ;acknowledge
          BSET SPCR,X,#SPE ;enable SPI
          BRCLR PORTD,X,#MISO,*
          STAA SPDR,X             ;master starts transfer
          BRCLR SPSR,X,#SPIE,*    ;wait for Tx to
                                  ;complete

          LDAA SPDR,X
          RTS


          ;M6805L3 Routine
          ;EQUATES follow
```

```
PORTA  EQU $0
PORTD  EQU $3
DDRA   EQU $4
DDRD   EQU $7
TADAT  EQU $8
TACR   EQU $9
MISC   EQU $A
TBDAT  EQU $C
TBCR   EQU $D
SPIDAT EQU $E
SPICR  EQU $F
PRESCL EQU $10

SS     EQU $0
SCK    EQU $1
SDA    EQU $2
CLAMP  EQU $3
SPE    EQU $4
SPIF   EQU $7
```

```
              ORG $20

BYTCNT RMB 1
DATLEN RMB 1             ;data length
DATA   RMB 40           ;reserve for 40 data bytes

              ORG $80

          ;SPI initialization

          LDA #$48
          STA MISC       ;inhibit INT2
          LDA #$D
          STA PORTD      ;set data,SS,CLAMP
          LDA #9
          STA DDRD       ;clock slave,D2input,CLAMP,SS
                         ;output
          LDA #$44
```

```
                      STA SPICR       ;SPI disabled,sample on negative
                                      ;clock
                      BRSET SCK,PORTD,*;wait for master to gain
                                      ;control
                      CLRA
SLAVE                 LDX  #1          ;1st receive command byte
                      BSR  XFR_SLAV
                      CMP  #1          ;if slave listen mode,read data
                      BEQ  RD_SLAV

                      CMP  #2          ;if slave talk mode,send data
                      BEQ  SND_SLAV
                      BRA  SLAVE

RD_SLAV               LDX  #1
                      BSR  XFR_SLAV
                      STA  BYTCNT
                      STA  DATLEN      ;store data length
NXT_SLAV              LDX  #1
                      BSR  XFR_SLAV
                      LDX  BYTCNT
                      STA  DATA,X      ;store received data in
                                      ;buffer
                      DEC  BYTCNT      ;all done yet?
                      BNE  NXT_SLAV         ;no,continue
                      BRA  SLAVE

SND_SLAV              LDA  DATLEN
                      STA  BYTCNT      ;store length in counter
                      LDX  #5
                      BSR  XFR_SLAV    ;send it to master
NXT_SLAV              LDX  BYTCNT      ;get next byte
                      LDA  DATA,X
                      LDX #5
                      BSR  XFR_SLAV
                      DEC  BYTCNT
                      BNE  NXT_SLAV
                      BRA  SLAVE

XFR_SLAV              BRCLR SCK,PORTD,*;wait for master to
                                      ;acknowledge on clock line
                      BCLR CLAMP,PORTD  ;send acknowledge to master
                      BRSET SCK,PORTD,* ;wait for master to enable
                                      ;its SPI
                      STA  SPIDAT
                      BSET SPE,SPICR
                      STX  DDRD
                      LDA  #$44        ;disable SPI
                      LDX  #$D
                      BRCLR SPIF,SPICR,* ;wait for data to arrive
                      STA  SPICR            ;disable SPI
```

```
LDA   SPIDAT          ;read data
STX   PORTD           ;force data pin
                      ;high,release clamp
STX   DDRD
RTS
END
```

Exercises

1. Develop a protocol to transfer data between two M68HC11 microcontrollers using the full handshaking I/O. Write the steps of the handshaking events, then develop an assembly language routine to accomplish that.

2. Repeat Exercise 1 to transfer the data using the SCI. You can select either polling or interrupt for all operations.

10 Programming Microcontrollers in C

This chapter is not intended to be a complete C programming text. You should be familiar with the basic concepts of C programming, besides having gained some experience in assembly language from this book. This chapter will demonstrate how to use C in real-time programming with microcontrollers—specifically the M68HC11.

10.1   Introductory Remarks

The high-level language C is gradually replacing assembly language because it has several advantages over assembly language, as noted in Chapter 3. The most important is that the programmer does not have to know the hardware or the programming model of a micro-controller. This makes the language universal and portable between different machines. For example, a C program written for the MCU M6805 can be run on the MCU M6811 with little modification using different compilers and header files. For the M6805, a 6805 compiler is used, and for the M6811, a 6811 compiler is employed, and so on.

C programs are divided into functions. A function in C is similar to a subroutine in assembly language. The *main( )* function is the one to which control is passed when the program is executed, and it is the first function to call.

#include Directive and Header Files

The stdio.h file stands for *standard input/output header.* The header file is a collection of information that goes at the top of a file with an extension ".h". When the C file is compiled, the compiler "sees" the text of that header file as if it were part of the source file.

The effect of the #include <stdio.h> or the #include <any header file> is the same as if we had typed the entire contents of this file into our source file. This is a smart way to save a lot of typing time.

#define Directive

The #define directive has two primary uses. The first is to define symbolic contents. The second is to create macros.

10.2   C Program Structure and Data Types

A typical C program is as follows:

```
#include < >
main()
  statements
function a()
  statements
function b()
  statements
```

The types of statements could be declaration, assignment, function, control, and/or null.

Table 10.1 summarizes the data type variables and the number of memory bits used to store each type.

Table 10.1

Type	Subtype	Size
char	unsigned char	8
	char	8
int	unsigned short	16
	short	16
	unsigned int	16
	int	16
	unsigned long	32
	long	32
float	float	32
	double	64
	long double	80
ptr	near	16
	far	32

10.3   Arithmetic Operators

The following operators are used in arithmetic operations:

Operation	Symbol
addition	+
subtraction	−
multiplication	*
division	/
modulus	%
increment	++
decrement	−−

Bitwise Operators

The following operators correspond to assembly language instructions such as AND, OR, XOR, NOT, ROL, and ROR. They perform specific operation on a bit-by-bit basis. They are used most frequently in embedded systems control.

Operation	Symbol
AND	&
OR	|
XOR	^
NOT	~
shift right	>>
shift left	<<

Bitfields

The |= operator means left variable = left variable OR right expression. It is used in the BSET instruction. The &= operator means left variable = left variable AND right expression. It is used in the BCLR instruction.

Assignment Operator

The assignment operator in C is the "=" sign. This sign means to evaluate the expression to the right of this sign and write the result to the variable to the left of the sign. For example, count = 10 means that the value 10 is assigned to "count".

Relational Operators

Relational operators are used in expressions to compare the values of two operands. If the result of the comparison is true, then the value of the expression is 1. If the result is false, then the value of the expression is 0. Here are the relational operators:

Operator	Symbol
equal to	==
not equal to	!=
greater than	>
greater than or equal to	>=
less than	<
less than or equal to	<=

Logical Operators

The C logical operators allow you to include two or more conditions in a decision.

Operator	Symbol
AND	&&
OR	‖
NOT	!

Do not confuse the bitwise operators and the logical operators. The bitwise operators, as mentioned earlier, manipulate individual bit(s) of a variable, while the logical operators deal with certain conditions in an *if* statement. As an example:

```
if(condition1 && condition2)
    then (decision)
```

or

```
if(condition1 ‖ condition2)
    then (decision)
```

C Library Functions

The C packages come with a library containing a huge number of predefined built-in functions and macros. The library functions allow the user to perform I/O operations with a variety of devices, dynamically allocating memory in a program. Examples of these functions include printf(), scan(), getche(), and so on.

Prefixes

	Example
Decimal is assumed	245
Octal uses a leading 0	0245
Hexadecimal uses leading 0x	0x245
ASCII characters	"ABC"

10.4 Control Statements and Looping

for Loop

```
for(expression)
  body of the loop;
    |
```

Example:

```
main()
{
  int count;
  for(count=0;count<8;count++)
    printf("count=%d",count);
}
```

while Loop

```
while(expression)
  body of the loop;
    |
```

Example:

```
main()
{
  int count=0;
  int sum=0;
  while(count<9)
    {
      sum+=count;
      printf("count=%d,sum=%d",count++,sum);
    }
}
```

if/else Statement

```
if(expression)
   statement1;
else
   statement2;
```

Example:

```
main()
{
  int temp;
```

```
printf("please type in the temperature: ");
scanf("%d",&temp);
if(temp < 90)
    if(temp > 65)
        printf("nice day!");
    else
        printf("it is chilly!");
else
    printf("it is hot!");
}
```

Switch with Break

```
switch(n)
{
  case 1:
     statement 1;
     break;
  case 2:
     statement 2;
     break;
  case 3:
     statement 3;
     break;
   default:
     statement 4;
  }
```

10.5   Arrays, Pointers, and Structures

Arrays

An array is a collection of like types of data (called *elements*) stored in consecutive memory locations.

Example:

```
1. short array1[8];              /* one dimensional array with 8
                                    elements */
2. static short array2[2][3];    /* 2 dimensional array with 2 rows
                                 /* of 3 elements each */
```

Pointers

A pointer is a variable that points to another variable.

Example:

```
*px = 6;        /* px points to the address of 6 */
*py = 9;        /* py points to the address of 9 */
```

Structures

A structure is a collection of one or more variables identified by a single name.

Example:

```
struct person
{
  char name[15];
  int height;
  int weight;
  char address[20];
  char city[20];
  char state[2];
  char zip[5];
};
```

10.6   Header File for the MC68HC11E9

Appendix C contains the C header file hc11e9.h. This file must be included in any program used on the MCU M68HC11E9. The header file includes:

```
#ifndef HC11E9
#define HC11E9
```

The first instruction determines that if HC11E9 is not defined, it will be defined in the second line.

Next is a definition of a structure called *Register*. The structure contains 8 bits B0 to B7. The second part is the definition of *all of the I/O registers*. The global "Register_set" is given a value 0x1000 (hex). The various register addresses are each cast onto the proper type and then dereferenced to receive a value in the register definition macros. For example, #define DDRC (*(Register *)(Register_set + 7) casts the number Register_set+7 (0x1007) to a pointer type Register. Then the assignment DDR7 = 0f means that the value 0f is assigned to the address 0x1007. The last entry in the file is:

```
#endif
```

The user should put the following include statement at the beginning of each program module: #include <hc11e9.h>.

10.7   Example of Using C Compilers

The Whitesmiths M68HC11 Toolkit is a fully integrated software development system consisting of the following components:

- Optimizing ANSI C compiler
- Macro Assembler
- Run-time library
- Programming utilities
- CXDB source level debugged

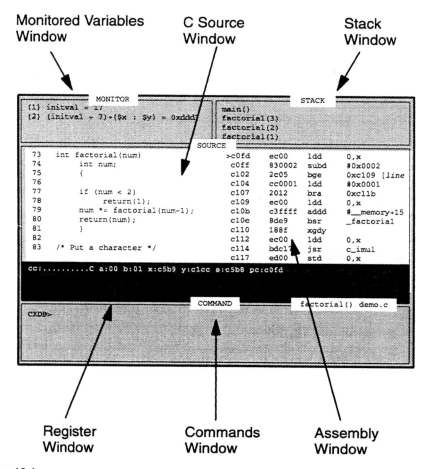

Figure 10.1

From *Motorola M68HC11 C Compiler and Source Level Debugger.* Courtesy of Intermetrics Microsystems Software, Inc.

The M68HC11 C compiler supports ANSI features important for embedded systems such as language extensions, function prototypes, "const" and "volatile" qualifiers, and "void" date types. The M68HC11 Assembler supports full modern macro features and listing generators.

The programming utilities include the linker, librarian, data initializer, file inspector, and hex file generators.

The CXDB is a true C source-level debugged, multiwindow interface that can display C and assembly source code, commands, M68H11 registers, monitored variables , and stack lends.

The CXDB command set allows you to set code and data breakpoints, and simulate I/0 channels, interrupts and timers, and CPU cycles. (See Figure 10.1.)

CXDB is available in three configurations: CXDBsim, CXDBice, and CXDBrom. The CXDBsim is a simulation/debugged. The CXDBice has the features of in-circuit emulators. The CXDBrom is used for debugging employing evaluation boards such as EVM and EVS.

10.8   ZAP V2.3a Simulator/Debugger

ZAP version 2.3a is a window driver simulator/debugged tool created by COSMIC Co. for the 68HC11. It has a toolbar and a pull-down menu for its features (Figure 10.2).

After compiling the C program, ZAP can display the equivalent assembly program. The following are the capabilities and features of this program:

1. You can compile C code from the utility menu.

2. You can load the C code only or the equivalent assembly language code only or both side by side as shown in Figure 10.3.

3. It has the capability of displaying all function names in the C code from the pop-up browse menu, as Figure 10.3 illustrates.

4. It has the capability of single-stepping each line in the C program and highlighting the equivalent instruction(s) in the assembly language program using the "debug" menu.

5. It has the capability of creating and browsing the breakpoints from the breakpoint menu.

6. ZAP can display the memory addresses and their contents and the 68HC11 registers, as shown in Figure 10.3.

7. It can display all the variables in the current function or in the current file from the view menu.

8. You can execute the program using the "Go," "Go till," or "Go from reset" commands from the debug menu.

9. You can stop program execution any time by clicking on the stop menu.

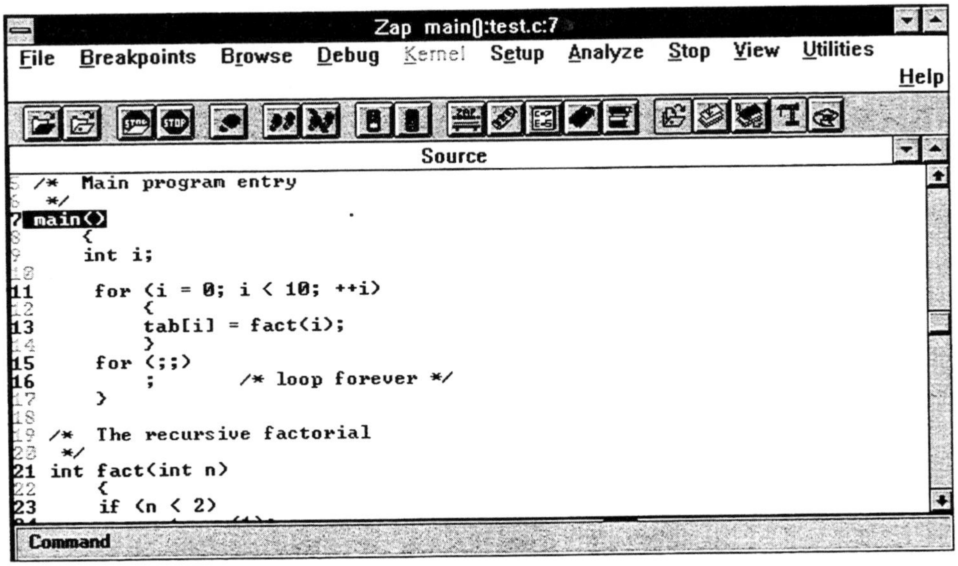

Figure 10.2

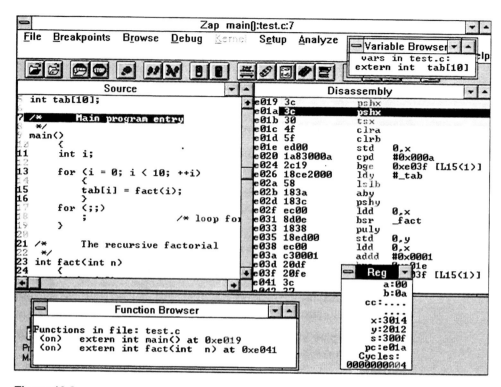

Figure 10.3

After compiling the C code (filename.c), two new files are generated: relocatable object (filename.o) and the absolute file (filename.h11). When we use ZAP, the absolute file (.h11) is loaded to the simulator/debugger and the C program is displayed.

Let's discuss a program to calculate a factorial of a number using recursion.

10.9 Recursion

A function that calls itself is said to be *recursive*. If a function f1 calls another function f2, which under certain circumstances calls f1, then f1 is called recursive. Recursion uses the stack to pass parameters. Both the calling function and the called function use a portion of the stack as a frame for accessing variables. A called function sets up its own frame by using the stack pointer (SP) to set up a frame pointer. It uses an index register (X reg, for example) as a frame pointer to address the variables stored in the frame. Recursion has advantages and disadvantages. The advantage is that it provides a simple solution to a problem that can be difficult to solve. The disadvantage is that it takes more time in execution due to the extra function calls.

Generally, a stack frame set up during a function call looks like the one shown in Figure 10.4.

The following is a program that uses a recursive function to calculate the factorial of numbers from 0 to 9. (Also see Figure 10.5.) The factorial of *n* is:

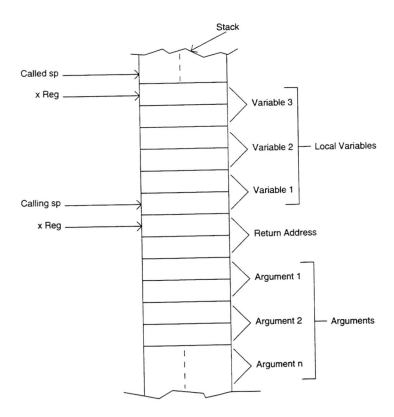

Figure 10.4

$$n! = n \times (n-1) \times (n-2) \times \ldots \times 2 \times 1$$

where
$$0! = 1$$
$$3! = 3 \times 2 \times 1 = 6$$

Here is the program:

```
/* Factorial program */

int tab [10];

/* Main program  entry */

main( )
      {
       int i;
       for (i =0; i<10; ++i)
            {
            tab [i] = fact (i);
            }
       for (;;)
            ;                /* loop forever */
```

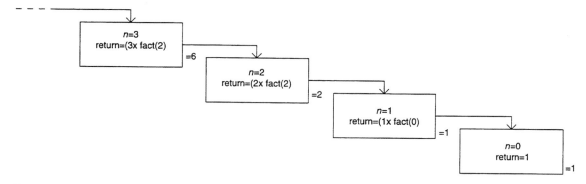

Figure 10.5

```
/* The recursive factorial */

int fact(int n)
        {
        if (n < 2)
                return (1);
        else
                return (n * fact(n - 1));
        }
```

From the program, if n is less than 2 (0 or 1), then the factorial is 1. If n is higher than 1, then the factorial of n is calculated as follows:

$$n! = n * (n–1)!$$

You conclude that the function "fact(n-1)" always calls itself over and over. That is the reason this function is recursive.

The resultant assembly code is the result of compiling the program by ZAP as follows:

```
e000        clra
            ldx     #_tab
            bra     0xe009
e006        staa    0,x
            inx
e009        cpx     #_memory
            bne     0xe006
            xgdx
            addd
            xgdx
            txs
            jsr     _main
            bra     _exit
            pshx
            pshx
            tsx
            clra
            clrb
```

```
        std   0,x
        cpd   #0x000a
        bge   0xe03f
        ldy   #_tab
        lslb
        aby
        pshy
        ldd   0,x
        bsr   _fact
        puly
        std   0,y
        ldd   0,x
        addd  #0x0001
        bra   0xe01e
e03f    bra   0xe03f
        pshx
        pshb
        psha
        tsx
        ldd   0,x
        cpd   #0x0002
        bge   0xe052
        ldd   #0x0001
        bra   0xe05d
e052    addd  #0xffff
        bsr   _fact
        ldy   0,x
        jsr   c_imul
e05d    pulx
        pulx
        rts
```

Understanding the idea behind using the stack to store function variables as explained above, you can walk through the preceding assembly code and try to grasp how it works.

If we single-step through the C code, starting at the main program entry as shown in Figure 10.6, the assembly equivalent start-up instructions are:

```
    2 pshx's to create a local variable space
    frame pointer = X reg = SP + 1, as follows:


    pshx
    pshx
    tsx
```

This will push the contents of X reg into the stack, then transfer the stack pointer to X reg. At the next step shown in Figure 10.7, the equivalent instructions are:

```
    clra
    clrb
    std   0,x
```

This will clear the address pointed by X reg.

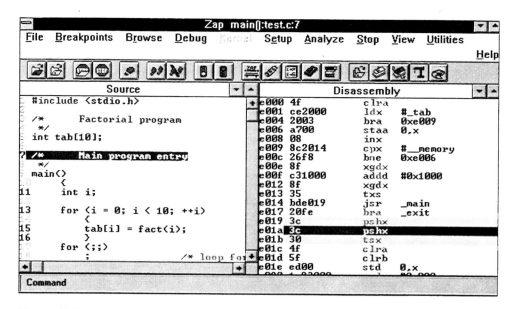

Figure 10.6

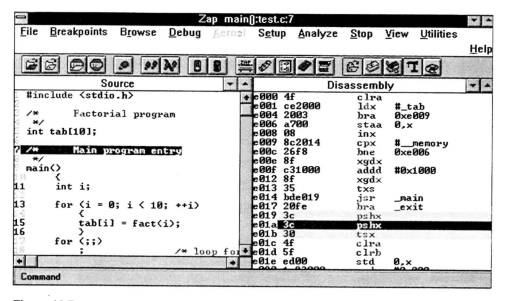

Figure 10.7

At the next single-step, $i = 0$ and the program will compare this value with 10. If it equals 10 or higher, execution will stop (branch to itself); otherwise it will proceed (Figure 10.8). The equivalent assembly instructions are:

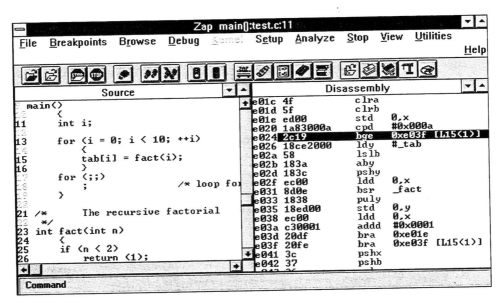

Figure 10.8

```
cpd    #0x000a
bge    0xe03f
```

The "0x" means that the number is a hex number.

In the next single-step, $i = 1$ and the assembly lines are as shown in Figure 10.9.

```
ldy    #_tab
lslb
aby
pshy
ldd    0,x
bsr    _fact
puly
std    0,y
```

And the program execution continues.

Program 1: Motor Start/Stop

This program shows how to start/stop a DC motor. The start push button PB is connected to PC0 while the stop is connected to PC1. The motor is interfaced to a driver connected to PB0, as Figure 10.10 indicates. (Example taken from *Programming Microcontrollers in C*, Ted Van Sickle, copyright 1994, HighText Publications, Inc. Reprinted with permission.)

```
#include <hc11e9.h>
#define TRUE 1
#define ON 1
main()
{
```

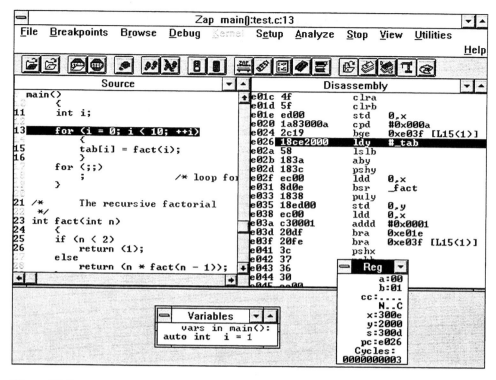

Figure 10.9

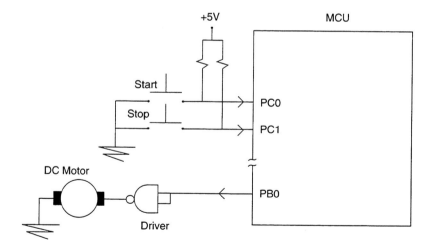

Figure 10.10

```
PORTB.PB0 &=~on;                    /* turn off motor initially */
DDRC = 0;                           /* port C is input */
while(true)                         /* repeat */
   {
```

```
        if(PORTC.PC0 == 0)              /* is start PB pressed */
            PORTB.PB0 |=on;             /* yes,turn on motor */
        else if(PORTC.PC1 ==0)          /* is stop PB pressed */
            PORTB.PB0 &=~on;            /* yes,turn off motor */
    }
}
```

Exercise

Assume that there are two stations apart from each other to control the above motor. Each station has start and stop pushbuttons. The start pushbuttons are usually connected in parallel, while the stop pushbuttons are connected in series. Write C code to control the motor from either station.

Program 2: Generating PWM Square Wave Polling

In this program, we want to generate a PWM square wave on the OC2 pin. This square wave has a duty cycle (DC) of 80%. The wave period value is stored as a variable called *period* and the positive portion is stored in another variable called *on_time,* as shown in Figure 10.11. OC1 is used to control OC2, as explained in Chapter 5. (Example taken from *Programming Microcontrollers in C,* Ted Van Sickle, copyright 1994, HighText Publications, Inc. Reprinted with permission.)

```c
#include <hc11e9.h>

period = 0x1000;
on_time = 0x800;

main()
{
  OC1M.OC1M7 = 1;                /* oc1 is affected */
  OC1M.OC1M6 = 1;                /* oc2 is affected */
  OC1D.OC1D6 = 1;                /* oc2 is high when oc1 occurs */
  TCTL1.OL2 = 1;                 /* toggle oc2 */
  PACTL.DDRA7 = 1;               /* oc1 is an output */
  TOC1 = TCNT + period;          /* set oc1 to period */
  TOC2 = TOC1 + on_time;         /* set oc2 to on_time */
  FOREVER
  {
    if(TFLG1&OC1F)               /* if oc1f is set */
    {
      TFLG1 = OC1F;              /* reset oc1f */
```

Figure 10.11

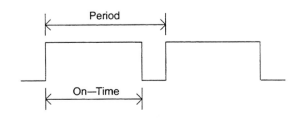

```
        TOC1+ = period;
    if(OC1D.OC1D7 == 1)
        OC1D.OC1D7 = 0;            /* oc1 toggles */
    else
        OC1D.OC1D7 = 1;
    }
    if(TFLG1&OC2F)                 /* if oc2f is set */
    {
      TFLG1 = OC2F;                /* reset oc2f */
      TOC2 = TOC1 + on_time;
    }
  }
}
```

Program 3:   Generating PWM Square Wave Using Interrupt

This program is similar to the previous program except that it uses an interrupt service routine (ISROC2) to generate the square wave. The ISROC2 is set up by using the @port construct. (Example taken from *Programming Microcontrollers in C,* Ted Van Sickle, copyright 1994, HighText Publications, Inc. Reprinted with permission.)

```
#include <hc11e9.h>

period = 0x1000;
on_time = 0x800;

@port void ISROC2(void);            /* prototype */

main()
{
  OC1M.OC1M7 = 1;                   /* oc1 is affected */
  OC1M.OC1M6 = 1;                   /* oc2 is affected */
  TMSK1.OC2I = 1;                   /* enable oc2 interrupt */
  OC1D.OC1D6 = 1;                   /* oc2 is high when oc1 occurs  */
  TCTL1.OL2 = 1;                    /* toggle oc2 */
  PACTL.DDRA7 = 1;                  /* oc1 is an output */
  TOC1 = TCNT + period;             /* set oc1 to period */
  TOC2 = TOC1 + on_time;            /* set oc2 to on_time */
  cli();                            /* enable system interrupt */
  FOREVER
{                                   /* wait for interrupt */
  }
}
@port void ISROC2(void)             /* interrupt service routine */
{
    TFLG1 = OC1F;                   /* reset oc1f */
    TOC1+ = period;
    if(OC1D.OC1D7 == 1)
      OC1D.OC1D7 = 0;               /* oc1 toggles */
    else
```

```
        OC1D.OC1D7 = 1;
    TFLG1 = OC2F;                    /* reset oc2f */
    TOC2 = TOC1 + on_time;
}
```

Program 4:   An SCI Application

This routine will use the SCI system to read the required motor speed and echo the same
data back to the originating source to provide full-duplex operation.

The number is any character that lies between 0 and 9. It is converted to a number
when the character 0 is subtracted from it. The value is added to 10 times the value stored
in new_speed, and the result is saved back to this location. The baud rate is chosen as
9600. Refer to Chapter 5 for more detail on SCI. (Example taken from *Programming
Microcontrollers in C,* Ted Van Sickle, copyright 1994, HighText Publications, Inc.
Reprinted with permission.)

```
#include <hc11e9.h>

int new_character;

WORD new_speed;
main()
{
                          /* init routine */
  BAUD.SCP1 = 1;
  BAUD.SCP0 = 0;
  BAUD.SCR2 = 0;
  BAUD.SDR2 = 0;
  BAUD.SCR0 = 0;              /* baud rate is 9600 */
  SCCR2.TE = 1;              /* trans is enabled */
  SCCR2.RE = 1;              /* recv is enabled */

  if(SCSR.RDRF==1)           /* read data if flag is set */
   {
      new_character=SCDR;      /* store data and reset flag*/
      while(SCSR.TDRE==0);     /* wait till trans buff is empty */
      SCDR = new_character;    /* trans data and reset flag */
      if(new_character>='0' && new_character <='9')
        new_speed = 10*new_speed + new_character - '0';
      else
         if(new_character=='\r')
       {
         if(new_speed>=1000 && new_speed<=10000)
            motor_speed=new_speed;
         new_speed=0;
       }
       else
          new_speed=0;
   }
```

Repeated execution of this code sequence will convert a series of ASCII character numbers to an appropriate unsigned integer value. If the character is a carriage return, the new_speed value is put into motor_speed to change the motor speed and the new_speed is then set to zero. If any other character is received, the data saved in new_speed is lost and the whole entry of a new speed into the system must be repeated from the beginning.

Exercises

1. Write C code to measure a period length of a square wave using polling.
2. Repeat Exercise 1 using interrupts.

11 Development and Debugging Tools

Software development can start as soon as the development of the hardware prototype is completed. However, some software routines can be written and tested even before the system hardware is ready. Examples of such routines would be mathematical routines (addition, multiplication, and so on) or keyscan routines if the keypad type is known. Other examples would include display routines if the display type and its driver are known in advance. These routines could be written in advance and when the hardware is ready, they could be tested. Some minor changes may be needed.

Writing parts of the program in advance save a great deal of time, and the programmer does not have to wait until the hardware is completed. As portions of the target-system hardware become ready, the software that interacts with each portion can be developed and tested. Also, we need to check the worst-case timing considerations for I/O devices when they request services through interrupts. We have to determine the following for each interrupt service routine:

1. How long it takes to execute
2. The maximum rate at which it will request service
3. How long it will leave interrupts disabled for higher-priority devices

By having all this information, we can determine whether a worst-case sequence of interrupts will result in an interrupting device not being served within an acceptable time.

As each software task is completed, its performance in conjunction with the prototype target-system hardware can be checked. In this way, the software can be built up and tested task by task until every task has been completed.

In "task-by-task" software development, we still need some kind of "development tools" to assist us in writing the code and debugging it. Most of the time, when we test our task using these tools, we find out that a lot of things go wrong or behave differently from what we had expected.

As mentioned in Chapter 3, assemblers, compilers, and linkers will generate syntax errors such as undefined variables, constants, or multiple declarations. After these errors are corrected, there is no guarantee that the code will run successfully. This means that the program instructions are executed with no problems but the program task is not achieved yet.

11.1   Types of Development and Debugging Tools

System developing and debugging verifies that the program performs correctly and the hardware is under the control of the software. The popular development and debugging tools are divided into three categories:

1. Emulators
2. Simulators
3. Logic Analyzers

Emulators

Emulators are the most popular developing and debugging tool on the market these days. Many manufacturers can provide software developers with emulators; some of the most popular emulator manufacturers include Motorola, Nohau, and others for Motorola microcontroller products. These emulators can provide us with the following:

1. RAM memory for the microcontrollers to use as data and program memory
2. Communications between the monitor program of the MCU and a personal computer such as an IBM PC
3. Ability to download the object code from the PC to the emulator's RAM
4. Displaying and changing of RAM variables and CPU registers, timer registers, pointers, condition code registers, and I/O ports
5. Ability to assign break points at various points in the program to execute it in parts for the purpose of debugging (some emulators even have their own "in-line assembler")
6. Ability to step the code (single-step) one instruction at a time and observe the contents of memory and registers

Evaluation Support

The programmer writes the code for the target in the host computer, then the code is assembled, compiled, and linked in the same host computer. Next the code is downloaded to the development board through either a serial or a parallel link.

The development board has the microcontroller that is to be emulated on board. This MCU is operating in a nonuser mode that allows internal bus access. A second computer on the development board controls the operation of the MCU. Code delivered from the host is put into memory accessed by the MCU, and the MCU can operate as if the code was contained within the internal memory.

All the I/O lines associated with the MCU are brought to a header on the development board, and a cable can be attached from this header to a plug-in device that plugs into a target board (Figure 11.1). This target system then operates as if it had a programmed MCU plugged into its socket. The MCU on the evaluation board has a complete monitor system in its firmware.

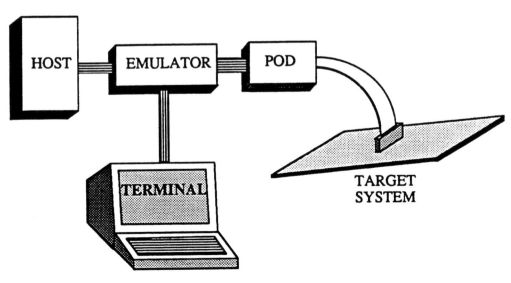

Figure 11.1
Copyright of Motorola. Used by permission.

MC68HC11EVB (Evaluation Board)

The EVB is a low-cost tool for debugging and evaluating the MC68HC11 system. The block diagram is shown in Figure 11.2. Its features are:

1. MC68HC11XX MCU
2. 8K of EPROM
3. 8K of RAM
4. MC68HC24 Port Replacement Unit (PRU)
5. Serial and parallel ports
6. Download capability from the host computer
7. Single line assembler/disassembler

The EVB uses a monitor/debugger program called *Bit User Fast Friendly Aid to Logical Operations (BUFFALO)*. It contains commands that allow the user to write programs, check the contents of memory, registers, and I/O ports and modify them as needed.

The EVB has two modes of operation: debugging and evaluation (emulation).

Debugging Mode. This mode allows the user to debug the code under the control of BUFFALO in two ways:

1. The code could be assembled on the EVB using the line assembler in the BUFFALO monitor.
2. The code could be assembled on the host computer (PC in our case) and then downloaded to the EVB in the form of an S-record. This method is more popular and can be used with

EVB BLOCK DIAGRAM

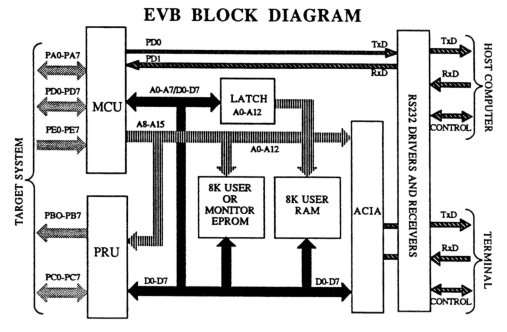

Figure 11.2
Copyright of Motorola. Used by permission.

any other debugging tool (such as EVM or CDS8). The communication between the PC and the EVB can be done by either Kermit or the EVM program. Kermit is a public domain terminal emulator and file transfer program supporting a wide range of computer systems. The EVM program is written by the P&E Company for Motorola products.

Evaluation Mode. This mode allows the user to evaluate the code in a target-system environment utilizing the memory of the MCU 68HC11.

Many BUFFALO commands are used to display and modify memory contents and CPU registers. These commands can also single-step through the program line by line, or assign breakpoints at different points in the program to execute a certain part of the code, or execute the entire program at once. They can change the disassembly lines, too. The EVB and its BUFFALO commands are explained in detail in the lab manual.

MC68HC11EVM (Evaluation Module)

The block diagram is shown in Figure 11.3. The features of this module are:

1. It emulates both single-chip and expanded-multiplexed modes for 48-pin and 52-pin packages.
2. A dual memory map scheme is used to allow unrestricted user access to the full 64K memory map.

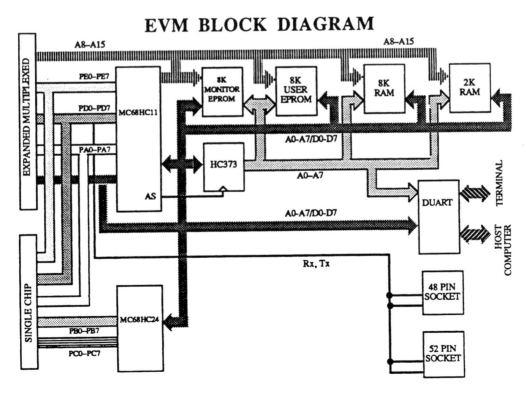

Figure 11.3
Copyright of Motorola. Used by permission.

3. Download capability is provided for the host computer and for user-supplied 8K EPROM.

4. There is target MC68HC11 EEPROM programming capability.

The EVM uses a monitor/debugger program called *EVMBUG,* which contains many commands to allow the user to write the code and check and modify the contents of memory, the CPU registers, and the I/O ports.

CDS8 Emulator

The Compact Development System for an 8-bit MCU (CDS8) is an emulator, bus analyzer, and control station for debugging hardware and software during development of the MC68HC11 family-based system. EPROM and EEPROM versions of the MC68HC11 family and conventional EPROMS can be programmed with the CDS8. The main boards in this emulator are a station module, which contains an emulator board and bus analyzer board; EPROM programmer board; and EPROM adapter board.

Advantages. The CDS8 is considered the most sophisticated emulator/analyzer unit for MC68HC11 development, for the following reasons:

1. It is window driven.

2. You can replace any microcontroller family member following certain procedures.

3. The most important capability of the CDS8 is a means of analyzing program execution activities on the target MCU bus. This ability allows the user to determine what is going on in a system without actually disturbing it.

4. On the CDS8 screen, there are three lines of concern: command line, status line, and error line.

5. The CDS8 has the other capabilities of other emulators, such as assembling/disassembling, breakpoints, single-stepping, program execution, memory display, memory modify, register display, register modify, and so on.

CDS8 Bus Analyzer. The various modes of the bus analyzer make it possible to choose the actions to take when a certain event(s) appears on the bus. Triggering of the bus analyzer is specified by defining the desired bus state(s) as terms and selecting the desired sequences of terms as a trigger event.

When the bus analyzer is "armed" and the emulator is running, a frame of the selected type is strobed into the trace buffer at each clock (Figure 11.4).

Simulators

Simulators are programs that run on the IBM PC and simulate the operation of the microcontroller. No target system hardware exists as in emulators.

Simulators can provide us with the following:

1. Simulation of timer registers, I/O ports, PC counter, pointers, and accumulators

2. Simulation of memory variable display and capability of changing these variables

3. Simulation of downloading from the PC to the target system

4. Simulation of single-stepping through the program

5. Simulation of assigning breakpoints

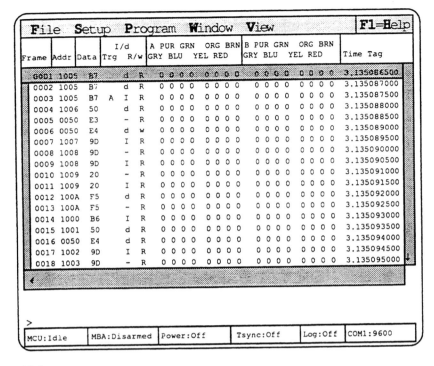

Figure 11.4
Copyright of Motorola. Used by permission.

Disadvantages of Simulators

1. They do not run in real time. The program execution time does not depend on the microcontroller system clock but on the PC system clock.
2. We cannot check hardware performance with the software developed through simulators because no hardware system is involved at all. This is a major disadvantage.

The best application for simulators, in our opinion, is developing the mathematical routines that do not require any hardware.

Logic Analyzers

The logic analyzer is used only in debugging stages. It is basically a RAM that is continually storing data from the microcontroller buses. The memory is First In First Out (FIFO) memory; as each new word is stored, an old word is discarded.

Data and control information are captured. Breakpoints are assigned to different addresses in the program, then the user runs it. At the breakpoints, the user can see the contents of memory. The user can also trace the problem before and after triggering.

Logic analyzers have many nice features, such as displaying the execution time or cycles of each instruction and the absolute or relative addresses. The data stored by the logic analyzer may be displayed in a variety of formats:

1. *Waveform display* makes the analyzer look like a multichannel oscilloscope.
2. *Binary display* represents both address and data in binary form.
3. *Hex display* shows both address and data in Hex, octal, ASCII, and decimal form.

Logic analyzers have some advantages—for example, there is no in-line-editor feature as in evaluation or emulation boards, and the user has no access to the CPU internal registers.

Host Computer Communications

Communication between emulators or simulators with the host computer is through an RS-232c serial link from the board. The object files can be downloaded using software tools such as PROCOMM or KERMIT created by Motorola, or using EVM software created by P&E Microcomputer Systems, Inc.

11.2   Debugging Techniques

Debugging tools employ some common techniques in debugging and testing:

1. Breakpoints, used by emulators, evaluation boards, and simulators
2. Single-stepping, used by emulators, evaluation boards, and simulators
3. Trace buffer, used by some emulators and logic analyzers

Breakpoints

Breakpoints are the most useful method in debugging. The user assigns breakpoints at different addresses in the program where code is to be tested. The breakpoint routines temporarily insert a software interrupt (SWI) instruction in place of a user's program instruction. In BUFFALO, up to four breakpoints can be stored at one time. The program is executed until the breakpoint is reached, then it stops. At this specific breakpoint, the user checks memory contents and CPU registers. Sometimes the breakpoint cannot be reached because the program flow bypasses this particular piece of code under certain conditions. The program also can crash for different reasons. In these cases, the user may employ the next technique, single-stepping.

Single-Stepping

Single-stepping (also called the *TRACE* or *T* command) is the execution of a piece of code, instruction by instruction. The user assigns a breakpoint before the location of the trouble, then executes the program from the beginning until the breakpoint. Next, the user begins single-stepping line by line. In each single-step, the data and the flow of the program are checked. The user continues doing this until the problem is located.

Note that single-stepping cannot be used in debugging time loops.

Trace Buffers

Some emulators and most logic analyzers can provide a real-time trace capability. A real-time trace buffer is a large FIFO memory that records bus activity while the target system is running at full speed. The trace captures and displays information, which enables the user to track down bugs without having to stop the program execution with a breakpoint.

Below is an example of a real-time trace using an emulator/logic analyzer. The first column shows the time or number of cycles before and after the trigger has occured. Negative numbers indicate cycles before the trigger, while positive numbers indicate cycles after the trigger. The second and third columns indicate the address and data of each instruction. The fourth column shows the disassembly or the code in mnemonic form.

Cycle #			Address	Data	Disassembly	
-9			C000	CE100	LDX	#BASE
-6	START		C003	86F0	LDAA	#$F0
-4	REPEAT		C005	A704	STAA	PORTB,X
→ TRIG			C00A	4C	INCA	
6			C00B	BDC015	JSR	DLY
8			C00E	81FA	CMPA	#$FA
11			C010	27F1	BEQ	START
14			C012	20F1	BRA	REPEAT

The trace shows the history of the instructions and the bus activities (read/write). When the bug appears, the recording of bus cycles (time) stops and we can examine the program flow before and after the bug has occurred.

Automatic Trace Comparisons

The intermittent bug is the nastiest bug we can find. To capture this bug, an automatic trace comparison feature offered by some emulators/analyzers can be used. The procedure looks like this:

1. Repeatedly capture trace buffers in a synchronized manner on the part of code where we think that a bug occurs.
2. Save these traces into a file.
3. Use the comparison feature to gather these traces.
4. Automatically compare these traces with the bug-free trace previously saved.
5. If there are differences, a file with these differences will be created. By examining this file, we can detemine where the problem is.

11.3   Testing Phase

We mentioned in Chapter 3 that testing and debugging is one of the three phases of software development. In this chapter, we will explain this phase.

Software Testing Techniques

There are three levels of testing: black-box, white-box, and integration tests.

Black-Box Testing

Testing of each module (or function) individually is called *unit-level testing*. Black-box testing is usually performed during later phases of the software project. The test enables the software engineer to derive sets of input conditions and verify the outputs. This technique is used to detect missing functions, initialization errors, and interface errors. The disadvantage of this test is that dead or unreachable code cannot be detected.

White-Box Testing (Unit-Level Testing)

This test is conducted to be sure that each module functions properly. Steps for proper unit-level testing are:

1. Understand the main purpose of the module (or function).
2. Recognize the I/Os that need to be tested.
3. Recognize the range, resolution, and units of the variables.
4. Understand the flow of the program from entry to exit points.
5. Write down the number of different independent control paths the program can take. This is called *path testing*.
6. Scrutinize all logical decisions on their true and false sides.
7. Test all loops at their boundaries and within their operational bonds.

White-box testing solves the problem of the black-box testing by testing all of the flow-of-control paths through the module (unit).

EXAMPLE: Path Testing

Assume a flowgraph of a module is as shown in Figure 11.5a. The paths of this figure depend on the value of two variables, A and B. The paths of the flowgraph can be expanded as shown in Figure 11.5b to display each path separately.

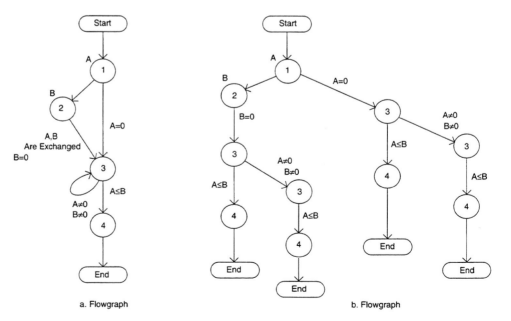

a. Flowgraph

b. Flowgraph

Figure 11.5

Table 11.1

A	B	Path #
4	0	1
0	3	3
6	8	4

The paths are as follows:

Path 1:   Start -1-2-3-4-end
                   B=0, A≠0
Path 2:   Start -1-2-3-3-4-end
                   0<A≤B
Path 3:   Start -1-3-4-end
                   A=0 , B is any value
Path 4:   Start -1-3-3-4-end
                   0<A≤B

A table should be created with different values (simple numbers) for A and B and the resulting paths checked. The table paths can be compared with the expected paths in the program (see Table 11.1).

Integration Testing

Integration testing is conducted only if all the software modules have successfully passed the unit test. This test is conducted because modules may work alone but not work together, or because the integrated software does not work with the hardware. Steps in integration testing are:

1. The modules have to be linked without any errors.
2. Development tools such as emulators, logic analyzers, simulators, and oscilloscopes are used to debug the integrated software.
3. As with unit testing, in integration testing each control path has to be checked and the results of each path have to be verified.
4. Any incompatibilities with the hardware have to be resolved by changing either the hardware or the software to achieve the desired results.

Top-Down Integration. Modules are integrated by moving downward from the main control module through the control hierarchy. As show in Figure 11.6, the integration could be "depth-first" or "breadth-first."

Bottom-Up Integration. Testing begins with the modules at the lowest levels in the program, moving upward to the main modules (Figure 11.7). Modules are combined to form clusters. Each cluster is tested using a driver.

11.4   Other 68HC11 Family Members

There are at least 32 members of the 68HC11 family at the time of writing. Some M68HC11 members have "special" features that are not applicable to the M68HC11E9.

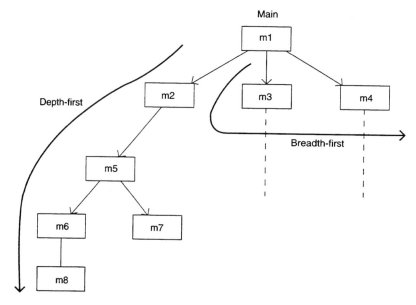

Figure 11.6

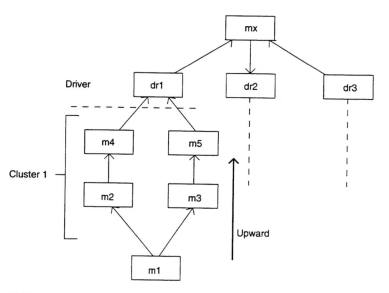

Figure 11.7

Special Features

Pulse-Width Modulation (PWM)

Some M68HC11 family members have up to six channels of 8-bit PWM outputs. At a 4-MHz bus frequency, signals can be produced from 40 KHz to less than 10 Hz. These family members are:

```
68HC11C0:   2 channels
68HC11G0:   4 channels
68HC11G5:   4 channels
68HC11G7:   4 channels
 68HC11J6:  4 channels
68HC11K0:   4 channels
68HC11KA0:  4 channels
68HC11K1:   4 channels
68HC11KA1:  4 channels
68HC11K3:   4 channels
68HC11KA3:  4 channels
68HC11K4:   4 channels
68HC11KA4:  4 channels
68HC11N4:   6 channels
```

Direct Memory Address (DMA)

An M68HC11 family member includes a four-channel DMA and a Memory Management Unit (MMU). The DMA provides fast data transfer between memories and registers and includes externally mapped memory in the expanded mode. The MMU allows up to 1 MB of address space in a physical 64-KB allocation, and integrated chip selects help reduce or eliminate glue logic. This member is the M68HC11M2, with four channels.

Math Coprocessors

Other family members offer a 16-bit on-chip math coprocessor that accelerates math operations by as much as 10 times. Located as a peripheral on the CPU bus, the math coprocessor requires no special instructions and is ideal for low-bandwidth DSP-type functions such as closed-loop control, servo positioning, and signal conditioning. These members are the 68HC11M2 and 68HC11N4.

Digital-to-Analog Conversion (DAC)

There is only one family member that has two 8-bit channels of DAC: the 68HC11N4. No one member contains all the features just mentioned, partly because the chip would be prohibitively expensive and also because no application would require all these features. So choosing the member depends on the application itself.

MCU 68HC11K4

The M68HC11Kx series has four members: 68HC11K0, 68HC11K1, 68HC11K3, and 68HC11K4. The M68HC11K4 is considered the most advanced member of the 68HC11 family for general applications. Figure 11.8 shows the 68HC11K4 block diagram. This MCU, with a nonmultiplexed expanded bus, is characterized by high speed, low power consumption, and an operating frequency from 4 MHz to DC.

The features of the 68HC11K4 are:

- Power-saving STOP and WAIT modes
- 24 KB of ROM/EPROM
- 640 bytes of EEPROM
- 768 bytes of RAM

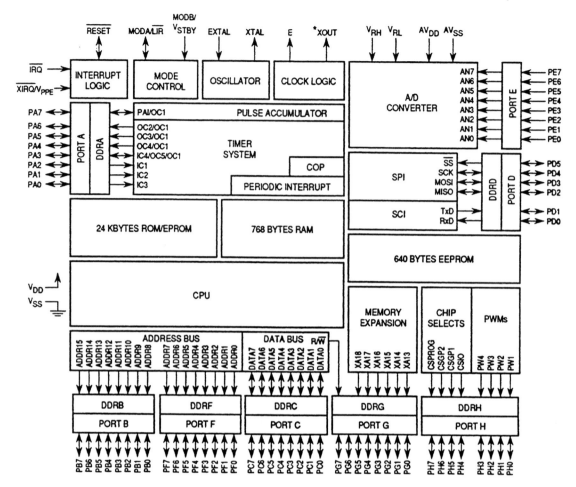

Figure 11.8
Copyright of Motorola. Used by permission.

- 1 MB of address space using mapping logic
- Nonmultiplexed address and data buses
- Four programmable chip selects
- Enhanced 16-bit timer
- 8-bit pulse accumulator
- Four 8-bit PWM channels
- Real-time interrupt (RTI) circuit
- Cop watchdog
- Enhanced SCI and SPI systems
- Eight 8-bit A/D channels
- Available in 84-pin PLCC, 84-pin windowed Arquad, or 80-pin Quad Flat Pack (QFP)
- Seven I/O ports: A to H (56 I/O lines), all programmable in either direction; port D (6 I/O lines) is input only

In the 84-pin PLCC, 64 pins are arranged in seven 8-bit ports: A, B, C, E, F, G, and H and one 6-bit port D. The lines of ports A, B, C, D, F, G, and H are fully bi-directional. Each of these seven ports serves a purpose other than I/O, depending on the operating mode or the peripheral functions selected. Port D is an input-only port. It is used as a general-purpose I/O or as an analog input port for the D/A converter. Some of the new features of the 68HC11K4 are explained below.

PWM Timer

The PWM timer subsystem provides up to four 8-bit modulated waveforms on port H. These outputs can be separated or concatenated to create 16-bit PWM outputs. The block diagram in Figure 11.9 shows three different clock sources that provide inputs to the control registers. These clock sources include clock A, clock B, and clock S (scaled). Clock A can be software selected to be $E/2n$ where $n = 0$ to 3. Clock B can be software selected to be $E/2m$ where $m = 0$ to 7. The scaled clock S uses clock A as an input and divides it with a reloadable counter. The rates available are software selectable to be clock A divided by 2 down to clock A divided by 5/2. Each PWM timer channel has the capability of selecting either of two clocks.

 Each of the four channels has a counter, a period register, and a duty register, as shown in Figure 11.9. The waveform output is the result of a match between the period register and the value in the counter. After specifying the period (in clocks) in PWPER 1-4 registers and the duty cycle (in clocks) in PWDTY 1-4 registers, the generated waveform is output automatically by the hardware without CPU intervention.

11.5 Motorola MC68HC16 Family (16-Bit)

The MC68HC16 is a high-speed, highly integrated 16-bit MCU designed using fully static HCMOS technology. It is upward source-code compatible with the MC68HC11. At the time of writing, there are seven members of the MCU 68HC16 family (Table 11.2).

M68HC16Z2

Figure 11.10 shows the block diagram of the M68HC16Z2. It consists of seven modules. These modules are interconnected by the intermodule bus (IMB).

System Integration Module (SIM)

- External bus support
- Programmable chip-select output
- System protection logic
- Watchdog timer, clock monitor, and bus monitor
- One 8-bit dual-function port
- One 7-bit output port
- One 7-bit dual-function port
- Phase-locked loop (PLL) clock system

CPU16

- 16-bit architecture
- Full 16-bit instruction set

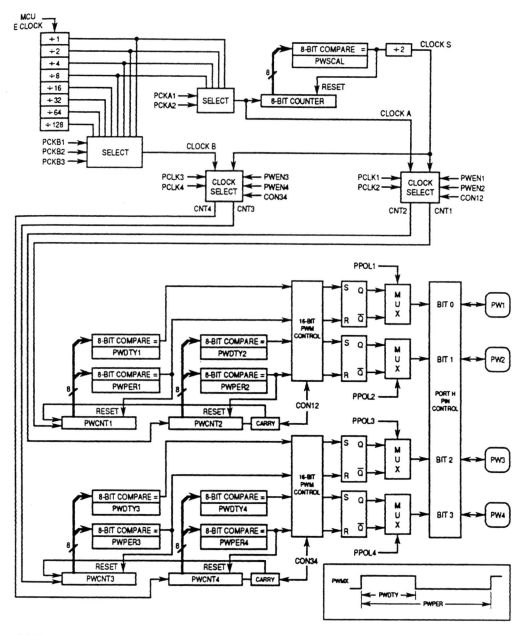

Figure 11.9
Copyright of Motorola. Used by permission.

- Three 16-bit index registers (IX,IY,IZ)
- Two 16-bit accumulators (D,E)
- Control-oriented DSP capability
- 1 MB of program memory and 1 MB of data memory
- High-level language support
- Fast interrupt response time
- Background debugging mode
- Fully static operation

Table 11.2

M68HC16 Family

Part Number	ROM	SRAM	EEPROM	Timer	I/O	Serial Communication	ADC	Peripheral	Package Integration	Comments Option
68HC16Z1	—	1K	—	GPT	46	QSM	8 Ch, 10-Bit	SIM	132-FC 132-FD 144-FM 144-FV	20 Address Lines, 12 Chip Selects, Synthesized Clock
68HC16Z2	8K	2K	—	GPT	46	QSM	8 Ch, 10-Bit	SIM	132-FC 132-FD	20 Address Lines, 12 Chip Selects, Synthesized Clock
68HC16Y1	48K	2K	—	TPU + GPT	95	MCCI	8 CH, 10-Bit	SCIM	160-FT 160-FM	20 Address Lines, 9 Chip Selects, Single Chip or Expanded Mode
68HC916Y1	—	4K	48K Flash	TPU + GPT	95	MCCI	8 Ch, 10-Bit	SCIM	160-FT 160-FM	20 Address Lines, 9 Chip Selects, Single Chip or Expanded Mode
68HC16X2	32K	1K	2K BEFlash	GPT	66	QSM	8 Ch, 10-Bit	RPSCIM	120-TH	20 Address Lines, 5 Chip Selects, Single Chip or Expanded Mode
68HC916X1	—	2K	32K Flash 16K Flash 2K BEFlash	GPT	66	QSM	8 Ch, 10-Bit 10-Bit	RPSCIM	120-TH	20 Address Lines, 5 Chip Selects, Single Chip or Expanded Mode
68HC16W1	—	3.5K	—	TPU + CTM	71	QSM + ABC	Queued 8 Ch, 10-Bit	SIM	160-FT 160-FM	20 Address Lines, 12 Chip Selects, Synthesized Clock

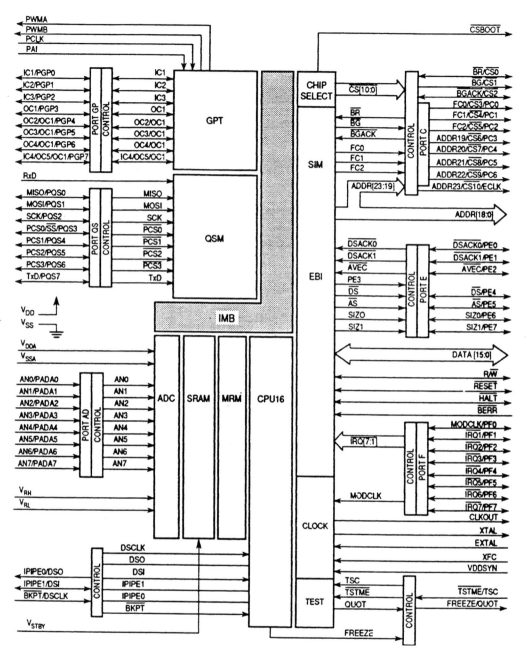

Figure 11.10
Copyright of Motorola. Used by permission.

Analog-to-Digital Converter (ADC)

- 10-bit resolution
- Eight channels and eight result registers
- Eight automated conversion modes

348

- Three result alignment modes
- One 8-bit digital port

Queued Serial Module (QSM)

- Enhanced SCI
- Queued SPI
- One 8-bit dual function port

General-Purpose Timer (GPT)

- Two 16-bit free-running counters with prescalar
- Three input capture channels
- Four output compare channels
- One input capture/output compare
- One pulse accumulator/event counter input
- Two pulse-width modulation outputs
- One 8-bit dual-function port
- Two optional discrete inputs
- An optional external clock input

Standby RAM (SRAM)

- 1024-byte Ram
- Extended standby voltage supply

Masked RAM Module (MRM)

- 8-byte array, accessible as bytes or words
- User-selectable default base address
- User-selectable bootstrap RAM function
- User-selectable RAM verification code

A phase-locked loop circuit synthesizes the system clock from a frequency (21,768 KHz) reference. Either a crystal with a nominal frequency of 4.19 MHz or an externally generated signal can be used.

11.6   MC68HC332

The MC68332 combines high-performance data manipulation capabilities with powerful peripheral subsystems. The MCU is built up from standard modules that interface through a common IMB. Figure 11.11 shows the M68HC332 block diagram.

The MCU incorporates a 32-bit CPU (CPU32), a system integration module (SIM), a time processor unit (TPU), a queued serial module (QSM), and a 2-KB static RAM with TPU emulation capability (TPU RAM).

The MCU can either synthesize an internal clock signal from an external reference or use an external clock input directly. A phase-locked loop circuit synthesizes the system clock from that frequency reference. Operation with 32,768 KHz reference frequency is standard.

350 DEVELOPMENT AND DEBUGGING TOOLS

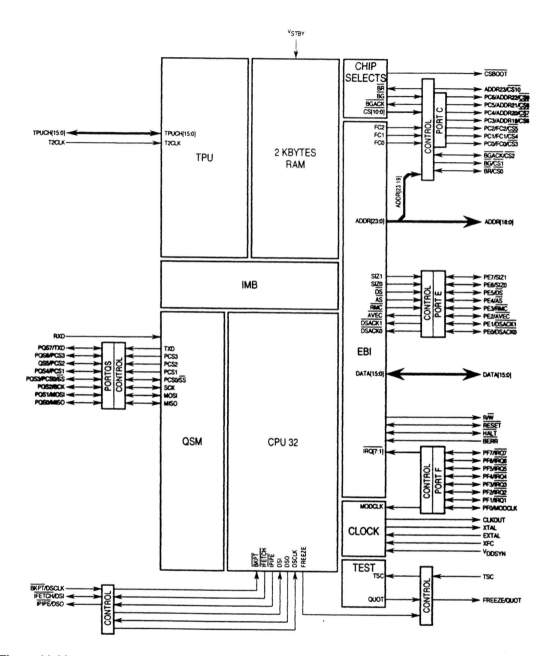

Figure 11.11
Copyright of Motorola. Used by permission.

The features of the MCU's modules are the system integration module, central processing unit, timer processor unit, queued serial module, and static RAM module with TPU emulation capability.

System Integration Module (SIM)

- External bus support
- Programmable chip-select outputs
- System protection logic
- Watchdog timer, clock monitor, and bus monitor
- System clock based on 32,768 KHz crystal
- Test/debug submodule for factory/user test and development

Central Processing Unit (CPU)

- Upward object-code compatible
- New instructions for controller applications
- 32-bit architecture
- Virtual memory implementation
- Loop mode of instruction execution
- Table look-up and interpolate instruction
- Improved exception handling for controller applications
- Trace on change of flow
- Hardware breakpoint signal, background mode
- Fully static operation

Time Processor Unit (TPU)

- Dedicated microengine operating independently of CPU32
- 16 independent, programmable channels and pins
- Capability of any channel to perform any time function
- Two timer count registers with programmable prescalars
- Selectable channel priority levels

Queued Serial Module (QSM)

- Enhanced SCI, modulus baud rate, parity
- Queued SPI, 80-byte RAM, up to 16 automatic transfers
- Dual-function I/O ports
- Continuous cycling, 8–16 bits per transfer

Static RAM Module with TPU Emulation Capability (TPU RAM)

- 2 KB of static RAM
- May be used as normal RAM or TPU microcode Emulation RAM

The CPU32 has eight 32-bit data registers, seven 32-bit address registers, a 32-bit program counter, separate 32-bit supervisor and user stack pointer, a 16-bit status register, two alternate function code registers, and a 32-bit vector base register.

Exercises

1. Make up a table to show similarities and differences between the features of the 68HC11E9 and 68HC11K4.

2. List a minimum set of tools with which every programmer should be equipped.

3. List three popular development and debugging tools for Motorola MCUs. What are advantages and disadvantages of each?

4. Explain the differences between the black-box and white-box tests.

5. Why do we call the M68HC16 a 16-bit MCU and the M68HC332 a 32-bit MCU?

6. In Program 2 of Chapter 6, assign a breakpoint and use a TRACE *(T)* command to single-step through the code to confirm that the program runs as intended. Show where you will assign the breakpoint. Repeat that debugging using both method 1 and method 2 of the program.

7. Explain the difference of generating a square waveform using M68HC11E9 versus M68HC11K4.

8. Explain the difference between unit testing and integration testing.

9. Describe the advantages of M68HC16 over M68HC11.

Practical Projects

These selected projects include many applications that require the student to design both the hardware and the software. Some of these projects require more research in technical libraries. Students are encouraged to develop and add more features to the projects. The instructor may assign more than one project at the end of the semester. Students have the option to select the M68HC11 version of their choice that suits a specific application. The software of each project should be designed using flowcharts or pseudocode before coding. The code should start with an initialization routine. The initialization routine clears memory RAM and initializes all variables.

Here are some suggested supply companies for parts needed in these projects:

- Transducer Techniques (909) 676-3965   strain gauges
- Digi-Key (800) 344-4539   electronic components
- Jameco (415) 592-8097   electronic components
- Newark Electronics (508) 683-0913   electronic components
- Pioneer Electronics (617) 861-9200   electronic components
- Micro Mo Electronics (813) 822-2529   small motors
- BEI (805) 968-0782   shaft encoders
- Detection Systems (716) 223-4060   detection systems

Project 1: A Mobile Robot Control Circuit

Design the hardware and the software of a mobile robot control circuit using MC68HC11xx. A block diagram of a mobile robot is shown in Figure 1. Because we are concerned with the control portion only, the mechnical parts and the motor gears are not covered here.

The block diagram includes:

1. An MCU MC68HC11xx, the system controller

2. The following sensors:

Figure 1

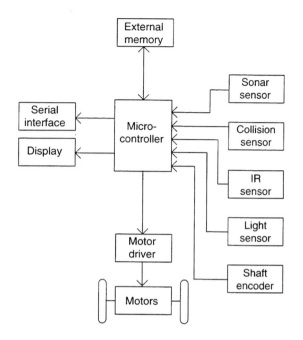

- *Light sensors* include photo-resistor, photo-diodes, and photo-transistor. The light can be used to make the robot trace it or hide from it.
- *Infrared (IR) sensors* consist of IR transmitter and receiver. They are used to detect a presence or absence of an object to avoid it.
- *Collision sensors* could be microswitches, force sensors, or others. They determine when the robot has run into an obstacle.
- *Sonar sensors* can provide the distance information of an object.
- *Shaft encoders* are covered in detail in chapter 7. They are mainly used to measure the speed and position of the motor's shaft.

3. Motor-driver ICs, which were discussed in chapter 7. You can choose one of the H-bridge IC drivers such as MPC1710A, made by Motorola; L293D, made by SGS Thompson; or UDN2993B, made by Allegro Microsystems.

4. LEDs or LCDs displays, used to display the RPM.

Operation

1. The right and left motors should run using PI compensator algorithm. The command should be entered either by keypad or by a host computer via the serial port.

2. If there is a light, follow the light.

3. If the robot has a collision in front, then back up.

4. If the robot has a collision in rear, turn left.

5. If the robot has a collision on left side, turn right.

6. If the robot has a collision on right side, turn left.

Diagnostics

The control software shall include a self-test routine to monitor the integrity of some sensors such as the light or IR sensors. The self-test shall include open-circuit and short-circuit tests. If the sensor is a short circuit to the ground, then the drop voltage across this sensor is zero volt. On the other hand, if the sensor is an open circuit, then the drop voltage is five volt. This integrity test should be conducted every 500 mS. If the number of fault occurrences reaches 10 in five seconds, then the fault is considered a true fault and should be recorded in EEPROM.

Project 2:   **A Remote Car Starter**

With a remote car starter, you can start your car from your home or office. In a cold winter, you can start your heater to warm up your car and drive it off immediately. In a hot summer, you can start your car's air conditioning to cool off the inside dead heat. You accomplish all this by simply leaving the heater or the air conditioning on before you turn off the ignition and leave the car.

Design a remote car starter that has other features besides starting and stopping the car's engine. It should also monitor the engine's rpm for stalling and overspeed.

Hardware Description

The hardware design should include these four major components:

1. A radio transmitter with two push-buttons: engine start and engine stop.

2. A radio receiver.

3. A microcontroller MC68HC11xx.

4. An interfacing circuit between the receiver and the MCU that includes three electronic switches (transistors). The first switches the +12 volts to the ignition. The second switches this voltage to the horn. The third switches the battery voltage to the starter. The horn beeps twice if the engine is started successfully and it beeps once if the engine shuts down or stalls.

Software Description

The software should include these five stages for this engine starter.

Stage 1

All relays are de-energized. When the start button is pressed, the program verifies that the engine is not running and there is a minimum of 9 volts across the battery. If these conditions are met, the software advances to Stage 2. If the engine is currently running, the horn beeps twice. If the stop button is pressed, the horn beeps once.

Stage 2

The ignition relay is energized. There is a waiting time of 3 seconds to be sure that the engine is running and the car's other electronics are online. If the engine is already running, the start sequence is canceled and the program goes back to Stage 1. After the 3-second time has elapsed with no interruption, the program advances to Stage 3.

Stage 3

If the engine is not running, the starter relay is energized. The logic monitors the engine's rpm and if it starts successfully, the starter relay is de-energized and the horn beeps twice. The program then advances to Stage 4. If the engine did not start after 3 seconds, the starter relay de-energizes and the horn beeps once. The program then advances to Stage 5 (engine stalled).

Stage 4

The engine will idle for 12 minutes. The program monitors rpm and the stop button. If the engine exceeds 3000 rpm (overspeed), the engine is shut down. If the stop button is pressed, the program returns to Stage 1. If the engine is stalled the first time, it goes to Stage 5. If it is stalled the second time, it returns to Stage 1.

Stage 5

The engine has to wait about 4 seconds in this stage, and then the program jumps to Stage 2 to start over again.

Implement this software algorithm using a flowchart or pseudocode method before coding.

Project 3:   Smart Scale

Design a weight measurement system with a load cell (such as the foil-type strain gauge) using an M68HC11. The load cell has a rated capacity of 5 pounds.

Hardware Description

The hardware circuit includes the following:

1. A wheatstone bridge circuit with a temperature compensation.
2. A signal conditioning circuit with a good common mode rejection (CMR) to amplify the bridge output. This output is fed to one of the MCU's A/D inputs.
3. Display digits of your choice to display different tasks as explained below. See Figure 2.

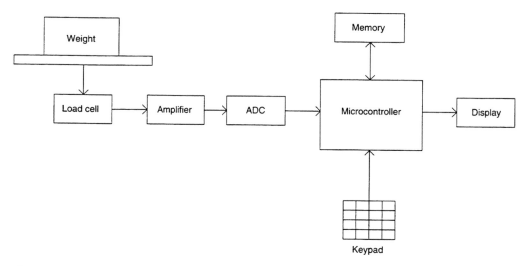

Figure 2

Software Description

The program of this project should accomplish the following:

1. Identify and display a valid part number (1 to 5).
2. Read the unit's price and display it.
3. Measure the weight in pounds and display it.
4. Calculate the cost, round it off, and then display it.

If the salesperson enters a nonvalid part number, the program should display the message "invalid part number." The student is free to use some of the EVB utility routines to save time.

Project 4:   Coffee Machine Controller

Operation

An operator deposits 45 cents for a cup of coffee and has the following possible varieties of coffee:

- Coffee black
- Coffee with sugar
- Coffee with cream
- Coffee with sugar and cream

The operator can cancel the selection and take his money back before the machine starts making coffee.

Inputs

1. Three coin detectors for nickles, dimes, and quarters
2. A keypad for making the coffee selection

Outputs

1. Cup drop solenoid
2. Coffee on solenoid
3. Sugar on solenoid
4. Cream on solenoid
5. Coin return solenoid

Timing

1. The cup has 1.5 seconds to get to place.
2. Powders will be released for different times depending on the selection:

• Black coffee	1.5 seconds
• Coffee with sugar	3 seconds
• Coffee with cream	3 seconds
• Coffee with cream and sugar	4.5 seconds

Develop the hardware circuit and the software of the control circuit using the MC68HC11xx.

Project 5:   **Real-Time Clock/Calendar**

The device used for this project is the Motorola MC68HC68T1 real-time clock (RTC), which is an HCMOS peripheral chip containing a real-time clock/calendar and a 32-byte static RAM array. A synchronous, serial, three-wire interface is provided for communication with a microcomputer.

The MC68HC68T1 can use a crystal or the 50/60 hertz line frequency as its timebase. During a normal operation, the timebase is provided by the power mains (50 or 60 hertz). When the main power is lost, a 3.6-volt NiCad backup battery is used to supply power to the chip and a crystal oscillator provides the timebase.

The RTC has full clock features: seconds, minutes, hours (AM/PM), day of the week, date, month, year (0-99). The MCU MC68HC11xx communicates with the RTC chip via the SPI port. More information on the MC68HC68T1 can be found in the Motorola Application-Specific Standard ICs data book "DL130 REV 1."

Design the hardware and the software of a clock/calendar using this chip to display the time and date. Use two buttons: one to set the clock and the other to set the alarm.

Project 6:   **A Single-Chip Telephone**

The MC34010 Electronic Telephone Circuit (ETC) provides all the necessary elements of a tone dialing telephone in a single IC. The block diagram shown in Figure 3 includes the dual-tone multiple frequency (DTMF) dialer, speech network, tone ringer, DC line interface circuit, and a microprocessor interface port that facilitates automatic dialing features. More information on this chip can be found in Motorola's communications data document "DL136/D REV 3."

The addition of an MCU to the ETC does nothing to enhance the performance of the telephone circuit. MCUs can offer additional functions such as visual digital display, visual clock/calendar, automated dialing, and so on.

Using the MC68HC11xx, design the hardware and the software to display the pressed telephone numbers on a display of your choice.

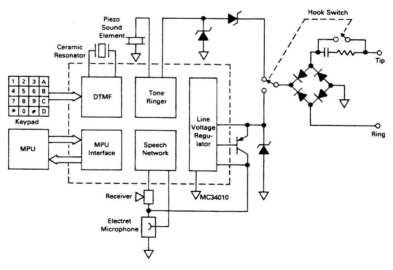

Figure 3
Copyright of Motorola. Used by permission.

Appendix A
Instruction Set Summary

Data Book Table 10–1:[1] MC68HC11A8
Instructions, Addressing Modes, and Execution Times

---

[1] *Data Book* means Motorola Document MC68HC11A8/D, *HCMOS Single-Chip Microcontroller 68HC11 Data Book.* Copyright of Motorola. Used by permission.

TABLE A.1 (*DATA BOOK* TABLE 10-1)

Source Form(s)	Operation	Boolean Expression	Addressing Mode for Operand	Opcode	Operand(s)	Bytes	Cycle	Cycle by Cycle*	S	X	H	I	N	Z	V	C
ABA	Add Accumulators	A + B → A	INH	1B		1	2	2-1	-	-	↕	-	↕	↕	↕	↕
ABX	Add B to X	IX + 00:B → IX	INH	3A		1	3	2-2	-	-	-	-	-	-	-	-
ABY	Add B to Y	IY + 00:B → IY	INH	18 3A		2	4	2-4	-	-	-	-	-	-	-	-
ADCA (opr)	Add with Carry to A	A + M + C → A	A IMM	89	ii	2	2	3-1	-	-	↕	-	↕	↕	↕	↕
			A DIR	99	dd	2	3	4-1								
			A EXT	B9	hh ll	3	4	5-2								
			A IND,X	A9	ff	2	4	6-2								
			A IND,Y	18 A9	ff	3	5	7-2								
ADCB (opr)	Add with Carry to B	B + M + C → B	B IMM	C9	ii	2	2	3-1	-	-	↕	-	↕	↕	↕	↕
			B DIR	D9	dd	2	3	4-1								
			B EXT	F9	hh ll	3	4	5-2								
			B IND,X	E9	ff	2	4	6-2								
			B IND,Y	18 E9	ff	3	5	7-2								
ADDA (opr)	Add Memory to A	A + M → A	A IMM	8B	ii	2	2	3-1	-	-	↕	-	↕	↕	↕	↕
			A DIR	9B	dd	2	3	4-1								
			A EXT	BB	hh ll	3	4	5-2								
			A IND,X	AB	ff	2	4	6-2								
			A IND,Y	18 AB	ff	3	5	7-2								
ADDB (opr)	Add Memory to B	B + M → B	B IMM	CB	ii	2	2	3-1	-	-	↕	-	↕	↕	↕	↕
			B DIR	DB	dd	2	3	4-1								
			B EXT	FB	hh ll	3	4	5-2								
			B IND,X	EB	ff	2	4	6-2								
			B IND,Y	18 EB	ff	3	5	7-2								
ADDD (opr)	Add 16-Bit to D	D + M:M + 1 → D	IMM	C3	jj kk	3	4	3-3	-	-	-	-	↕	↕	↕	↕
			DIR	D3	dd	2	5	4-7								
			EXT	F3	hh ll	3	6	5-10								
			IND,X	E3	ff	2	6	6-10								
			IND,Y	18 E3	ff	3	7	7-8								
ANDA (opr)	AND A with Memory	A•M → A	A IMM	84	ii	2	2	3-1	-	-	-	-	↕	↕	0	-
			A DIR	94	dd	2	3	4-1								
			A EXT	B4	hh ll	3	4	5-2								
			A IND,X	A4	ff	2	4	6-2								
			A IND,Y	18 A4	ff	3	5	7-2								
ANDB (opr)	AND B with Memory	B•M → B	B IMM	C4	ii	2	2	3-1	-	-	-	-	↕	↕	0	-
			B DIR	D4	dd	2	3	4-1								
			B EXT	F4	hh ll	3	4	5-2								
			B IND,X	E4	ff	2	4	6-2								
			B IND,Y	18 E4	ff	3	5	7-2								
ASL (opr)	Arithmetic Shift Left	← □—[ ][ ][ ][ ]— 0 C b7 b0	EXT	78	hh ll	3	6	5-8	-	-	-	-	↕	↕	↕	↕
			IND,X	68	ff	2	6	6-3								
			IND,Y	18 68	ff	3	7	7-3								
ASLA			A INH	48		1	2	2-1								
ASLB			B INH	58		1	2	2-1								
ASLD	Arithmetic Shift Left Double	□←[ ]- -[ ]← 0 C b15 b0	INH	05		1	3	2-2	-	-	-	-	↕	↕	↕	↕
ASR (opr)	Arithmetic Shift Right	→ [ ][ ][ ][ ]—□ b7 b0 C	EXT	77	hh ll	3	6	5-8	-	-	-	-	↕	↕	↕	↕
			IND,X	67	ff	2	6	6-3								
			IND,Y	18 67	ff	3	7	7-3								
ASRA			A INH	47		1	2	2-1								
ASRB			B INH	57		1	2	2-1								
BCC (rel)	Branch if Carry Clear	? C = 0	REL	24	rr	2	3	8-1	-	-	-	-	-	-	-	-
BCLR (opr) (msk)	Clear Bit(s)	M•(m̄m̄) → M	DIR	15	dd mm	3	6	4-10	-	-	-	-	↕	↕	0	-
			IND,X	1D	ff mm	3	7	6-13								
			IND,Y	18 1D	ff mm	4	8	7-10								
BCS (rel)	Branch if Carry Set	? C = 1	REL	25	rr	2	3	8-1	-	-	-	-	-	-	-	-
BEQ (rel)	Branch if = Zero	? Z = 1	REL	27	rr	2	3	8-1	-	-	-	-	-	-	-	-

*Cycle-by-cycle number provides a reference to Tables 10-2 through 10-8 which detail cycle-by-cycle operation.
    Example: Table 10-1 Cycle-by-Cycle column reference number 2-4 equals Table 10-2 line item 2-4.

TABLE A.1 (*DATA BOOK* TABLE 10-1) continued

Source Form(s)	Operation	Boolean Expression	Addressing Mode for Operand	Opcode	Operand(s)	Bytes	Cycle	Cycle by Cycle*	S	X	H	I	N	Z	V	C
BGE (rel)	Branch if ≥ Zero	? N ⊕ V = 0	REL	2C	rr	2	3	8-1	-	-	-	-	-	-	-	-
BGT (rel)	Branch if > Zero	? Z + (N ⊕ V) = 0	REL	2E	rr	2	3	8-1	-	-	-	-	-	-	-	-
BHI (rel)	Branch if Higher	? C + Z = 0	REL	22	rr	2	3	8-1	-	-	-	-	-	-	-	-
BHS (rel)	Branch if Higher or Same	? C = 0	REL	24	rr	2	3	8-1	-	-	-	-	-	-	-	-
BITA (opr)	Bit(s) Test A with Memory	A•M	A IMM	85	ii	2	2	3-1	-	-	-	-	↕	↕	0	-
			A DIR	95	dd	2	3	4-1								
			A EXT	B5	hh ll	3	4	5-2								
			A IND,X	A5	ff	2	4	6-2								
			A IND,Y	18 A5	ff	3	5	7-2								
BITB (opr)	Bit(s) Test B with Memory	B•M	B IMM	C5	ii	2	2	3-1	-	-	-	-	↕	↕	0	-
			B DIR	D5	dd	2	3	4-1								
			B EXT	F5	hh ll	3	4	5-2								
			B IND,X	E5	ff	2	4	6-2								
			B IND,Y	18 E5	ff	3	5	7-2								
BLE (rel)	Branch if ≤ Zero	? Z + (N ⊕ V) = 1	REL	2F	rr	2	3	8-1	-	-	-	-	-	-	-	-
BLO (rel)	Branch if Lower	? C = 1	REL	25	rr	2	3	8-1	-	-	-	-	-	-	-	-
BLS (rel)	Branch if Lower or Same	? C + Z = 1	REL	23	rr	2	3	8-1	-	-	-	-	-	-	-	-
BLT (rel)	Branch If < Zero	? N ⊕ V = 1	REL	2D	rr	2	3	8-1	-	-	-	-	-	-	-	-
BMI (rel)	Branch if Minus	? N = 1	REL	2B	rr	2	3	8-1	-	-	-	-	-	-	-	-
BNE (rel)	Branch if Not = Zero	? Z = 0	REL	26	rr	2	3	8-1	-	-	-	-	-	-	-	-
BPL (rel)	Branch if Plus	? N = 0	REL	2A	rr	2	3	8-1	-	-	-	-	-	-	-	-
BRA (rel)	Branch Always	? 1 = 1	REL	20	rr	2	3	8-1	-	-	-	-	-	-	-	-
BRCLR(opr) (msk) (rel)	Branch if Bit(s) Clear	? M• mm = 0	DIR	13	dd mm rr	4	6	4-11	-	-	-	-	-	-	-	-
			IND,X	1F	ff mm rr	4	7	6-14								
			IND,Y	18 1F	ff mm rr	5	8	7-11								
BRN (rel)	Branch Never	? 1 = 0	REL	21	rr	2	3	8-1	-	-	-	-	-	-	-	-
BRSET(opr) (msk) (rel)	Branch if Bit(s) Set	? (M̄)•mm = 0	DIR	12	dd mm rr	4	6	4-11	-	-	-	-	-	-	-	-
			IND,X	1E	ff mm rr	4	7	6-14								
			IND,Y	18 1E	ff mm rr	5	8	7-11								
BSET(opr) (msk)	Set Bit(s)	M + mm → M	DIR	14	dd mm	3	6	4-10	-	-	-	-	↕	↕	0	-
			IND,X	1C	ff mm	3	7	6-13								
			IND,Y	18 1C	ff mm	4	8	7-10								
BSR (rel)	Branch to Subroutine	See Special Ops	REL	8D	rr	2	6	8-2	-	-	-	-	-	-	-	-
BVC (rel)	Branch if Overflow Clear	? V = 0	REL	28	rr	2	3	8-1	-	-	-	-	-	-	-	-
BVS (rel)	Branch if Overflow Set	? V = 1	REL	29	rr	2	3	8-1	-	-	-	-	-	-	-	-
CBA	Compare A to B	A - B	INH	11		1	2	2-1	-	-	-	-	↕	↕	↕	↕
CLC	Clear Carry Bit	0 → C	INH	0C		1	2	2-1	-	-	-	-	-	-	-	0
CLI	Clear Interrupt Mask	0 → I	INH	0E		1	2	2-1	-	-	-	0	-	-	-	-
CLR (opr)	Clear Memory Byte	0 → M	EXT	7F	hh ll	3	6	5-8	-	-	-	-	0	1	0	0
			IND,X	6F	ff	2	6	6-3								
			IND,Y	18 6F	ff	3	7	7-3								
CLRA	Clear Accumulator A	0 → A	A INH	4F		1	2	2-1	-	-	-	-	0	1	0	0
CLRB	Clear Accumulator B	0 → B	B INH	5F		1	2	2-1	-	-	-	-	0	1	0	0
CLV	Clear Overflow Flag	0 → V	INH	0A		1	2	2-1	-	-	-	-	-	-	0	-
CMPA (opr)	Compare A to Memory	A - M	A IMM	81	ii	2	2	3-1	-	-	-	-	↕	↕	↕	↕
			A DIR	91	dd	2	3	4-1								
			A EXT	B1	hh ll	3	4	5-2								
			A IND,X	A1	ff	2	4	6-2								
			A IND,Y	18 A1	ff	3	5	7-2								

*Cycle-by-cycle number provides a reference to Tables 10-2 through 10-8 which detail cycle-by-cycle operation.
  Example: Table 10-1 Cycle-by-Cycle column reference number 2-4 equals Table 10-2 line item 2-4.

TABLE A.1 (*DATA BOOK* TABLE 10-1) continued

Source Form(s)	Operation	Boolean Expression	Addressing Mode for Operand	Opcode	Operand(s)	Bytes	Cycle	Cycle by Cycle*	S	X	H	I	N	Z	V	C
CMPB (opr)	Compare B to Memory	B – M	B IMM	C1	ii	2	2	3-1	-	-	-	-	↕	↕	↕	↕
			B DIR	D1	dd	2	3	4-1								
			B EXT	F1	hh ll	3	4	5-2								
			B IND,X	E1	ff	2	4	6-2								
			B IND,Y	18 E1	ff	3	5	7-2								
COM (opr)	1's Complement Memory Byte	$FF – M → M	EXT	73	hh ll	3	6	5-8	-	-	-	-	↕	↕	0	1
			IND,X	63	ff	2	6	6-3								
			IND,Y	18 63	ff	3	7	7-3								
COMA	1's Complement A	$FF – A → A	A INH	43		1	2	2-1	-	-	-	-	↕	↕	0	1
COMB	1's Complement B	$FF – B → B	B INH	53		1	2	2-1	-	-	-	-	↕	↕	0	1
CPD (opr)	Compare D to Memory 16-Bit	D – M:M + 1	IMM	1A 83	jj kk	4	5	3-5	-	-	-	-	↕	↕	↕	↕
			DIR	1A 93	dd	3	6	4-9								
			EXT	1A B3	hh ll	4	7	5-11								
			IND,X	1A A3	ff	3	7	6-11								
			IND,Y	CD A3	ff	3	7	7-8								
CPX (opr)	Compare X to Memory 16-Bit	IX – M:M + 1	IMM	8C	jj kk	3	4	3-3	-	-	-	-	↕	↕	↕	↕
			DIR	9C	dd	2	5	4-7								
			EXT	BC	hh ll	3	6	5-10								
			IND,X	AC	ff	2	6	6-10								
			IND,Y	CD AC	ff	3	7	7-8								
CPY (opr)	Compare Y to Memory 16-Bit	IY – M:M + 1	IMM	18 8C	jj kk	4	5	3-5	-	-	-	-	↕	↕	↕	↕
			DIR	18 9C	dd	3	6	4-9								
			EXT	18 BC	hh ll	4	7	5-11								
			IND,X	1A AC	ff	3	7	6-11								
			IND,Y	18 AC	ff	3	7	7-8								
DAA	Decimal Adjust A	Adjust Sum to BCD	INH	19		1	2	2-1	-	-	-	-	↕	↕	↕	↕
DEC (opr)	Decrement Memory Byte	M – 1 → M	EXT	7A	hh ll	3	6	5-8	-	-	-	-	↕	↕	↕	-
			IND,X	6A	ff	2	6	6-3								
			IND,Y	18 6A	ff	3	7	7-3								
DECA	Decrement Accumulator A	A – 1 → A	A INH	4A		1	2	2-1	-	-	-	-	↕	↕	↕	-
DECB	Decrement Accumulator B	B – 1 → B	B INH	5A		1	2	2-1	-	-	-	-	↕	↕	↕	-
DES	Decrement Stack Pointer	SP – 1 → SP	INH	34		1	3	2-3	-	-	-	-	-	-	-	-
DEX	Decrement Index Register X	IX – 1 → IX	INH	09		1	3	2-2	-	-	-	-	-	↕	-	-
DEY	Decrement Index Register Y	IY – 1 → IY	INH	18 09		2	4	2-4	-	-	-	-	-	↕	-	-
EORA (opr)	Exclusive OR A with Memory	A ⊕ M → A	A IMM	88	ii	2	2	3-1	-	-	-	-	↕	↕	0	-
			A DIR	98	dd	2	3	4-1								
			A EXT	B8	hh ll	3	4	5-2								
			A IND,X	A8	ff	2	4	6-2								
			A IND,Y	18 A8	ff	3	5	7-2								
EORB (opr)	Exclusive OR B with Memory	B ⊕ M → B	B IMM	C8	ii	2	2	3-1	-	-	-	-	↕	↕	0	-
			B DIR	D8	dd	2	3	4-1								
			B EXT	F8	hh ll	3	4	5-2								
			B IND,X	E8	ff	2	4	6-2								
			B IND,Y	18 E8	ff	3	5	7-2								
FDIV	Fractional Divide 16 by 16	D/IX → IX; r → D	INH	03		1	41	2-17	-	-	-	-	-	↕	↕	↕
IDIV	Integer Divide 16 by 16	D/IX → IX; r → D	INH	02		1	41	2-17	-	-	-	-	-	↕	0	↕
INC (opr)	Increment Memory Byte	M + 1 → M	EXT	7C	hh ll	3	6	5-8	-	-	-	-	↕	↕	↕	-
			IND,X	6C	ff	2	6	6-3								
			IND,Y	18 6C	ff	3	7	7-3								
INCA	Increment Accumulator A	A + 1 → A	A INH	4C		1	2	2-1	-	-	-	-	↕	↕	↕	-
INCB	Increment Accumulator B	B + 1 → B	B INH	5C		1	2	2-1	-	-	-	-	↕	↕	↕	-
INS	Increment Stack Pointer	SP + 1 → SP	INH	31		1	3	2-3	-	-	-	-	-	-	-	-

*Cycle-by-cycle number provides a reference to Tables 10-2 through 10-8 which detail cycle-by-cycle operation.
   Example: Table 10-1 Cycle-by-Cycle column reference number 2-4 equals Table 10-2 line item 2-4.

TABLE A.1 (*DATA BOOK* TABLE 10-1) continued

Source Form(s)	Operation	Boolean Expression	Addressing Mode for Operand	Opcode	Operand(s)	Bytes	Cycle	Cycle by Cycle*	S	X	H	I	N	Z	V	C
INX	Increment Index Register X	IX + 1 → IX	INH	08		1	3	2-2	-	-	-	-	-	$\updownarrow$	-	-
INY	Increment Index Register Y	IY + 1 → IY	INH	18 08		2	4	2-4	-	-	-	-	-	$\updownarrow$	-	-
JMP (opr)	Jump	See Special Ops	EXT	7E	hh ll	3	3	5-1	-	-	-	-	-	-	-	-
			IND,X	6E	ff	2	3	6-1								
			IND,Y	18 6E	ff	3	4	7-1								
JSR (opr)	Jump to Subroutine	See Special Ops	DIR	9D	dd	2	5	4-8	-	-	-	-	-	-	-	-
			EXT	BD	hh ll	3	6	5-12								
			IND,X	AD	ff	2	6	6-12								
			IND,Y	18 AD	ff	3	7	7-9								
LDAA (opr)	Load Accumulator A	M → A	A IMM	86	ii	2	2	3-1	-	-	-	-	$\updownarrow$	$\updownarrow$	0	-
			A DIR	96	dd	2	3	4-1								
			A EXT	B6	hh ll	3	4	5-2								
			A IND,X	A6	ff	2	4	6-2								
			A IND,Y	18 A6	ff	3	5	7-2								
LDAB (opr)	Load Accumulator B	M → B	B IMM	C6	ii	2	2	3-1	-	-	-	-	$\updownarrow$	$\updownarrow$	0	-
			B DIR	D6	dd	2	3	4-1								
			B EXT	F6	hh ll	3	4	5-2								
			B IND,X	E6	ff	2	4	6-2								
			B IND,Y	18 E6	ff	3	5	7-2								
LDD (opr)	Load Double Accumulator D	M → A, M + 1 → B	IMM	CC	jj kk	3	3	3-2	-	-	-	-	$\updownarrow$	$\updownarrow$	0	-
			DIR	DC	dd	2	4	4-3								
			EXT	FC	hh ll	3	5	5-4								
			IND,X	EC	ff	2	5	6-6								
			IND,Y	18 EC	ff	3	6	7-6								
LDS (opr)	Load Stack Pointer	M:M + 1 → SP	IMM	8E	jj kk	3	3	3-2	-	-	-	-	$\updownarrow$	$\updownarrow$	0	-
			DIR	9E	dd	2	4	4-3								
			EXT	BE	hh ll	3	5	5-4								
			IND,X	AE	ff	2	5	6-6								
			IND,Y	18 AE	ff	3	6	7-6								
LDX (opr)	Load Index Register X	M:M + 1 → IX	IMM	CE	jj kk	3	3	3-2	-	-	-	-	$\updownarrow$	$\updownarrow$	0	-
			DIR	DE	dd	2	4	4-3								
			EXT	FE	hh ll	3	5	5-4								
			IND,X	EE	ff	2	5	6-6								
			IND,Y	CD EE	ff	3	6	7-6								
LDY (opr)	Load Index Register Y	M:M + 1 → IY	IMM	18 CE	jj kk	4	4	3-4	-	-	-	-	$\updownarrow$	$\updownarrow$	0	-
			DIR	18 DE	dd	3	5	4-5								
			EXT	18 FE	hh ll	4	6	5-6								
			IND,X	1A EE	ff	3	6	6-7								
			IND,Y	18 EE	ff	3	6	7-6								
LSL (opr)	Logical Shift Left		EXT	78	hh ll	3	6	5-8	-	-	-	-	$\updownarrow$	$\updownarrow$	$\updownarrow$	$\updownarrow$
			IND,X	68	ff	2	6	6-3								
			IND,Y	18 68	ff	3	7	7-3								
LSLA			A INH	48		1	2	2-1								
LSLB			B INH	58		1	2	2-1								
LSLD	Logical Shift Left Double		INH	05		1	3	2-2	-	-	-	-	$\updownarrow$	$\updownarrow$	$\updownarrow$	$\updownarrow$
LSR (opr)	Logical Shift Right		EXT	74	hh ll	3	6	5-8	-	-	-	-	0	$\updownarrow$	$\updownarrow$	$\updownarrow$
			IND,X	64	ff	2	6	6-3								
			IND,Y	18 64	ff	3	7	7-3								
LSRA			A INH	44		1	2	2-1								
LSRB			B INH	54		1	2	2-1								
LSRD	Logical Shift Right Double		INH	04		1	3	2-2	-	-	-	-	0	$\updownarrow$	$\updownarrow$	$\updownarrow$
MUL	Multiply 8 by 8	A x B → D	INH	3D		1	10	2-13	-	-	-	-	-	-	-	$\updownarrow$

*Cycle-by-cycle number provides a reference to Tables 10-2 through 10-8 which detail cycle-by-cycle operation.
  Example: Table 10-1 Cycle-by-Cycle column reference number 2-4 equals Table 10-2 line item 2-4.

TABLE A.1 (*DATA BOOK* TABLE 10-1) continued

Source Form(s)	Operation	Boolean Expression	Addressing Mode for Operand	Opcode	Operand(s)	Bytes	Cycle	Cycle by Cycle*	S	X	H	I	N	Z	V	C
NEG (opr)	2's Complement Memory Byte	0 − M → M	EXT	70	hh ll	3	6	5-8	-	-	-	-	↕	↕	↕	↕
			IND,X	60	ff	2	6	6-3								
			IND,Y	18 60	ff	3	7	7-3								
NEGA	2's Complement A	0 − A → A	A INH	40		1	2	2-1	-	-	-	-	↕	↕	↕	↕
NEGB	2's Complement B	0 − B → B	B INH	50		1	2	2-1	-	-	-	-	↕	↕	↕	↕
NOP	No Operation	No Operation	INH	01		1	2	2-1	-	-	-	-	-	-	-	-
ORAA (opr)	OR Accumulator A (Inclusive)	A + M → A	A IMM	8A	ii	2	2	3-1	-	-	-	-	↕	↕	0	-
			A DIR	9A	dd	2	3	4-1								
			A EXT	BA	hh ll	3	4	5-2								
			A IND,X	AA	ff	2	4	6-2								
			A IND,Y	18 AA	ff	3	5	7-2								
ORAB (opr)	OR Accumulator B (Inclusive)	B + M → B	B IMM	CA	ii	2	2	3-1	-	-	-	-	↕	↕	0	-
			B DIR	DA	dd	2	3	4-1								
			B EXT	FA	hh ll	3	4	5-2								
			B IND,X	EA	ff	2	4	6-2								
			B IND,Y	18 EA	ff	3	5	7-2								
PSHA	Push A onto Stack	A → Stk, SP=SP−1	A INH	36		1	3	2-6	-	-	-	-	-	-	-	-
PSHB	Push B onto Stack	B → Stk, SP=SP−1	B INH	37		1	3	2-6	-	-	-	-	-	-	-	-
PSHX	Push X onto Stack (Lo First)	IX → Stk, SP=SP−2	INH	3C		1	4	2-7	-	-	-	-	-	-	-	-
PSHY	Push Y onto Stack (Lo First)	IY → Stk, SP=SP−2	INH	18 3C		2	5	2-8	-	-	-	-	-	-	-	-
PULA	Pull A from Stack	SP=SP+1, A ← Stk	A INH	32		1	4	2-9	-	-	-	-	-	-	-	-
PULB	Pull B from Stack	SP=SP+1, B ← Stk	B INH	33		1	4	2-9	-	-	-	-	-	-	-	-
PULX	Pull X from Stack (Hi First)	SP=SP+2, IX ← Stk	INH	38		1	5	2-10	-	-	-	-	-	-	-	-
PULY	Pull Y from Stack (Hi First)	SP=SP+2, IY ← Stk	INH	18 38		2	6	2-11	-	-	-	-	-	-	-	-
ROL (opr)	Rotate Left	C b7 ← b0 C	EXT	79	hh ll	3	6	5-8	-	-	-	-	↕	↕	↕	↕
			IND,X	69	ff	2	6	6-3								
			IND,Y	18 69	ff	3	7	7-3								
ROLA			A INH	49		1	2	2-1								
ROLB			B INH	59		1	2	2-1								
ROR (opr)	Rotate Right	C b7 → b0 C	EXT	76	hh ll	3	6	5-8	-	-	-	-	↕	↕	↕	↕
			IND,X	66	ff	2	6	6-3								
			IND,Y	18 66	ff	3	7	7-3								
RORA			A INH	46		1	2	2-1								
RORB			B INH	56		1	2	2-1								
RTI	Return from Interrupt	See Special Ops	INH	3B		1	12	2-14	↕	↕	↕	↕	↕	↕	↕	↕
RTS	Return from Subroutine	See Special Ops	INH	39		1	5	2-12	-	-	-	-	-	-	-	-
SBA	Subtract B from A	A − B → A	INH	10		1	2	2-1	-	-	-	-	↕	↕	↕	↕
SBCA (opr)	Subtract with Carry from A	A − M − C → A	A IMM	82	ii	2	2	3-1	-	-	-	-	↕	↕	↕	↕
			A DIR	92	dd	2	3	4-1								
			A EXT	B2	hh ll	3	4	5-2								
			A IND,X	A2	ff	2	4	6-2								
			A IND,Y	18 A2	ff	3	5	7-2								
SBCB (opr)	Subtract with Carry from B	B − M − C → B	B IMM	C2	ii	2	2	3-1	-	-	-	-	↕	↕	↕	↕
			B DIR	D2	dd	2	3	4-1								
			B EXT	F2	hh ll	3	4	5-2								
			B IND,X	E2	ff	2	4	6-2								
			B IND,Y	18 E2	ff	3	5	7-2								
SEC	Set Carry	1 → C	INH	0D		1	2	2-1	-	-	-	-	-	-	-	1
SEI	Set Interrupt Mask	1 → I	INH	0F		1	2	2-1	-	-	-	1	-	-	-	-
SEV	Set Overflow Flag	1 → V	INH	0B		1	2	2-1	-	-	-	-	-	-	1	-

*Cycle-by-cycle number provides a reference to Tables 10-2 through 10-8 which detail cycle-by-cycle operation.
   Example: Table 10-1 Cycle-by-Cycle column reference number 2-4 equals Table 10-2 line item 2-4.

TABLE A.1 (*DATA BOOK* TABLE 10-1) continued

Source Form(s)	Operation	Boolean Expression	Addressing Mode for Operand	Opcode	Operand(s)	Bytes	Cycle	Cycle by Cycle*	S	X	H	I	N	Z	V	C
STAA (opr)	Store Accumulator A	A → M	A DIR	97	dd	2	3	4-2	-	-	-	-	↕	↕	0	-
			A EXT	B7	hh ll	3	4	5-3								
			A IND,X	A7	ff	2	4	6-5								
			A IND,Y	18 A7	ff	3	5	7-5								
STAB (opr)	Store Accumulator B	B → M	B DIR	D7	dd	2	3	4-2	-	-	-	-	↕	↕	0	-
			B EXT	F7	hh ll	3	4	5-3								
			B IND,X	E7	ff	2	4	6-5								
			B IND,Y	18 E7	ff	3	5	7-5								
STD (opr)	Store Accumulator D	A → M, B → M + 1	DIR	DD	dd	2	4	4-4	-	-	-	-	↕	↕	0	-
			EXT	FD	hh ll	3	5	5-5								
			IND,X	ED	ff	2	5	6-8								
			IND,Y	18 ED	ff	3	6	7-7								
STOP	Stop Internal Clocks		INH	CF		1	2	2-1	-	-	-	-	-	-	-	-
STS (opr)	Store Stack Pointer	SP → M:M + 1	DIR	9F	dd	2	4	4-4	-	-	-	-	↕	↕	0	-
			EXT	BF	hh ll	3	5	5-5								
			IND,X	AF	ff	2	5	6-8								
			IND,Y	18 AF	ff	3	6	7-7								
STX (opr)	Store Index Register X	IX → M:M + 1	DIR	DF	dd	2	4	4-4	-	-	-	-	↕	↕	0	-
			EXT	FF	hh ll	3	5	5-5								
			IND,X	EF	ff	2	5	6-8								
			IND,Y	CD EF	ff	3	6	7-7								
STY (opr)	Store Index Register Y	IY → M:M + 1	DIR	18 DF	dd	3	5	4-6	-	-	-	-	↕	↕	0	-
			EXT	18 FF	hh ll	4	6	5-7								
			IND,X	1A EF	ff	3	6	6-9								
			IND,Y	18 EF	ff	3	6	7-7								
SUBA (opr)	Subtract Memory from A	A − M → A	A IMM	80	ii	2	2	3-1	-	-	-	-	↕	↕	↕	↕
			A DIR	90	dd	2	3	4-1								
			A EXT	B0	hh ll	3	4	5-2								
			A IND,X	A0	ff	2	4	6-2								
			A IND,Y	18 A0	ff	3	5	7-2								
SUBB (opr)	Subtract Memory from B	B − M → B	B IMM	C0	ii	2	2	3-1	-	-	-	-	↕	↕	↕	↕
			B DIR	D0	dd	2	3	4-1								
			B EXT	F0	hh ll	3	4	5-2								
			B IND,X	E0	ff	2	4	6-2								
			B IND,Y	18 E0	ff	3	5	7-2								
SUBD (opr)	Subtract Memory from D	D − M:M + 1 → D	IMM	83	jj kk	3	4	3-3	-	-	-	-	↕	↕	↕	↕
			DIR	93	dd	2	5	4-7								
			EXT	B3	hh ll	3	6	5-10								
			IND,X	A3	ff	2	6	6-10								
			IND,Y	18 A3	ff	3	7	7-8								
SWI	Software Interrupt	See Special Ops	INH	3F		1	14	2-15	-	-	-	1	-	-	-	-
TAB	Transfer A to B	A → B	INH	16		1	2	2-1	-	-	-	-	↕	↕	0	-
TAP	Transfer A to CC Register	A → CCR	INH	06		1	2	2-1	↕	↕	↕	↕	↕	↕	↕	↕
TBA	Transfer B to A	B → A	INH	17		1	2	2-1	-	-	-	-	↕	↕	0	-
TEST	TEST (Only in Test Modes)	Address Bus Counts	INH	00		1	**	2-20	-	-	-	-	-	-	-	-
TPA	Transfer CC Register to A	CCR → A	INH	07		1	2	2-1	-	-	-	-	-	-	-	-
TST (opr)	Test for Zero or Minus	M − 0	EXT	7D	hh ll	3	6	5-9	-	-	-	-	↕	↕	0	0
			IND,X	6D	ff	2	6	6-4								
			IND,Y	18 6D	ff	3	7	7-4								
TSTA		A − 0	A INH	4D		1	2	2-1	-	-	-	-	↕	↕	0	0
TSTB		B − 0	B INH	5D		1	2	2-1	-	-	-	-	↕	↕	0	0
TSX	Transfer Stack Pointer to X	SP + 1 → IX	INH	30		1	3	2-3	-	-	-	-	-	-	-	-
TSY	Transfer Stack Pointer to Y	SP + 1 → IY	INH	18 30		2	4	2-5	-	-	-	-	-	-	-	-

*Cycle-by-cycle number provides a reference to Tables 10-2 through 10-8 which detail cycle-by-cycle operation.

   Example: Table 10-1 Cycle-by-Cycle column reference number 2-4 equals Table 10-2 line item 2-4.

TABLE A.1   (*DATA BOOK* TABLE 10-1) continued

Source Form(s)	Operation	Boolean Expression	Addressing Mode for Operand	Machine Coding (Hexadecimal)		Bytes	Cycle	Cycle by Cycle*	Condition Codes							
				Opcode	Operand(s)				S	X	H	I	N	Z	V	C
TXS	Transfer X to Stack Pointer	IX − 1 → SP	INH	35		1	3	2-2	-	-	-	-	-	-	-	-
TYS	Transfer Y to Stack Pointer	IY − 1 → SP	INH	18 35		2	4	2-4	-	-	-	-	-	-	-	-
WAI	Wait for Interrupt	Stack Regs & WAIT	INH	3E		1	***	2-16	-	-	-	-	-	-	-	-
XGDX	Exchange D with X	IX → D, D → IX	INH	8F		1	3	2-2	-	-	-	-	-	-	-	-
XGDY	Exchange D with Y	IY → D, D → IY	INH	18 8F		2	4	2-4	-	-	-	-	-	-	-	-

*Cycle-by-cycle number provides a reference to Tables 10-2 through 10-8 which detail cycle-by-cycle operation.
    Example: Table 10-1 Cycle-by-Cycle column reference number 2-4 equals Table 10-2 line item 2-4.
**Infinity or Until Reset Occurs
***12 Cycles are used beginning with the opcode fetch. A wait state is entered which remains in effect for an integer number of MPU E-clock cycles (n) until an interrupt is recognized. Finally, two additional cycles are used to fetch the appropriate interrupt vector (14 + n total).

dd = 8-Bit Direct Address ($0000 – $00FF) (High Byte Assumed to be $00)
ff = 8-Bit Positive Offset $00 (0) to $FF (255) (Is Added to Index)
hh = High Order Byte of 16-Bit Extended Address
ii = One Byte of Immediate Data
jj = High Order Byte of 16-Bit Immediate Data
kk = Low Order Byte of 16-Bit Immediate Data
ll = Low Order Byte of 16-Bit Extended Address
mm = 8-Bit Bit Mask (Set Bits to be Affected)
rr = Signed Relative Offset $80 (− 128) to $7F (+ 127)
        (Offset Relative to the Address Following the Machine Code Offset Byte)

Appendix B
Selected Sensors/Drivers[1]

[1] Pages 368–372 are from *Integrated and Discrete Semiconductors Databook.* Copyright 1993 by Allegro MicroSystems, Inc. Reprinted with permission.

Pages 373–375 are from *Optoelectronics Designer's Catalog.* Copyright 1993 by Hewlett-Packard Company. Reproduced with permission of Hewlett-Packard Company.

Pages 376–377 are from *Production Data Documents.* Copyright 1983 by Texas Instruments, Inc. Reprinted with permission.

Pages 378–380 are from *National Application Specific Analog Products.* Copyright 1995 by National Semiconductor Company. Reprinted with permission.

Pages 381–386 are from *Pressure Sensor Device Data.* Copyright 1994 by Motorola, Inc. Used by permission.

2993

DUAL H-BRIDGE MOTOR DRIVERS

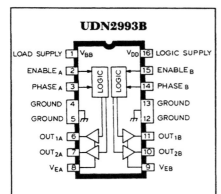

UDN2993B

LOAD SUPPLY	1	V$_{BB}$	V$_{DD}$	16	LOGIC SUPPLY
ENABLE$_A$	2			15	ENABLE$_B$
PHASE$_A$	3	LOGIC LOGIC	14	PHASE$_B$
GROUND	4			13	GROUND
GROUND	5			12	GROUND
OUT$_{1A}$	6			11	OUT$_{1B}$
OUT$_{2A}$	7			10	OUT$_{2B}$
V$_{EA}$	8			9	V$_{EB}$

Dwg. No. A-12,455

ABSOLUTE MAXIMUM RATINGS
at T$_J$ ≤ +150°C

Load Supply Voltage, V$_{BB}$ . . . . 40 V
Logic Supply Voltage, V$_{DD}$ . . . . 7.0 V
Logic Input Voltage Range, V$_{PHASE}$ or
  V$_{ENABLE}$ . . . . -0.3 V to V$_{DD}$ + 0.3 V
Output Current, I$_{OUT}$ . . . . ±600 mA
Sink Driver Emitter Voltage,
  V$_E$ . . . . 1.5 V
Package Power Dissipation,
  P$_D$ . . . . **See Graph**
Operating Temperature Range,
  T$_A$ . . . . -20°C to +85°C
Storage Temperature Range,
  T$_S$ . . . . -55°C to +150°C

IMPORTANT: Load supply voltage must never be applied without logic supply voltage present.

NOTE: Output current rating may be limited by chopping frequency, ambient temperature, airflow, and heat sinking. Under any set of conditions, do not exceed the specified maximum current and a junction temperature of +150°C.

Cost-effective monolithic drive electronics for bipolar stepper and dc (brush) servo motors to 40 V and 500 mA is very practical with the UDN2993B and UDN2993LB. These dual full-bridge motion control ICs integrate separate inputs, level shifting for upper power outputs, control logic, integral inductive transient protection, and source (upper) and sink (lower) drivers in an H-bridge configuration. The single-chip power IC provides improved space utilization and reliability unmatched by discrete component circuitry.

Excepting the power supply connections, the two H-bridges are independent. An ENABLE input is provided for each bridge and permits pulse-width modulation (PWM) through the use of external circuitry. PWM drive techniques provide the benefits of reduced power dissipation, improved motor performance (especially torque), and positively affect system efficiency. Separate PHASE inputs for each bridge determine the direction of current flow in the load. Additionally, each pair of (sink) emitters are terminated to package connections. This allows the use of current-sensing circuitry. Both devices incorporate an intrinsic "dead time" to preclude high crossover (or cross-conduction) currents during changes in direction (phase).

These devices are packaged in plastic DIPs (suffix B) or surface-mountable wide-body SOICs (suffix LB) with copper lead frames for optimum power dissipation without heat sinks. The lead configurations allow automatic insertion, fit standard IC sockets or printed wiring board layouts, and enable easy attachment of a heat sink for maximum power-handling capability. The heat-sink tabs are at ground potential and require no insulation.

Dual full-bridge drivers with peak current ratings of ±3 A are supplied as the UDN2998W.

FEATURES

- ±600 mA Output Current
- Output Voltage to 40 V
- Crossover Current Protection
- TTL/NMOS/CMOS Compatible Inputs
- Low Input Current
- Internal Clamp Diodes
- DIP or SOIC Packaging

Always order by complete part number:

Part Number	Package
UDN2993B	16-Pin DIP
UDN2993LB	20-Lead Wide-Body SOIC

2993
DUAL H-BRIDGE MOTOR DRIVERS

TYPICAL APPLICATION
2-PHASE BIPOLAR STEPPER MOTOR DRIVE
(Chopper Mode)

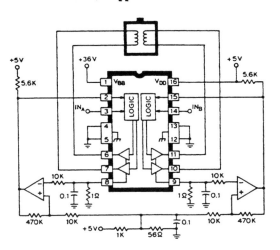

Dwg. No. A-12,453

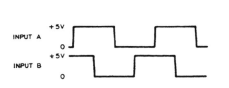

Dwg. No. A-12,454

TEST FIGURES

FIGURE 1

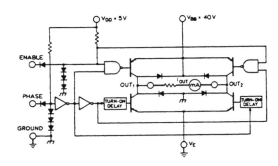

Dwg. No. A-12,449

FIGURE 2

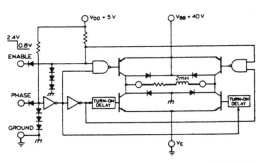

Dwg. No. A-12,450

5804

BiMOS II UNIPOLAR STEPPER-MOTOR TRANSLATOR/DRIVER

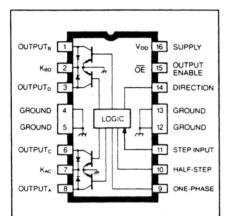

Dwg. No. W-194

Combining low-power CMOS logic with high-current and high-voltage bipolar outputs, the UCN5804B BiMOS II translator/driver provides complete control and drive for a four-phase unipolar stepper-motor with continuous output current ratings to 1.25 A per phase (1.5 A startup) and 35 V.

The CMOS logic section provides the sequencing logic, DIRECTION and OUTPUT ENABLE control, and a power-ON reset function. Three stepper-motor drive formats, wave-drive (one-phase), two-phase, and half-step are externally selectable. The inputs are compatible with standard CMOS, PMOS, and NMOS circuits. TTL or LSTTL may require the use of appropriate pull-up resistors to ensure a proper input-logic high.

The wave-drive format consists of energizing one motor phase at a time in an A-B-C-D (or D-C-B-A) sequence. This excitation mode consumes the least power and assures positional accuracy regardless of any winding inbalance in the motor. Two-phase drive energizes two adjacent phases in each detent position (AB-BC-CD-DA). This sequence mode offers an improved torque-speed product, greater detent torque, and is less susceptible to motor resonance. Half-step excitation alternates between the one-phase and two-phase modes (A-AB-B-BC-C-CD-D-DA), providing an eight-step sequence.

The bipolar outputs are capable of sinking up to 1.5 A and withstanding 50 V in the OFF state (sustaining voltages up to 35 V). Ground clamp and flyback diodes provide protection against inductive transients. Thermal protection circuitry disables the outputs when the chip temperature is excessive.

The UCN5804B is rated for operation over the temperature range of -20°C to +85°C. It is supplied in a 16-pin dual in-line plastic batwing package with a copper lead frame and heat-sinkable tabs for improved power dissipation capabilities.

FEATURES

- 1.5 A Maximum Output Current
- 35 V Output Sustaining Voltage
- Wave-Drive, Two-Phase, and Half-Step Drive Formats
- Internal Clamp Diodes
- Output Enable and Direction Control
- Power-ON Reset
- Internal Thermal Shutdown Circuitry

ABSOLUTE MAXIMUM RATINGS

Output Voltage, V_{CE} . . . . 50 V
Output Sustaining Voltage,
    $V_{CE\,(sus)}$ . . . . 35 V
Output Sink Current, I_{OUT} . . . . 1.5 A
Logic Supply Voltage, V_{DD} . . . . 7.0 V
Input Voltage, V_{IN} . . . . 7.0 V
Package Power Dissipation,
    P_D . . . . See Graph
Operating Temperature Range,
    T_A . . . . -20°C to +85°C
Storage Temperature Range,
    T_S . . . . -55°C to +150°C

Always order by complete part number: UCN5804B .

5804
BiMOS II UNIPOLAR STEPPER-MOTOR TRANSLATOR/DRIVER

APPLICATIONS INFORMATION

Internal power-ON reset (POR) circuitry resets OUTPUT$_A$ (and OUTPUT$_D$ in the two-phase drive format) to the ON state with initial application of the logic supply voltage. After reset, the circuit then steps according to the tables.

The outputs will advance one sequence position on the high-to-low transition of the STEP INPUT pulse. Logic levels on the HALF-STEP and ONE-PHASE inputs will determine the drive format (one-phase, two-phase, or half-step). The DIRECTION pin determines the rotation sequence of the outputs. Note that the STEP INPUT must be in the low state when changing the state of ONE-PHASE, HALFSTEP, or DIRECTION to prevent erroneous stepping.

All outputs are disabled (OFF) when OUTPUT ENABLE is at a logic high. If the function is not required, OUTPUT ENABLE should be tied low. In that condition, all outputs depend only on the state of the step logic.

During normal commutation of a unipolar stepper motor, mutual coupling between the motor windings can force the outputs of the UCN5804B below ground. This condition will cause forward biasing of the collector-to-substrate junction and source current from the output. For many L/R applications, this substrate current is high enough to adversely affect the logic circuitry and cause misstepping. External series diodes (Schottky are recommended for increased efficiency at low voltage operation) will prevent substrate current from being sourced through the outputs. Alternatively, external ground clamp diodes will provide a preferred current path from ground when the outputs are pulled below ground.

Internal thermal protection circuitry disables all outputs when the junction temperature reaches approximately 165°C. The outputs are enabled again when the junction cools down to approximately 145°C.

WAVE-DRIVE SEQUENCE

DIRECTION = L ↓ (left side) / DIRECTION = H ↑ (right side)

Half Step = L, One Phase = H				
Step	A	B	C	D
POR	ON	OFF	OFF	OFF
1	ON	OFF	OFF	OFF
2	OFF	ON	OFF	OFF
3	OFF	OFF	ON	OFF
4	OFF	OFF	OFF	ON

TWO-PHASE DRIVE SEQUENCE

DIRECTION = L ↓ (left side) / DIRECTION = H ↑ (right side)

Half Step = L, One Phase = L				
Step	A	B	C	D
POR	ON	OFF	OFF	ON
1	ON	OFF	OFF	ON
2	ON	ON	OFF	OFF
3	OFF	ON	ON	OFF
4	OFF	OFF	ON	ON

HALF-STEP DRIVE SEQUENCE

DIRECTION = L ↓ (left side) / DIRECTION = H ↑ (right side)

Half Step = H, One Phase = L				
Step	A	B	C	D
POR	ON	OFF	OFF	OFF
1	ON	OFF	OFF	OFF
2	ON	ON	OFF	OFF
3	OFF	ON	OFF	OFF
4	OFF	ON	ON	OFF
5	OFF	OFF	ON	OFF
6	OFF	OFF	ON	ON
7	OFF	OFF	OFF	ON
8	ON	OFF	OFF	ON

5804
BiMOS II UNIPOLAR STEPPER-MOTOR TRANSLATOR/DRIVER

TYPICAL APPLICATION
L/R STEPPER-MOTOR DRIVE

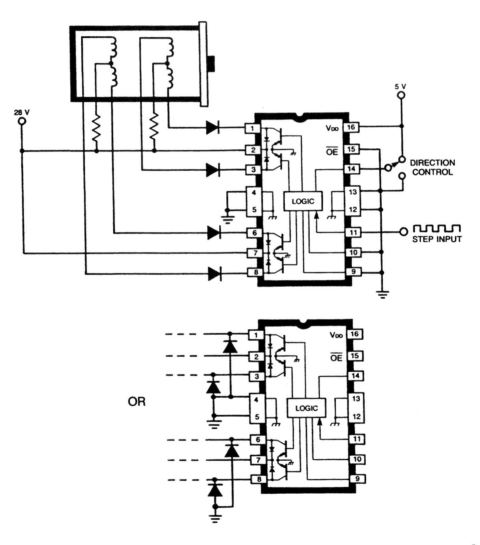

Dwg. No. EP-029A

Four Character Smart Alphanumeric Displays

Technical Data

HPDL-1414
HPDL-2416

Features
- **Smart Alphanumeric Display**
  Built-in RAM, ASCII Decoder and LED Drive Circuitry
- **Wide Operating Temperature Range**
  -40°C to +85°C
- **Fast Access Time**
  160 ns
- **Excellent ESD Protection**
  Built-in Input Protection Diodes
- **CMOS IC for Low Power Consumption**
- **Full TTL Compatibility Over Operating Temperature Range**
  $V_{IL} = 0.8$ V
  $V_{IH} = 2.0$ V
- **Wave Solderable**
- **Rugged Package Construction**
- **End-Stackable**
- **Wide Viewing Angle**

Description
The HPDL-1414 and 2416 are smart, four character, sixteen-segment, red GaAsP displays. The HPDL-1414 has a character height of 2.85 mm (0.112"). The HPDL-2416 has a character height of 4.10 mm (0.160"). The on-board CMOS IC contains memory, ASCII decoder, multiplexing circuitry and drivers. The monolithic LED characters are magnified by an immersion lens which increases both character size and luminous intensity. The encapsulated dual-in-line package provides a rugged, environmentally sealed unit.

The HPDL-1414 and 2416 incorporate many improvements over competitive products. They have a wide operating temperature range, very fast IC access time, and improved ESD protection. The displays are also fully TTL compatible, wave solderable, and highly reliable. These displays are ideally suited for industrial and commercial applications where a good-looking, easy-to-use alphanumeric display is required.

Typical Applications
- **Portable Data Entry Devices**
- **Medical Equipment**
- **Process Control Equipment**
- **Test Equipment**
- **Industrial Instrumentation**
- **Computer Peripherals**
- **Telecommunication Instrumentation**

ESD WARNING: STANDARD CMOS HANDLING PRECAUTIONS SHOULD BE OBSERVED WITH THE HPDL-1414 AND HPDL-2416.

HPDL-2416

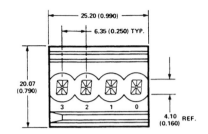

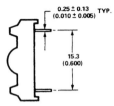

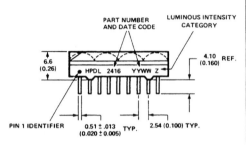

PIN NO.	FUNCTION	PIN NO.	FUNCTION
1	$\overline{CE}_1$ CHIP ENABLE	10	GND
2	$\overline{CE}_2$ CHIP ENABLE	11	D_0 DATA INPUT
3	$\overline{CLR}$ CLEAR	12	D_1 DATA INPUT
4	CUE CURSOR ENABLE	13	D_2 DATA INPUT
5	$\overline{CU}$ CURSOR SELECT	14	D_3 DATA INPUT
6	$\overline{WR}$ WRITE	15	D_6 DATA INPUT
7	ADDRESS INPUT A_1	16	D_5 DATA INPUT
8	ADDRESS INPUT A_0	17	D_4 DATA INPUT
9	V_{DD}	18	$\overline{BL}$ DISPLAY BLANK

NOTES:
1. UNLESS OTHERWISE SPECIFIED, THE TOLERANCE ON ALL DIMENSIONS IS 0.254 mm (0.010 in.).
2. DIMENSIONS IN mm (inches).

Recommended Operating Conditions

Parameter	Sym.	Min.	Nom.	Max.	Units
Supply Voltage	V_{DD}	4.5	5.0	5.5	V

Character Set

BITS	D_3 D_2 D_1 D_0	0 0 0 0	0 0 0 1	0 0 1 0	0 0 1 1	0 1 0 0	0 1 0 1	0 1 1 0	0 1 1 1	1 0 0 0	1 0 0 1	1 0 1 0	1 0 1 1	1 1 0 0	1 1 0 1	1 1 1 0	1 1 1 1
$D_6\ D_5\ D_4$	HEX	0	1	2	3	4	5	6	7	8	9	A	B	C	D	E	F
0 1 0	2	(space)	!	"	#	$	%	&	'	<	>	*	+	,	—	.	/
0 1 1	3	0	1	2	3	4	5	6	7	8	9	:	;	<	=	>	?
1 0 0	4	@	A	B	C	D	E	F	G	H	I	J	K	L	M	N	O
1 0 1	5	P	Q	R	S	T	U	V	W	X	Y	Z	[	\	]	^	_

Magnified Character Font Description

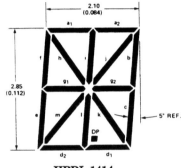

HPDL-1414

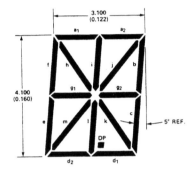

HPDL-2416

Relative Luminous Intensity vs. Temperature

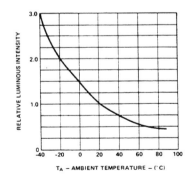

$\overline{WR}$	A_1	A_0	D_6	D_5	D_4	D_3	D_2	D_1	D_0	DIG_3	DIG_2	DIG_1	DIG_0
L	L	L	a	a	a	a	a	a	a	NC	NC	NC	A
L	L	H	b	b	b	b	b	b	b	NC	NC	B	NC
L	H	L	c	c	c	c	c	c	c	NC	C	NC	NC
L	H	H	d	d	d	d	d	d	d	D	NC	NC	NC
H	X	X	X	X	X	X	X	X	X	Previously Written Data			

L = LOGIC LOW INPUT
H = LOGIC HIGH INPUT
X = DON'T CARE
"a" = ASCII CODE CORRESPONDING TO SYMBOL "A"
NC = NO CHANGE

Figure 2. HPDL-1414 Write Truth Table.

TIL305
5 × 7 ALPHANUMERIC DISPLAY
D1033, MAY 1971 REVISED MARCH 1983

SOLID-STATE DISPLAY WITH RED TRANSPARENT PLASTIC ENCAPSULATION

- 7,62-mm (0.300-inch) Character Height
- High Luminous Intensity
- Low Power Requirements
- Wide Viewing Angle
- 5 X 7 Array with X-Y Select and Decimal
- Compatible with USASCII and EBCDIC Codes

mechanical data

This assembly consists of a display chip mounted on a printed circuit board with a red molded plastic body. Multiple displays may be mounted on 11,43-mm (0.450-inch) centers.

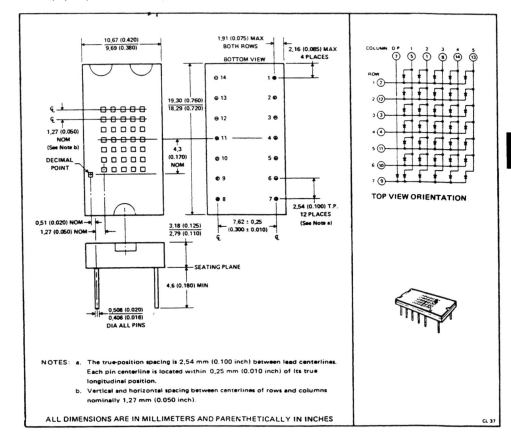

NOTES: a. The true-position spacing is 2,54 mm (0.100 inch) between lead centerlines. Each pin centerline is located within 0,25 mm (0.010 inch) of its true longitudinal position.
b. Vertical and horizontal spacing between centerlines of rows and columns nominally 1,27 mm (0.050 inch).

ALL DIMENSIONS ARE IN MILLIMETERS AND PARENTHETICALLY IN INCHES   CL 37

Intelligent LED Displays

4

TEXAS INSTRUMENTS
POST OFFICE BOX 655303 · DALLAS, TEXAS 75265

TIL305
5 × 7 ALPHANUMERIC DISPLAY

absolute maximum ratings over operating free-air temperature range (unless otherwise noted)

Reverse Voltage at 25°C Free-Air Temperature . . . . 3 V
Peak Forward Current, Each Diode . . . . 100 mA
Average Forward Current (see Note 1):
    Each Diode . . . . 10 mA
    Total . . . . 200 mA
Operating Free-Air Temperature Range . . . . 0° to 70°C
Storage Temperature Range . . . . −25°C to 85°C

operating characteristics of each diode at 25°C free-air temperature (unless otherwise noted)

	PARAMETER	TEST CONDITIONS	MIN	TYP	MAX	UNIT
I_v	Luminous Intensity (see Note 2)		40	110		μcd
λ_p	Wavelength at Peak Emission	I_F = 10 mA		660		nm
Δ'	Spectral Bandwidth			20		nm
V_F	Static Forward Voltage		1.5	1.65	2	V
α_{VF}	Average Temperature Coefficient of Static Forward Voltage	I_F = 10 mA, T_A = 0°C to 70°C		−1.4		mV/°C
I_R	Static Reverse Current	V_R = 3 V			10	μA
C	Anode-to-Cathode Capacitance	V_R = 0, f = 1 MHz			80	pF

NOTES: 1. This average value applies for any 1-ms period.
          2. Luminous intensity is measured with a light sensor and filter combination that approximates the CIE (International Commission on Illumination) eye-response curve.

TYPICAL CHARACTERISTICS

4

Intelligent LED Displays

RELATIVE LUMINOUS INTENSITY
vs
FREE-AIR TEMPERATURE

RELATIVE LUMINOUS INTENSITY
vs
FORWARD CURRENT

FORWARD CONDUCTION
CHARACTERISTICS

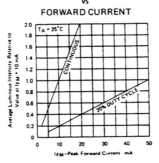

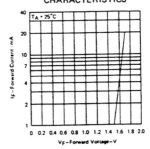

FIGURE 1

FIGURE 2

FIGURE 3

TEXAS
INSTRUMENTS
POST OFFICE BOX 655303 · DALLAS, TEXAS 75265

LM34/LM35 Precision Monolithic Temperature Sensors

National Semiconductor
Application Note 460

INTRODUCTION

Most commonly-used electrical temperature sensors are difficult to apply. For example, thermocouples have low output levels and require cold junction compensation. Thermistors are nonlinear. In addition, the outputs of these sensors are not linearly proportional to any temperature scale. Early monolithic sensors, such as the LM3911, LM134 and LM135, overcame many of these difficulties, but their outputs are related to the Kelvin temperature scale rather than the more popular Celsius and Fahrenheit scales. Fortunately, in 1983 two I.C.'s, the LM34 Precision Fahrenheit Temperature Sensor and the LM35 Precision Celsius Temperature Sensor, were introduced. This application note will discuss the LM34, but with the proper scaling factors can easily be adapted to the LM35.

The LM34 has an output of 10 mV/°F with a typical nonlinearity of only ±0.35°F over a −50 to +300°F temperature range, and is accurate to within ±0.4°F typically at room temperature (77°F). The LM34's low output impedance and linear output characteristic make interfacing with readout or control circuitry easy. An inherent strength of the LM34 over other currently available temperature sensors is that it is not as susceptible to large errors in its output from low level leakage currents. For instance, many monolithic temperature sensors have an output of only 1 μA/°K. This leads to a 1°K error for only 1 μ-Ampere of leakage current. On the other hand, the LM34 may be operated as a current mode device providing 20 μA/°F of output current. The same 1 μA of leakage current will cause an error in the LM34's output of only 0.05°F (or 0.03°K after scaling).

Low cost and high accuracy are maintained by performing trimming and calibration procedures at the wafer level. The device may be operated with either single or dual supplies. With less than 70 μA of current drain, the LM34 has very little self-heating (less than 0.2°F in still air), and comes in a TO-46 metal can package, a SO-8 small outline package and a TO-92 plastic package.

FORERUNNERS TO THE LM34

The making of a temperature sensor depends upon exploiting a property of some material which is a changing function of temperature. Preferably this function will be a linear function for the temperature range of interest. The base-emitter voltage (V_{BE}) of a silicon NPN transistor has such a temperature dependence over small ranges of temperature.

Unfortunately, the value of V_{BE} varies over a production range and thus the room temperature calibration error is not specified nor guaranteeable in production. Additionally, the temperature coefficient of about −2 mV/°C also has a tolerance and spread in production. Furthermore, while the tempo may appear linear over a narrow temperature, there is a definite nonlinearity as large as 3°C or 4°C over a full −55°C to +150°C temperature range.

Another approach has been developed where the difference in the base-emitter voltage of two transistors operated at different current densities is used as a measure of temperature. It can be shown that when two transistors, Q1 and Q2, are operated at different emitter current densities, the difference in their base-emitter voltages, ΔV_{BE}, is

$$\Delta V_{BE} = V_{BE1} - V_{BE2} = \frac{kT}{q} \ln \frac{(J_{E1})}{J_{E2}} \qquad (1)$$

where k is Boltzman's constant, q is the charge on an electron, T is absolute temperature in degrees Kelvin and J_{E1} and J_{E2} are the emitter current densities of Q1 and Q2 respectively. A circuit realizing this function is shown in *Figure 1*.

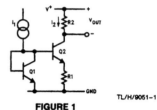

FIGURE 1

Equation 1 implies that as long as the ratio of I_{E1} to I_{E2} is held constant, then ΔV_{BE} is a linear function of temperature (this is not exactly true over the whole temperature range, but a correction circuit for the nonlinearity of V_{BE1} and V_{BE2} will be discussed later). The linearity of this ΔV_{BE} with temperature is good enough that most of today's monolithic temperature sensors are based upon this principle.

An early monolithic temperature sensor using the above principle is shown in *Figure 2*. This sensor outputs a voltage which is related to the absolute temperature scale by a factor of 10 mV per degree Kelvin and is known as the LM135. The circuit has a ΔV_{BE} of approximately

$$(0.2 \text{ mV/°K}) \times (T)$$

developed across resistor R. The amplifier acts as a servo to enforce this condition. The ΔV_{BE} appearing across resistor R is then multiplied by the resistor string consisting of R

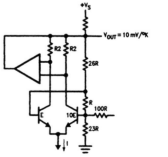

FIGURE 2

and the 26R and 23R resistors for an output voltage of (10 mV/°K) × (T). The resistor marked 100R is used for offset trimming. This circuit has been very popular, but such Kelvin temperature sensors have the disadvantage of a large constant output voltage of 2.73V which must be subtracted for use as a Celsius-scaled temperature sensor.

Various sensors have been developed with outputs which are proportional to the Celsius temperature scale, but are rather expensive and difficult to calibrate due to the large number of calibration steps which have to be performed. Gerard C.M. Meijer[4] has developed a circuit which claims to be inherently calibrated if properly trimmed at any one temperature. The basic structure of Meijer's circuit is shown in *Figure 3*. The output current has a temperature coefficient of 1 μA/°C. The circuit works as follows: a current which is proportional to absolute temperature, I_{PTAT}, is generated by a current source. Then a current which is proportional to the V_{BE} drop of transistor Q4 is subtracted from I_{PTAT} to get the output current, I_O. Transistor Q4 is biased by means of a PNP current mirror and transistor Q3, which is used as a feedback amplifier. In Meijer's paper it is claimed that the calibration procedure is straightforward and can be performed at any temperature by trimming resistor R4 to adjust the sensitivity, dI_O/dT, and then trimming a resistor in the PTAT current source to give the correct value of output current for the temperature at which the calibration is being performed.

Meijer's Celsius temperature sensor has problems due to its small output signal (i.e., the output may have errors caused by leakage currents). Another problem is the trim scheme requires the trimming of two resistors to a very high degree of accuracy. To overcome these problems the circuits of *Figure 4* (an LM34 Fahrenheit temperature sensor) and *Figure 5* (an LM35 Celsius temperature sensor) have been developed to have a simpler calibration procedure, an output voltage with a relatively large tempco, and a curvature compensation circuit to account for the non-linear characteristics of V_{BE} versus temperature. Basically, what happens is transistors Q1 and Q2 develop a ΔV_{BE} across resistor R1. This voltage is multiplied across resistor nR1. Thus at the

non-inverting input of amplifier A2 is a voltage two diode drops below the voltage across resistor nR1. This voltage is then amplified by amplifier A2 to give an output proportional to whichever temperature scale is desired by a factor of 10 mV per degree.

CIRCUIT OPERATION

Since the two circuits are very similar, only the LM34 Fahrenheit temperature sensor will be discussed in greater detail. The circuit operates as follows:

Transistor Q1 has 10 times the emitter area of transistor Q2, and therefore, one-tenth the current density. From *Figure 4*, it is seen that the difference in the current densities of Q1 and Q2 will develop a voltage which is proportional to absolute temperature across resistor R_1. At 77°F this voltage will be 60 mV. As in the Kelvin temperature sensor, an amplifier, A1, is used to insure that this is the case by servoing the base of transistor Q1 to a voltage level, V_{PTAT}, of $\Delta V_{BE} \times n$. The value of n will be trimmed during calibration of the device to give the correct output for any temperature.

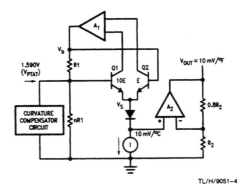

FIGURE 4

TL/H/9051–4

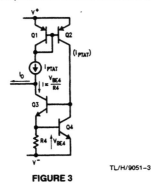

FIGURE 3

TL/H/9051–3

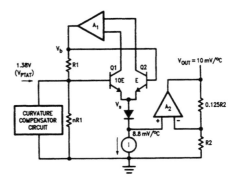

FIGURE 5

TL/H/9051–5

For purposes of discussion, suppose that a value of V_{PTAT} equal to 1.59V will give a correct output of 770 mV at 77°F. Then n will be equal to $V_{PTAT}/\Delta V_{BE}$ or 1.59V/60 mV = 26.5, and V_{PTAT} will have a temperature coefficient (tempco) of:

$$\frac{nk}{q} \ln \frac{I_1}{I_2} = 5.3 \text{ mV/°C}.$$

Subtracting two diode drops of 581 mV (at 77°F) with tempcos of −2.35 mV/°C each, will result in a voltage of 428 mV with a tempco of 10 mV/°C at the non-inverting input of amplifier A2. As shown, amplifier A2 has a gain of 1.8 which provides the necessary conversion to 770 mV at 77°F (25°C). A further example would be if the temperature were 32°F (0°C), then the voltage at the input of A2 would be 428 mV−(10 mV/°C) (25°C) = 0.178, which would give V_{OUT} = (0.178) (1.8) = 320 mV—the correct value for this temperature.

EASY CALIBRATION PROCEDURE

The circuit may be calibrated at any temperature by adjusting the value of the resistor ratio factor n. Note that the value of n is dependent on the actual value of the voltage drop from the two diodes since n is adjusted to give a correct value of voltage at the output and not to a theoretical value for PTAT. The calibration procedure is easily carried out by opening or shorting the links of a quasi-binary trim network like the one shown in *Figure 6*. The links may be opened to add resistance by blowing an aluminum fuse, or a resistor may be shorted out of the circuit by carrying out a "zener-zap". The analysis in the next section shows that when the circuit is calibrated at a given temperature, then the circuit will be accurate for the full temperature range.

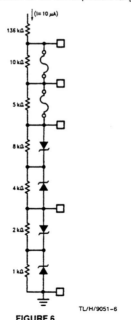

TL/H/9051–6

FIGURE 6

How The Calibration Procedure Works

Widlar[5] has shown that a good approximation for the base-emitter voltage of a transistor is:

$$V_{BE} = V_{G0}\left(1 - \frac{T}{T_0}\right) + V_{beo}\left(\frac{T}{T_0}\right) + \frac{nkT}{q}\ln\left(\frac{T_0}{T}\right) + \frac{kT}{q}\ln\frac{I_C}{I_{co}} \quad (1)$$

where T is the temperature in °Kelvin, T_0 is a reference temperature, V_{G0} is the bandgap of silicon, typically 1.22V, and V_{beo} is the transistor's base-emitter voltage at the reference temperature, T_0. The above equation can be re-written as

$$V_{BE} = \text{(sum of linear temp terms)} + \text{(sum of non-linear temp terms)} \quad (2)$$

where the first two terms of equation 1 are linear and the last two terms are non-linear. The non-linear terms were shown by Widlar to be relatively small and thus will be considered later.

Let us define a base voltage, V_b, which is a linear function of temperature as: $V_b = C_1 \cdot T$. This voltage may be represented by the circuit in *Figure 1*. The emitter voltage is $V_e = V_b - V_{be}$ which becomes:

$$V_e = C_1 T - V_{G0}\left(1 - \frac{T}{T_0}\right) - V_{beo}\left(\frac{T}{T_0}\right).$$

If V_e is defined as being equal to C_2 at $T = T_0$, then the above equation may be solved for C_1. Doing so gives:

$$C_1 = \frac{V_{beo} + C_2}{T_0} \quad (3)$$

Using this value for C_1 in the equation for V_e gives:

$$V_e = C_2\frac{T}{T_0} + V_{G0}\left(\frac{T - T_0}{T_0}\right) \quad (4)$$

If V_e is differentiated with respect to temperature, T, equation 4 becomes $dV_e/dT = (C_2 + V_{G0})/T_0$.

This equation shows that if V_b is adjusted at T_0 to give $V_e = C_2$, then the rate of change of V_e with respect to temperature will be a constant, independent of the value of V_b, the transistor's beta or V_{be}.

To proceed, consider the case where $V_e = C_2 = 0$ at $T_0 = 0$°C. Then $\frac{dV_e}{dT} = \frac{V_{G0}}{273.7} = 4.47$ mV/°C.

Therefore, if V_e is trimmed to be equal to (4.47 mV) T (in °C) for each degree of displacement from 0°C, then the trimming can be done at ambient temperatures.

In practice, the two non-linear terms in equation 1 are found to be quadratic for positive temperatures. Tsividis[6] showed that the bandgap voltage, V_0, is not linear with respect to temperature and causes nonlinear terms which become significant for negative temperatures (below 0°C). The sum of these errors causes an error term which has an approximately square-law characteristic and is thus compensated by the curvature compensation circuit of *Figure 7*.

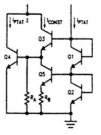

TL/H/9051–7

FIGURE 7

MOTOROLA
SEMICONDUCTOR ▰▰▰▰▰▰▰
TECHNICAL DATA

Order this document
by MTS102/D

Silicon Temperature Sensors

Designed for use in temperature sensing applications in automotive, consumer and industrial products requiring low cost and high accuracy.

- Precise Temperature Accuracy Over Extreme Temperature MTS102: ± 2°C from −40°C to +150°C
- Precise Temperature Coefficient
- Fast Thermal Time Constant
    - 3 Seconds — Liquid
    - 8 Seconds — Air
- Linear V_{BE} versus Temperature Curve Relationship
- Other Packages Available

MTS102
MTS103
MTS105

SILICON
TEMPERATURE
SENSORS

CASE 29-04, STYLE 1
TO-226AA
(TO-92)

Pin Number		
1	2	3
Emitter	Base	Collector

MAXIMUM RATINGS

Rating	Symbol	Value	Unit
Emitter-Base Voltage	V_{EB}	4.0	Vdc
Collector Current — Continuous*	I_C	100	mAdc
Operating and Storage Junction Temperature Range	T_J, T_{stg}	−55 to +150	°C

* See Note 5 on following page.

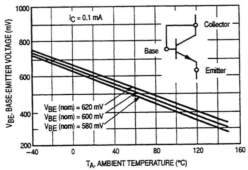

Figure 1. Base-Emitter Voltage versus Ambient Temperature

X-ducer is a trademark of Motorola Inc.

REV 2 1/94

MOTOROLA

© Motorola, Inc. 1994

ELECTRICAL CHARACTERISTICS (T_A = 25°C unless otherwise noted)

Characteristic		Symbol	Min	Typ	Max	Unit
Supply Voltage		V_S	−0.2	—	35	Vdc
Output Voltage		V_{out}	−1.0	—	6.0	Vdc
Output Current		I_O	—	—	10	mAdc
Emitter-Base Breakdown Voltage (I_E = 100 μAdc, I_C = 0)		$V_{(BR)EBO}$	4.0	—	—	Vdc
Base-Emitter Voltage (I_C = 0.1 mA)		V_{BE}	580	595	620	mV
Base-Emitter Voltage Matching, Note 1 (I_C = 0.1 mA, T_A = 25°C ±0.05°C)	MTS102 MTS103 MTS105	ΔV_{BE}	−3.0 −4.0 −7.0	— — —	3.0 4.0 7.0	mV
Temperature Matching Accuracy, Note 2 (T1 = 40°C, T2= +150°C, T_A = 25°C ± 0.05°C)	MTS102 MTS103 MTS105	ΔT	−3.0 −3.0 −5.0	— — —	3.0 3.0 5.0	°C
Temperature Coefficient, Notes 3 and 4 (V_{BE} = 595 mV, I_C = 0.1 mA)		T_C	−2.28	−2.265	−2.26	mV/°C
Thermal Time Constant Liquid Flowing Air		τ_{TH}	— —	3.0 8.0	— —	s
Dependence of T_C on V_{BE} @ 25°C (Note 4, Figure 3)		$\Delta T_C/\Delta V_{BE}$	—	0.0033	—	mV/°C mV

THERMAL CHARACTERISTICS

	Symbol	Min	Typ	Max	Unit
Thermal Resistance, Junction to Ambient	$R_{\theta JA}$	—	—	200	°C/W

MECHANICAL CHARACTERISTICS

Weight	—	—	87	—	Grams

NOTES:
1. All devices within any one group or package will be matched for V_{BE} to the tolerance identified in the electrical characteristics table. Each device will be labeled with the mean V_{BE} value for that group.
2. All devices within an individual group, as described in Note 1, will track within the specified temperature accuracy. This includes variations in T_C, V_{BE}, and nonlinearity in the range −40 to +150°C. Nonlinearity is typically less than ±1°C in this range. (See Figure 4)
3. The T_C as defined by a least-square linear regression for V_{BE} versus temperature over the range −40 to +150°C for a nominal V_{BE} of 595 mV at 25°C. For other nominal V_{BE} values the value of the T_C must be adjusted for the dependence of the T_C on V_{BE} (see Note 4).
4. For nominal VBE at 25°C other than 595 mV, the T_C must be corrected using the equation T_C = −2.265 + 0.003 (V_{BE} − 595) where V_{BE} is in mV and the T_C is in mV/°C. The accuracy of this T_C is typically ±0.01 mV/°C.
5. For maximum temperature accuracy, I_C should not exceed 2 mA. (See Figure 2)

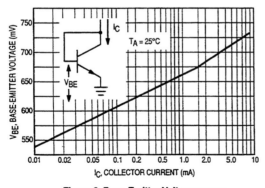

Figure 2. Base-Emitter Voltage versus Collector-Emitter Current

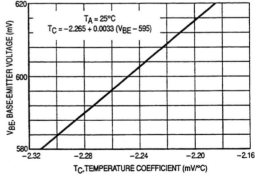

Figure 3. Temperature Coefficient versus Base-Emitter Voltage

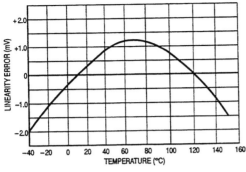

Figure 4. Linearity Error versus Temperature

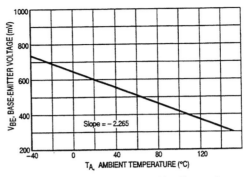

Figure 5. V_{BE} versus Ambient Temperature

APPLICATIONS INFORMATION

The base and collector leads of the device should be connected together in the operating circuit (pins 2 and 3). They are not internally connected.

The following example describes how to determine the V_{BE} versus temperature relationship for a typical shipment of various V_{BE} groups.

EXAMPLE:

Given — Customer receives a shipment of MTS102 devices. The shipment consists of three groups of different nominal V_{BE} values.

Group 1: V_{BE} (nom) = 595 mV
Group 2: V_{BE} (nom) = 580 mV
Group 3: V_{BE} (nom) = 620 mV

Find — V_{BE} versus temperature Relationship.

1. Determine value of T_C:
   a. If V_{BE} (nom) = 595 mV, T_C = −2.265 mV/°C from the Electrical Characteristics table.

   b. If V_{BE} (nom) is less than or greater than 595 mV determine T_C from the relationship described in Note 4.

   $T_C = -2.265 + 0.0033 (V_{BE} - 595)$ or see Figure 3.

2. Determine the V_{BE} value at extremes, − 40°C and +150°C:

   $V_{BE(T_A)} = V_{BE(25°C)} + (T_C)(T_A - 25°C)$ where $V_{BE(T_A)}$ = value of V_{BE} at desired temperature.

3. Plot the V_{BE} versus T_A curve using two V_{BE} values: V_{BE} (− 40°C), V_{BE}(25°C), or V_{BE}(+150°C)

4. Given any measured V_{BE}, the value of T_A (to the accuracy value specified: MTS102 − ±2°C, MTS103 − ±35°C, MTS105 − ± 5°C) can be read from Figure 5 or calculated from equation 2.

5. Higher temperature accuracies can be achieved if the collector current, I_C, is controlled to react in accordance with and to compensate for the linearity error. Using this concept, practical circuits have been built in which allow these sensors to yield accuracies within ± 0.1°C and ± 0.01°C.
   Reference: "Transistors − A Hot Tip for Accurate Temperature Sensing", Pat O'Neil and Carl Derrington, *Electronics 1979*.

MOTOROLA
SEMICONDUCTOR ▬▬▬▬▬
TECHNICAL DATA

0 to 7.3 PSI
On-Chip Temperature
Compensated & Calibrated,
Silicon Pressure Sensors

The MPX2050, MPX2051 and MPX2052 series device is a silicon piezoresistive pressure sensors providing a highly accurate and linear voltage output — directly proportional to the applied pressure. The sensor is a single, monolithic silicon diaphragm with the strain gauge and a thin-film resistor network integrated on-chip. The chip is laser trimmed for precise span and offset calibration and temperature compensation. This device was designed for use in applications such as pump/motor controllers, robotics, level indicators, medical diagnostics, pressure switching, and non-invasive blood pressure measurement.

- Temperature Compensated Over 0°C to +85°C
- Unique Silicon Shear Stress Strain Gauge
- Full Scale Span Calibrated to 40 mV (typical)
- Easy to Use Chip Carrier Package Options
- Ratiometric to Supply Voltage

MPX2050
MPX2051
MPX2052
SERIES
Motorola Preferred Devices

0–7.3 PSI
X-ducer™
SILICON
PRESSURE SENSORS

Pin Number			
1	2	3	4
Ground	+V_{out}	V_S	−V_{out}

**BASIC CHIP
CARRIER ELEMENT
CASE 344
Style 1**

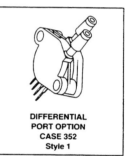

**DIFFERENTIAL
PORT OPTION
CASE 352
Style 1**

MAXIMUM RATINGS

Rating	Symbol	Value	Unit
Overpressure	P_{max}	200	kPa
Burst Pressure	P_{burst}	500	kPa
Supply Voltage (Note 11)	V_{Smax}	16	Vdc
Storage Temperature	T_{stg}	−50 to +150	°C
Operating Temperature	T_A	−40 to +125	°C

VOLTAGE OUTPUT versus APPLIED DIFFERENTIAL PRESSURE

The differential voltage output of the X-ducer is directly proportional to the differential pressure applied.

The output voltage of the differential element, differential ported and gauge ported sensors increases with increasing pressure applied to the pressure side relative to the vacuum side. Similarly, output voltage increases as increasing vacuum is applied to the vacuum side relative to the pressure side of the differential units.

The output voltage of the gauge vacuum ported sensor increases with increasing vacuum (decreasing pressure) applied to the vacuum side with the pressure side at ambient. Figure 1 shows the schematic diagram of the MPX2050 sensor circuit.

X-ducer is a trademark of Motorola Inc.

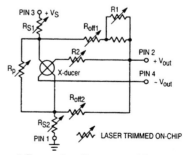

**Figure 1. Temperature Compensated Pressure
Sensor Schematic**

REV 3 1/94

MPX2100 • MPX2101 SERIES

OPERATING CHARACTERISTICS (V_S = 10 Vdc, T_A = 25°C unless otherwise noted)

Characteristic		Symbol	Min	Typ	Max	Unit
Pressure Range [1]		P_{OP}	0	—	100	kPa
Supply Voltage [11]		V_S	—	10	16	Vdc
Supply Current		I_o	—	6.0	—	mAdc
Full Scale Span [2], Figure 5	MPX2100A, MPX2100D, MPX2101D	V_{FSS}	38.5	40	41.5	mV
	MPX2101A		37.5	40	42.5	
Zero Pressure Offset, Figure 5	MPX2100D, MPX2101D	V_{off}	−1.0	—	1.0	mV
	MPX2100A		−2.0	—	2.0	
	MPX2101A		−3.0	—	3.0	
Sensitivity		$\Delta V/\Delta P$	—	0.4	—	mV/kPa
Linearity [3] Figure 2	MPX2100A, MPX2100D		—	−0.25	0.25	%FSS
	MPX2101A, MPX2101D		—	−0.5	0.5	
Pressure Hysteresis [4] (0 to 100 kPa)			—	−0.1	0.1	%FSS
Temperature Hysteresis [5] (−40°C to +125°C)			—	—	±0.5	—	%FSS
Temperature Effect on Full Scale Span [6] (0 to +85°C), Figure 6		TCV_{FSS}	−1.0	—	1.0	%FSS
Temperature Effect on Offset [7] (0 to +85°C), Figure 6		TCV_{off}	−1.0	—	1.0	mV
Input Impedance		Z_{in}	1000	—	2500	Ω
Output Impedance		Z_{out}	1400	—	3000	Ω
Response Time [8] (10% to 90%)		t_R	—	1.0	—	ms
Temperature Error Band, Figure 6			—	0	—	85	°C
Stability [9]			—	—	±0.5	—	%FSS

MECHANICAL CHARACTERISTICS

Characteristic	Symbol	Min	Typ	Max	Unit
Weight (Basic Element Case 344)	—	—	2.0	—	Grams
Warm-Up	—	—	15	—	Sec
Cavity Volume				0.01	IN^3
Volumetric Displacement	—	—	—	0.001	IN^3
Common Mode Line Pressure	—	—	—	690	kPA

NOTES:
1. 1 kPa (kiloPascal) equals 0.145 PSI.
2. Measured at 10 Vdc excitation for 100 kPa pressure differential. V_{FSS} and FSS are like terms representing the algebraic difference between full scale output and zero pressure offset.
3. Maximum deviation from end-point straight line fit at 0 and 100 kPa.
4. Maximum output difference at any pressure point within P_{OP} for increasing and decreasing pressures.
5. Maximum output difference at any pressure point within P_{OP} for increasing and decreasing temperatures in the range −40°C to +125°C.
6. Slope end-point straight line fit to full scale span at 0°C and +85°C relative to +25°C.
7. Slope end-point straight line fit to zero pressure offset at 0°C and +85°C relative to +25°C.
8. For a 0 to 100 kPa pressure step change.
9. Stability is defined as the maximum difference in output at any pressure within P_{OP} and temperature within +10°C to +85°C after:
   a. 1000 temperature cycles, −40°C to +125°C.
   b. 1.5 million pressure cycles, 0 to 100 kPa.
10. Operating characteristics based on positive pressure differential relative to the vacuum side (gauge/differential) or sealed reference (absolute).
11. Recommended voltage supply: 10 V ± 0.2 V, regulated. Sensor output is ratiometric to the voltage supply. Supply voltages above +16 V may induce additional error due to device self-heating.

MPX2100 • MPX2101 SERIES

LINEARITY

Linearity refers to how well a transducer's output follows the equation: $V_{out} = V_{off} +$ sensitivity x P over the operating pressure range. There are two basic methods for calculating nonlinearity: (1) end point straight line fit (see Figure 2) or (2) a least squares best line fit. While a least squares fit gives the "best case" linearity error (lower numerical value), the calculations required are burdensome.

Conversely, an end point fit will give the "worst case" error (often more desirable in error budget calculations) and the calculations are more straightforward for the user. Motorola's specified pressure sensor linearities are based on the end point straight line method measured at the midrange pressure.

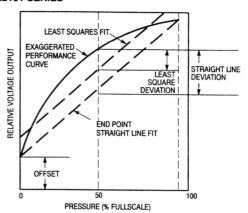

Figure 2. Linearity Specification Comparison

EXAMPLE INTERFACE CIRCUITS

The MPX2100 series sensors with on-chip compensation can be used individually or in multiples in research, design, or development projects to optimize a design. The small size and low cost of the compensated MPX2100 series sensors make these devices ideally suited for such applications.

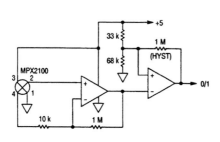

Output switches low at 55% full-scale input; switches high at 45% input.
1 M Hysteresis resistor may be removed or value changed according to user requirements.

Figure 3. Single-ended Supply, TTL or CMOS Logic Compatible Comparator

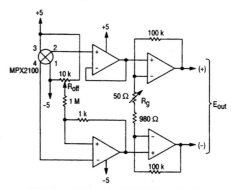

DVM µP compatible input. Set SPAN with R_g, the OFFSET with R_{off}.
Differential output is ±8 Vdc with full-scale pressure applied.

Figure 4. Precision Pressure-to-Voltage Converter using Quad Op Amp

These are offered as basic suggestions only: actual component selection and values are determined by the final circuit requirements.

Appendix C
MC68HC11E9 Header File[1]

```c
#ifndef HC11E9
#define HC11E9

typedef struct
{
     unsigned  char bit0 : 1;
     unsigned  char bit1 : 1;
     unsigned  char bit2 : 1;
     unsigned  char bit3 : 1;
     unsigned  char bit4 : 1;
     unsigned  char bit5 : 1;
     unsigned  char bit6 : 1;
     unsigned  char bit7 : 1;
} Register;

unsigned int Register_Set=0x1000 ;

#define PORTA  (*(volatile Register*)(Register_Set+0))
#define PIOC   (*(Register*)(Register_Set+2))
#define PORTC  (*(volatile Register*)(Register_Set+3))
#define PORTB  (*(volatile Register*)(Register_Set+4))
#define PORTCL (*(volatile Register*)(Register_Set+5))
#define DDRC   (*(Register*)(Register_Set+7))
#define PORTD  (*(volatile Register*)(Register_Set+8))
#define DDRD   (*(Register*)(Register_Set+9))
#define PORTE  (*(volatile Register*)(Register_Set+0xA))
```

---

[1] Example taken from *Programming Microcontrollers in C,* Ted Van Sickle, copyright 1994, HighText Publications, Inc. Reprinted with permission.

```
#define CFORC (*(Register*)(Register_Set+0xB))
#define OC1M  (*(Register*)(Register_Set+0xC))
#define OC1D  (*(Register*)(Register_Set+0xD))
#define TCNT  (*(volatile unsigned int *)(Register_Set+0xE))
#define TIC1  (*(volatile unsigned int *)(Register_Set+0x10))
#define TIC2  (*(volatile unsigned int *)(Register_Set+0x12))
#define TIC3  (*(volatile unsigned int *)(Register_Set+0x14))
#define TOC1  (*(volatile unsigned int *)(Register_Set+0x16))
#define TOC2  (*(volatile unsigned int *)(Register_Set+0x18))
#define TOC3  (*(volatile unsigned int *)(Register_Set+0x1A))
#define TOC4  (*(volatile unsigned int *)(Register_Set+0x1C))
#define TOC5  (*(volatile unsigned int *)(Register_Set+0x1E))
#define TCTL1 (*(volatile Register*)(Register_Set+0x20))
#define TCTL2 (*(volatile Register*)(Register_Set+0x21))
#define TMSK1 (*(volatile Register*)(Register_Set+0x22))
#define TFLG1 (*(volatile unsigned char*)(Register_Set+0x23))
#define TMSK2 (*(volatile Register*)(Register_Set+0x24))
#define TFLG2 (*(volatile unsigned char*)(Register_Set+0x25))
#define PACTL (*(Register*)(Register_Set+0x26))
#define PACNT (*(volatile Register*)(Register_Set+0x27))
#define SPCR  (*(Register*)(Register_Set+0x28))
#define SPSR  (*(volatile Register*)(Register_Set+0x29))
#define SPDR  (*(volatile Register*)(Register_Set+0x2A))
#define BAUD  (*(Register*)(Register_Set+0x2B))
#define SCCR1 (*(volatile Register*)(Register_Set+0x2C))
#define SCCR2 (*(Register*)(Register_Set+0x2D))
#define SCSR  (*(volatile Register*)(Register_Set+0x2E))
#define SCDR  (*(volatile unsigned char*)(Register_Set+0x2F))
#define ADCTL (*(volatile Register*)(Register_Set+0x30))
#define ADR1  (*(volatile unsigned char*)(Register_Set+0x31))
#define ADR2  (*(volatile unsigned char*)(Register_Set+0x32))
#define ADR3  (*(volatile unsigned char*)(Register_Set+0x33))
#define ADR4  (*(volatile unsigned char*)(Register_Set+0x34))
#define BPROT (*(Register*)(Register_Set+0x35))
#define OPTION (*(Register*)(Register_Set+0x39))
#define COPRSR (*(unsigned char*)(Register_Set+0x3A))
#define PPROG (*(Register*)(Register_Set+0x3B))
#define HPRIO (*(Register*)(Register_Set+0x3C))
#define INIT  (*(Register*)(Register_Set+0x3D))
#define TEST1 (*(Register*)(Register_Set+0x3E))
#define CONFIG (*(Register*)(Register_Set+0x3F))

/* Bit names for general use */

#define Bit_7 7
#define Bit_6 6
#define Bit_5 5
#define Bit_4 4
#define Bit_3 3
#define Bit_2 2
#define Bit_1 1
```

```
#define Bit_0 0
/* PIOC bit definitions 0x0*/

#define STAF bit7
#define STAI bit6
#define CWOM bit5
#define HNDS bit4
#define OIN  bit3
#define PLS  bit2
#define EGA  bit1
#define INVB bit0

/* CFORC bit definitions 0xB*/

#define FOC1 bit7
#define FOC2 bit6
#define FOC3 bit5
#define FOC4 bit4
#define FOC5 bit3

/* OC1M bit definitions 0xC*/

#define OC1M7 bit7
#define OC1M6 bit6
#define OC1M5 bit5
#define OC1M4 bit4
#define OCMM3 bit3

/* OC1D bit definitions 0xD*/

#define OC1D7 bit7
#define OC1D6 bit6
#define OC1D5 bit5
#define OC1D4 bit4
#define OCMD3 bit3

/* TCTL1 bit definition 0x20*/

#define OM2 bit7
#define OL2 bit6
#define OM3 bit5
#define OL3 bit4
#define OM4 bit3
#define OL4 bit2
#define OM5 bit1
#define OL5 bit0

/* TCTL2 bit definitions 0x21 */

#define EDG4B bit7
#define EDG4A bit6
```

```
#define EDG1B bit5
#define EDG1A bit4
#define EDG2B bit3
#define EDG2A bit2
#define EDG3B bit1
#define EDG3A bit0

/* TMSK1 bit definitions 0x22 */

#define OC1I bit7
#define OC2I bit6
#define OC3I bit5
#define OC4I bit4
#define I405I bit3
#define IC1I bit2
#define IC2I bit1
#define IC3I bit0

/* TFLG1 bit definitions 0x23 */

#define OC1F 0x80
#define OC2F 0x40
#define OC3F 0x20
#define OC4F 0x10
#define I405F 0x08
#define IC1F 0x04
#define IC2F 0x02
#define IC3F 0x01

/* TMSK2 bit definitions 0x24 */

#define TOI  bit7
#define RTII bit6
#define PAOVI bit5
#define PAII bit4
#define PR1 bit1
#define PR0 bit0

/* TFLG2 bit definitions 0x25 */

#define TOF  0x80
#define RTIF 0x40
#define PAOVF 0x20
#define PAIF 0x10

/* PACTL bit definitions 0x26 */

#define DDRA7  bit7
#define PAEN   bit6
#define PAMOD  bit5
```

```
#define PEDGE  bit4
#define DDRA3  bit3
#define I4O5   bit2
#define RTR1   bit1
#define RTR0   bit0
/* SPCR bit definitions 0x28 */

#define SPIE bit7
#define SPE  bit6
#define DWOM bit5
#define MSTR bit4
#define CPOL bit3
#define CPHA bit2
#define SPR1 bit1
#define SPR0 bit0

/* SPSR bit definitions 0x29 */

#define SPIF  bit7
#define WCOL  bit6
#define MODV  bit4

/* BAUD bit definitions 0x2B */

#define TCLR bit7
#define SCP1 bit5
#define SCP0 bit4
#define RCKB bit3
#define SCR2 bit2
#define SCR1 bit1
#define SCR0 bit0

/*SCCR1 bit definition 0x2C*/

#define R8 bit7
#define T8 bit6
#define M bit4
#define WAKE bit3

/* SCCR2 bit definitions 0x2D */

#define TIE bit7
#define TCIE bit6
#define RIE  bit5
#define ILIE bit4
#define TE   bit3
#define RE   bit2
#define RWU  bit1
#define SBK0 bit0
```

```
/* SCSR  bit definitions 0x2E */

#define TDRE bit7
#define TC   bit6
#define RDRF bit5
#define IDLE bit4
#define OR   bit3
#define NF   bit2
#define FE   bit1


/* ADCTL bit definitions 0x30 */

#define CCF  bit7
#define SCAN bit5
#define MULT bit4
#define CD   bit3
#define CC   bit2
#define CB   bit1
#define CA   bit0


/* BPROT bit definitions 0x35 */

#define PTCON bit4
#define BPRT3 bit3
#define BPRT2 bit2
#define BPRT1 bit1
#define BPRT0 bit0


/* OPTION bit definitions 0x39 */

#define ADPU bit7
#define CSEL bit6
#define IRQE bit5
#define DLY  bit4
#define CME  bit3
#define CR1  bit1
#define CR0  bit0


/* PPROG bit definitions 0x3B */

#define ODD  bit7
#define EVEN bit6
#define ELAT bit5
#define BYTE bit4
#define ROW  bit3
#define ERASE bit2
#define EELAT bit1
#define EEPGM bit0


/* HPRIO bit definitions 0x3C */
#define RBOOT bit7
```

```
#define SMOD  bit6
#define MDA   bit5
#define IRV   bit4
#define PSEL3 bit3
#define PSEL2 bit2
#define PSEL1 bit1
#define PSEL0 bit0

/* INIT  bit definitions 0x3D */

#define RAM3 bit7
#define RAM2 bit6
#define RAM1 bit5
#define RAM0 bit4
#define REG3 bit3
#define REG2 bit2
#define REG1 bit1
#define REG0 bit0

/* TEST1 bit definitions 0x3E */

#define TILOP bit7
#define OCCR bit5
#define CBYP bit4
#define DISR bit3
#define FCM  bit2
#define FCOP bit1
#define TCON bit0

/* CONFIG bit definitions 0x3F */

#define NOSEC bit3
#define NOCOP bit2
#define ROMON bit1
#define EEON  bit0

/*   Macros and function to permit interrupt service
     routine programming from C.
     To use the vector call, do vector(isr, vector_address)
     where isr is a pointer to the interrupt service
     routine, and vector_address is the vector address where
     the isr pointer must be stored.       */

#define vector(isr,vector_address) (*(void
**)(vector_address)=(isr))
#define cli() _asm("cli\n")
#define sei() _asm("sei\n")

#define TRUE 1
#define FALSE 0
#define NULL 0
```

```
#define FOREVER while(TRUE)
typedef unsigned int WORD;
typedef unsigned char BYTE;

#endif
```